PROBABILITY AND STATISTICS IN HYDROLOGY

by

VUJICA YEVJEVICH

WATER RESOURCES PUBLICATIONS
FORT COLLINS, COLORADO, U. S. A.
1972

This book can be purchased through

WATER RESOURCES PUBLI CATIONS,

P.O. Box 303, Fort Collins, Colorado 80521, U.S.A.

First Printing 1972
Second Printing 1978

PROBABILITY AND STATISTICS IN HYDROLOGY

ISBN-0-918334-00-4

Library of Congress Catalog Card Number: 74 — 168494.

Printed by LithoCrafters, Ann Arbor, Michigan, U.S.A.

PREFACE

The classical treatment of hydrologic data has been through the use of *descriptive statistics*, with properties of available sample of data derived and presented as various statistics and frequency distributions. Results from the descriptive statistics have then been used as if future samples will have properties identical to the original samples. Because this is not true more and more investigators in hydrology have been relying on *inferential statistics* for extracting information from available hydrologic data, even though the use of descriptive statistics is beneficial. This book is conceived as a tool for those interested in applying *inferential statistics* to *hydrology and water resources*. The substance of the text consists of mathematical description of properties of hydrologic random variables, and the use of *probability theory* and *mathematical statistics* in solving hydrologic problems.

Using inferential statistics means the maximum information about the populations of random variables, in its most condensed form, is extracted from the data. To accomplish this condensation, and use of data, mathematical models for describing hydrologic random variables must be postulated, the best techniques for estimating the parameters of the models used must be determined, and relevant hypotheses must be advanced and tested by the inferential statistics. The accomplishment of this condensation of information and use of hydrologic data through inferential statistics is the major objective of this book.

A book is sometimes described by stating what it is and what it is not. This book is not a conventional statistical treatment because no proofs of lemmas and no derivations of formulas of probability theory and mathematical statistics are presented, except in a few particular cases that have a hydrologic character. Certain concepts, however, are defined, mathematical expressions are given, techniques are explained, and their applications to hydrology, with examples, are presented. Nor is this book a manual or an exhaustive literature review of the various published contributions to the application of probability and statistics in hydrology. This book neither contains the computer programs necessary for solving problems presented in the text, or the classical statistical tables usually included in such a text; the statistical tables are already available in a large number of books on statistics. The selection of topics of probability theory and mathematical statistics that are presented are primarily based on the writer's experience and judgment. The book is also not a text on statistical hydrology or stochastic hydrology, but rather an illustration of how various topics of probability theory and mathematical statistics can be applied in hydrology. A book on statistical or stochastic hydrology would treat hydrologic topics of random nature by emphasizing the properties of these phenomena rather than the techniques used; this text emphasizes the techniques. This book is a self-contained text on various ways of applying probability theory and mathematical statistics to hydrology, provided the reader accepts certain statements and mathematical expressions.

Probability and Statistics in Hydrology is aimed at three audiences: (1) practicing hydrologists, water resources engineers and other specialists dealing with certain facets of hydrology, in short, for professionals in the field; (2) graduate students and participants of various post-graduate training courses, in short, for those who are entering or specializing in the field of hydrology; and (3) specialists others than hydrologists who are interested in probability and statistics as applied to hydrology, in short, for professionals in fields related to hydrology. Statisticians who want to increase their understanding of the ways in which probability theory and mathematical inferential statistics can be applied to hydrology may be also interested in this text.

Probability and Statistics in Hydrology is an outgrowth of notes prepared for a graduate course, *Statistics in Hydrology and Hydraulics*, developed and taught by the writer for the past decade in the Department of Civil Engineering of Colorado State University at Fort Collins. Through the years the course has emphasized the association between probability theory and mathematical statistics and hydrology, with hydrology considered both as an earth science and as an engineering and applied discipline. In this book this association is exemplified by numerous examples. At the end of each chapter some problems to discuss and solve are given for the reader to use for checking his understanding. Problems that require large numerical solutions should be calculated on a digital computer. Limited numbers of references are listed at the end of the book for each chapter. A list of symbols was not considered necessary because the symbols used are defined in the text whenever it was thought the reader may be in doubt of their meanings.

This book is related only to the application of probability theory and mathematical statistics to hydrology; the application of stochastic processes, in general, and time series analysis, in particular, to hydrology are the subjects of another book, *Stochastic Processes in Hydrology*, which is published simultaneously.

Chapter 1 of *Probability and Statistics in Hydrology* is an introduction to the character of hydrologic random phenomena. Chapters 2 and 3 represent a condensation of probability theory as viewed from hydrologic application. Chapter 4 is an introduction to the application of mathematical inferential statistics to hydrology. Chapters 5, 6, and 7 refer to frequency distributions, descriptor parameters, and probability functions. Chapters 8, 9, and 10 are related to estimation techniques,sampling theory and the testing of hypotheses. Chapter 11 treats correlation and regression, and Chapter 12 is a brief summary of the multivariate analysis looked at from hydrologic applicability.

For the last decade the writer has been involved in active basic and applied research in the application of probability theory, mathematical statistics and stochastic processes in hydrology and water resources, under grants awarded for research by the U. S. National Science Foundation. The NSF support is gratefully acknowledged. Most of the results of this research and the research of graduate students under the guidance of the writer, have been published in the Hydrology Papers of Colorado State University, presented at various national or international conferences, or published in professional journals. The writer also gratefully acknowledges the benefits of a conducive environment for teaching and research at Colorado State University.

Feedback the writer obtained from graduate students attending his classes during the 1969/70 and 1970/71 academic years regarding the distributed class notes that this text is based on was exceptionally helpful for the final shaping of this text and is acknoweldged. Mr. Leo R. Beard, Director of the Hydrologic Engineering Center, U. S. Corps of Engineers, Davis, California; Dr. Chester C. Kisiel, Professor of Hydrology and Water Resources, University of Arizona, Tucson, Arizona; Dr. Ulrich Maniak, Privatdozent, Technical University Braunschweig, Federal Republic of Germany, and Dr. Subin Pinkayan, Assistant Professor, Asian Institute of Technology, Bangkok, Thailand, have either reviewed this manuscript in general or read it in detail. Their suggestions have been very beneficial.The writer expresses his appreciation for these suggestions, even though some of them could not be incorporated in the text. The writer also acknowledges Mr. Stan Cortner, technical editor at the Engineering Research Center of Colorado State University, for his assistance in hammering the text into, hopefully, a readable form. The last but not the least acknowledgement goes to the writer's family for their endurance.

Vujica Yevjevich
Professor of Civil Engineering
and Professor-in-Charge of
Hydrology and Water Resources Program
Colorado State University

May 1971
Fort Collins, Colorado

TABLE OF CONTENTS

Chapter

CHAPTER 1

CHARACTERISTICS OF HYDROLOGIC PHENOMENA

An introductory chapter to a book on the application of probability theory and mathematical statistics to a natural science such as hydrology should describe the characteristic phenomena and the basic approaches to their analysis. Therefore, this chapter contains a brief review of the types of environments and processes in hydrology, data, phenomena, variables, parameters of these processes, cause-effect relations, and a brief discussion of the nature of water requirements, deliveries,and consumption.

1.1. Deterministic and Stochastic Processes of Hydrology.

Definition of hydrology. Hydrology is an earth science concerned with the distribution of water quantities and water qualities in time and space. It mainly investigates water in liquid, gaseous and solid states over continental land areas. When hydrologic processes and environments are studied by theoretical methods, the study may be called theoretical hydrology. When hydrologic principles are used, it is applied hydrology. Development of hydrology has a long history [1, 2]*.

Hydrologic environments. Those parts of the earth in which particular hydrologic phenomena of the water cycle occur are defined as hydrologic environments. Oceans and seas are primary water suppliers in the water cycle and receivers of continental water outflows, so they may be considered as hydrologic environments. Air, in the sense that it is the receiver of evaporated water, the carrier of water between areas of the earth, and the environment of water condensation and precipitation, is also a hydrologic environment. Continental solid and liquid surfaces as the receivers of precipitation are important hydrologic environments. Many other hydrologic phenomena occur on land surfaces, such as evaporation, infiltration, water retention, snow and ice accumulation and melting, and surface runoff. The upper layer of the earth's crust is the

* The symbol [ ] designates the reference number and sometimes the pages of references given at the end of the book for each chapter separately.

hydrologic environment where water retention occurs. In this environment, ground water accumulates and flows either in shallow or in deep pervious geologic formations. Lakes, swamps, artificial ponds and reservoirs, and streams represent hydrologic environments of water accumulation, evaporation, infiltration, and flow. Soils and plants are also environments in the water cycle with relatively complex hydrologic processes. These various environments affect the water quality in their particular ways. Each hydrologic environment may be considered as a hydrologic system with particular properties.

Of particular interest are water interchanges across various interfaces among hydrologic environments, such as the water flux at the ocean--atmosphere interface, between the air and solid and/or liquid continental areas, across the interface of continental bodies of water and underlying solid surfaces, and similar interfaces.

Example 1.1. Studying oceans and seas as the main source of continental precipitation is relevant for eventually predicting moisture supply over continental areas. Presently, this type of predicting water supply seems to be the only hope for a prediction longer than a couple of days, say for two or more weeks or longer. Studying the atmosphere is relevant for short-range predictions of precipitation and evaporation, say for several days, as well as for weather modification either as increases of some types of precipitation or as decreases or suppressions of other types of precipitation and evaporation, and other similar problems.

Hydrologic processes. Any hydrologic environment may be generalized as to consist of water inputs, environmental responses and water storage to these inputs, and water outputs. Usually output of an environment represents input into another environment. This trinity, input-response (storage)-output, may be described as the basic parts of a hydrologic system. Each of these parts represents a hydrologic process. The environmental storage of water quantity or quality variables is also a hydrologic process as a function of time and space.

Natural hydrologic processes are either stochastic or combined deterministic-stochastic processes. Pure deterministic processes with a negligible stochastic variation can be realized only under controlled, artificial conditions. Generally, they are a combination of various deterministic and stochastic processes.

Deterministic processes of hydrology. Deterministic processes of hydrology are the results of physical, chemical, and biological deterministic laws, but primarily the result of fluid mechanics and termodynamics laws and regularities. These processes relate to those hydrologic phenomena for which a limited number of variables may be uniquely related by functional expressions or experimental curves.

Example 1.2. The response of a well defined impervious surface to a specified input of rainfall in the form of a runoff hydrograph, with negligible evaporation and infiltration losses, is the most typical example of a deterministic process in hydrology. Under natural conditions these responses are affected by many random factors.

Example 1.3. A rating function of a unique discharge-stage relation at a river cross section with a fixed bed, including the two-stage case of unsteady flow, is another example of a deterministic relation. For rivers with movable beds, many random factors influence these relations.

Stochastic processes of hydrology. The majority of hydrologic phenomena in nature are stochastic processes, or processes governed by laws of chance. Examples of these processes are phenomena such as precipitation, evaporation, runoff, ground-water levels, sediment transport, lake levels, snow and ice accumulation and melt, water quality properties, properties of porous environments, river basin geomorphologic forms, etc. The outcomes of these phenomena are governed by laws of chance and the sequences of various variables that describe these random phenomena are stochastic processes.

It is important to stress that the terms, *chance, random, probabilistic* and *stochastic* are considered synonyms in this text; they all refer to phenomena subject to the laws of chance.

Strictly speaking, there are no pure deterministic hydrologic processes in nature. They are either a combination of deterministic processes of the periodic or transient phenomena, and of stochastic processes, or they are predominantly stochastic processes. Therefore, hydrology cannot be fully understood, described, and best applied to water resources conservation, development and control without the extensive use of methods of probability theory.

Example 1.4. The annual precipitation, at a point or over a defined area, the annual river runoff at a gauging station, and similar hydrologic processes of annual values are stochastic processes, provided there are no trends created by systematic errors or by man-made changes in nature as deterministic components of these processes.

Example 1.5. The monthly, daily or continuous sequences of precipitaton, runoff, evaporation, or sediment transport phenomena are considered as a combination of deterministic and stochastic processes. Usually, the periodicity of the day or the year in these sequences are deterministic processes, while the chance fluctuations about these periodic components are stochastic processs, thus comprising periodic-stochastic processes.

Joint application of deterministic and stochastic methods in hydrology. The world's leading hydrologists maintain that a simultaneous use of both deterministic and stochastic (or statistical) methods of analysis and description of hydrologic processes in nature is necessary for producing the best scientific and practical information for hydrology, in general, and for water resource development, conservation and control, in particular.

Treatment of data for the objectives of finding regularities in hydrology may be either by a deterministic approach, by a statistical approach, or by both simultaneously. The use of one approach independently of the other may lead to an analysis of data without a sound theoretical background or to a theoretical treatment without the proper verification of hypotheses and of discovered regularities. Future progress of hydrology may depend to a large extent on how these two approaches are combined for discovering, understanding, and generalizing hydrologic regularities of nature.

Often, if a deterministic relationship function or curve is developed between random variables, the process is mistaken as deterministic. This relation is a function of random variables. As it is shown in Chapter 2, the functions of random variables are also random variables, and their processes must be stochastic.

Example 1.6. A runoff to rainfall relationship is best understood by a simultaneous investigation of the deterministic river basin responses and of the statistical properties and relations of observed data on rainfall, runoff, climate, river basin characteristics, and random factors in these responses.

Example 1.7. To discover man-made non-homogeneity in hydrologic series in the form of trends or jumps in the mean or in any other hydrologic parameter, only an investigation of physical factors, that create the non-homogeneity, simultaneously with the use of statisitcal methods of identifying and testing it, can give completely reliable information about the existence, type and amount of non-homogeneity.

Example 1.8. To understand groundwater properties well, the analysis of physical and deterministic hydraulic properties of the field response to any withdrawal or recharge of water must be combined with the investigation of stochastic character of natural water recharge and outflow, and of stochastic properties of permeability and dimensions of a ground-water environment.

1.2. Probability, Determinism and Hydrology.

Applicability of probability theory in hydrology. Probability theory is that branch of mathematics that investigates random phenomena. It is part of a broader chapter of mathematics, the theory of measures. Because hydrology is basically a collection of phenomena related to distributions in time and space of random variables of water quantity and water quality, probability theory is the main branch of mathematics applicable to hydrologic phenomena.

Two offshoots of probability theory are mathematical statistics and stochastic processes. The first deals with observed samples and makes inferences about the phenomena with all its possible realizations, or about its population. The second deals with phenomena in the sense of sequential stochastic connections. This trinity, probability theory — mathematical statistics — stochastic processes, gives basic mathematical techniques for analyzing and describing hydrologic random variables and processes. These techniques are usually encompassed by practitioners into a unique concept of *statistics applied to hydrology*.

Determinism and probability. Outcomes in observation of a random variable are affected by an extremely large number of causative factors. Determinists in natural sciences sometimes contend that the search for deterministic relations will include with time more and more causative factors to explain effects on a random variable. Following this reasoning it could be expected with time to continuously extend the deterministic explanation and description of present-day hydrologic random phenomena and substantially restrict the application of probability theory in hydrology. This reasoning has a fundamental weakness. A finite number of causative factors taken into any deterministic equation could rarely exhaust all causative factors involved in the outcome of a hydrologic random phenomenon. If the causative factors are also random variables, which they are more often than not in hydrology, these cause-effect deterministic relationships, as stated previously, are nothing else than functions of random variables.

Though it can be expected for deterministic explanations of a cause-effect relationship to undoubtedly increase with time as research progresses and investigation techniques develop, two factors will limit this progress, the economy of analysis, and the large number of causative factors involved. The first limitation means that one will rarely expect to write equations with hundreds of variables involved, and one will rarely expect to measure all these variables in order to predict their effect. The second limitation means that any deterministic approach could not account for or exhaust all causative factors. As the discussion of types of relations among hydrologic variables in the next subchapter shows, deterministic relations may eventually reduce, in some particular cases, the level of randomness but they can never eliminate it completely. In other cases even this reduction becomes impossible.

Example 1.9. The variable of annual precipitation is close to being an independent stochastic process in time when regional studies are made with a large number of stations, [3]. It is unlikely that man will be able, at least in the foreseeable future, to develop a deterministic relation between the annual precipitation and the myriads of causative factors which affect annual precipitation in every possible way. Like the study of variables affecting deviations in shooting at and hitting a target, it is easy to show that hundreds of causative factors, most of them also random variables, affect the outcomes of annual precipitation amount.

1.3. Hydrologic Data.

Types of data. To investigate properties of random variables related to distributions of water quantity and water quality in time and space, four types of data are available.

(a) The first type is historic or chronologic data, or observations of processes in time, with the resulting continuous or discrete hydrologic time series. If not observed initially, this information is lost. A majority of present-day hydrologic data belong to this type.

(b) The second type is field data observations along lines, or surveys of hydrologic phenomena across areas or in space, like the determination of depths of groundwater porous strata, determination of sediment characteristics along a riverbed and similar surveys. Field data can be resurveyed when necessary; economic factors are usually the limiting condition. A substantial part of hydrologic data may be of this type.

(c) The third type is laboratory and field experimental data related to hydrology acquired by methods similar to data obtained in hydraulic experiments. This type of hydrologic data is mainly used in basic and applied research.

(d) The fourth type is the simultaneous measurements of two or more random variables in order to establish a relationship among these variables, mainly for the purpose of transferring statistical information among variables.

Quality of data. A distinction is often useful and necessary between a true, virgin, and observed value of any hydrologic variable. The true value of any observation is never known because the data obtained, through inevitable errors in observation, are not exact values. The virgin value is the value produced by unchanged conditions of an environment; data are not influenced by either unpredictable natural or man-made significant changes in hydrologic environments. The observed value is available as the result of various surveys, recordings, or experiments; this value is usually published by hydrologic services.

Observed data have errors in each of the four phases of sensing, transmitting, recording, and processing data. Computed values always contain errors that are either random or systematic. In this latter case the data are said to be inconsistent. Changes of many hydrologic environments occur intermittently by accident, by slow evolution in nature, or by various human activities. The data affected by these environmental changes become non-homogeneous data. In this case no difference is made between treating inconsistent and non-homogeneous data.

The three concepts of random errors, systematic errors (inconsistency, bias) and non-homogeneity are necessary and useful in practical hydrologic applications. Random errors are always in data. It is often either assumed or proven by repeated experiments that these random errors are approximately symmetrically distributed about the true value. In most cases it is supposed also that they follow a normal or Gaussian distribution. Standard deviation then is used as a measure of the magnitude of these errors. A typical random error may be the measurment errors of a hydrologic quantity. The order of magnitude of random errors should be determined for each measuring technique.

Inconsistency is defined as systematic errors in hydrologic data. Inconsistency is the difference between the observed values, with the inherent systematic error, and the true values. A typical case is a systematic positive error in observations for some time, and then by a change in observational technique, a systematic negative error, or values close to true values are produced. Another case is the effect of a change in environment around a precipitation gauge. Changes in time, such as a tree growing slowly around a gauge, influence rain measurements so that the rain gauged may show a systematic trend downward or upward that does not exist in the true rainfall. Various techniques exist for determining the type and the amount of inconsistency.

Non-homogeneity is the difference between the true values and the virgin values of a hydrologic variable. It is sometimes also called non--stationarity for the case of a time series. After a river is dammed, the variance of river flows below the dam may show a jump at the time it becomes operational. A large forest fire may produce sudden changes in several hydrologic parameters. An evaluation of the effects of a forest fire represents a slow natural change of the river basin environment, or a weak trend in many hydrologic parameters.

The analysis of hydrologic errors and changes by detecting magnitudes of random errors, inconsistency, and non-homogeneity, and their effects on time series properties [4] is necessary as soon as one requires reliability in hydrologic statistical information. The time sequence of hydrologic variables with regularities or irregularities involved is of great practical interest. Any projection into the future based on past properties of hydrologic variables requires that the conclusions are derived from data free of significant inconsistency and non-homogeneity, and with tolerable random errors.

Example 1.10. Figure 1.1 represents the time series of annual runoff of the Nile River at Aswan Dam for the period 1869—1955 [5], with V_i any annual value. Around 1899—1903 a significant negative jump of 29 percent of the general mean runoff is indicated. Probability is less than 0.01 percent that this jump resulted from sampling fluctuations. Four combined factors are explanations for this sudden change: (a) natural and sampling stochastic variation; (b) inconsistency in data brought about by the use of a rating curve after 1903 but applied to river gauging stages for the preceding and transitional period during the construction of the Aswan Dam; (c) non-homogeneity in data produced by the Aswan reservoir after 1903 in the form of increased evaporation and water percolation into banks, and (d) non-homogeneity in data produced by an increase in the upstream water consumption.

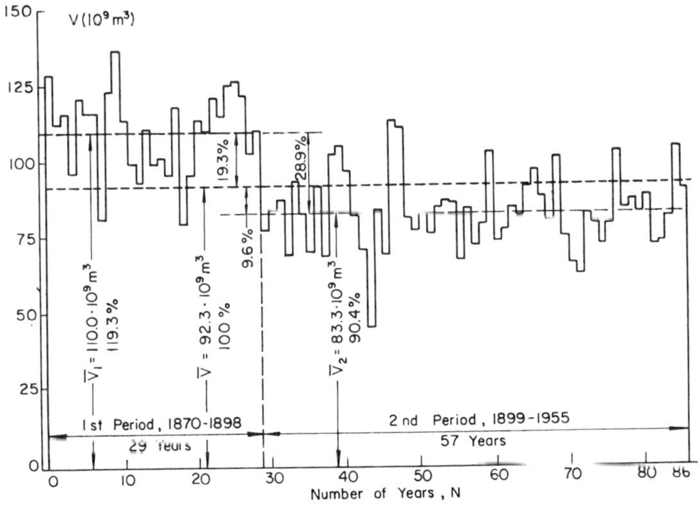

Figure 1.1. Fluctuations of annual flow of the River Nile as an example of inconsistency in data in time series.

Example 1.11. Figure 1.2 represents an example of non-homogeneity of maximum annual flood peaks of the Danube River at Orshova, Romania, for a period of 110 years. The significant upward trend can be explained by the continuous construction of levees for flood protection along the channels of the Danube and its major tributaries during this period. The use of this sample for the prediction of future events would underestimate floods for given probabilities of exceedence, without removing this non-homogeneity.

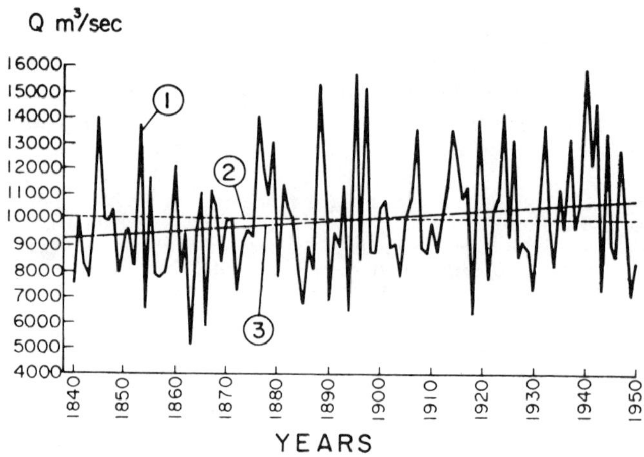

Figure 1.2. Series of the maximum annual flood discharge of the Danube River at Orshova in Romania, 1840 – 1950, with an upward trend: (1) series of peaks; (2) average flood peak, and (3) trend of the average flood.

1.4. Hydrologic Magnitudes and Cause-Effect Relations.

Hydrologic phenomena, variables and parameters. A distinction is made in this text between the term *hydrologic phenomena* such as precipitation, evaporation, runoff, sediment transport, etc.; *hydrologic variables* that describe these phenomena, such variables as instantaneous discharge, daily or monthly river flows, annual river runoff, for the case of runoff; and *hydrologic statistical parameters* or statistics that describe the probability distributions and stochastic processes of these hydrologic variables, such parameters as the mean, median, variance, coefficient of variation, skewness coefficient, serial correlation coefficients, and others.

Example 1.12. Precipitation is a general hydrologic phenomenon measured and recorded at various gauging stations. The precipitation intensity computed from the rainfall and snowfall recordings is the basic variable of precipitation. Its continuous sequence in time, including zero values, is a periodic-stochastic process. The hourly, daily, monthly, annual precipitation, the maximal values of each day of the year, and similar magnitudes are also precipitation variables. In describing these variables, by probability distribution functions or by various other descriptions, numbers in the form of statistical parameters of mathematical models are used.

Basic and derived variables. A distinction is also made in this text between a basic and a derived hydrologic variable. The instantaneous river discharge, the rainfall intensity, the instantaneous sediment discharge and so forth are conceived as basic hydrologic variables. The daily, monthly, seasonal, annual and similar values, the peak values, the lowest values, the total highest or lowest water volumes of a time unit in a given time interval, and similar, are variables derived from the basic variable. One of the fundamental objectives of modern application of probability theory, mathematical statistics, and stochastic processes in hydrology is the development or the use of methods and practical procedures of determining the characteristics of derived random variables from the characteristics of basic random variables. Basic hydrologic variables are usually given as continuous sequences, including zero values, while derived variables are primarily given in the form of discrete sequences.

Numerical-valued and ranking variables. Hydrologic variables are mainly numerical-valued or number-oriented magnitudes. Sometimes they may be variables with ranks, given as 1, 2, 3, 4, ..., n; or A, B, C, ..., and similar. The tendency in modern hydrology is toward using only numerical-valued variables.

Types of cause-effect relations among hydrologic variables. The relation of a hydrologic variable x_1 to many other variables, x_2, x_3, ..., may be either deterministic, usually represented by the relation of a finite number of functionally or experimentally related variables, or stochastic, with relations that involve small, large or infinite numbers of variables.

In the three graphs of Fig. 1.3, variables x_2, x_3, ..., are sorted according to the magnitude of their effect on variable x_1. In this case, the variance S_i^2 is used as a measure of these individual influences, which variance is the contribution of each variable x_2, x_3, ..., to explain the total variation of the variable x_1, with ΣS_i^2, $i = 2, 3, \ldots, \infty$, being equal to the variance of x_1.

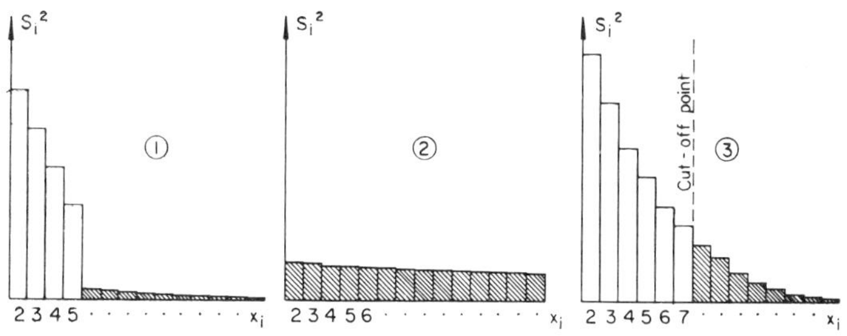

Figure 1.3. Basic effects of a set of variables x_i on the variable x_1, measured by S_i^2, the contribution to the explained variance of x_1 : (1) scheme of a functional relationship; (2) scheme of no functional relationship, with x_1 an independent random variable; and (3) scheme of a correlative relationship.

The first graph of Fig. 1.3 shows four deterministic variables that mainly influence the outcome of x_1. Effects of all other causative factors, if they exist, are assumed to be negligible. This case is considered as a deterministic relation between five variables. To find an equation that relates x_1 to x_2, ..., x_5, either analytical or experimental methods may be used. This deterministic case is rare in hydrology under natural conditions because either many factors of small individual effects are neglected, or measurement of variables $x_1, \ldots, x_5$ produce errors $e_1, \ldots, e_5$. These errors are of random nature and have similar effects as random causative factors. It is often difficult to distinguish in hydrology whether the small variation of x_1, unexplained by variables x_2 through x_5, is produced by the neglected small effects of variables $x_6, x_7, \ldots$, or by random measurement errors in variables x_1 through x_5. Usually both sources are responsible for this unexplained variation.

The second graph of Fig. 1.3 is the case for which observations of the variable x_1 are affected by a large or even infinite number of other variables. Usually, each of them has a small approximately equal but additive and independent effect on x_1. In this case, x_1 may be considered as independent or nearly independent of the variables $x_2, \ldots, x_\infty$ in the classical sense of functional or correlative dependence among variables. If the above conditions of $x_2, \ldots, x_\infty$ are satisfied, x_1 is a normally distributed variable or very close to it. This is the case where no practical relation exists among variables investigated, though the cause-effect association is present. It is the limiting case when the number of causative factors goes to infinity while the effect on x_1 of each of them individually goes to zero. This is another extreme or idealized case in hydrology, opposite to the case of the first graph for a pure functional relation. Strictly speaking, these two extreme and limiting cases of relations among variables are rare in hydrology, except in an approximate way. In this latter case of x_1, its relation to a multitude of causative factors is neglected and its properties are investigated by the methods of probability theory, mathematical statistics and stochastic processes.

The third graph of Fig. 1.3 represents the concept of typical hydrologic cause-effect relations. Some variables affect x_1 significantly. Practical considerations limit the relation equation to a cut-off point, with a finite number of variables. The association between selected variables is now in the form of multiple regression or multivariate equations. A normal occurrence in hydrology is that variables $x_2, x_3, \ldots$, are also dependent among themselves, besides being related to x_1. The basic dilemma is the selection of a cut-off point in the third graph. It is usually a compromise between the accuracy of a relation, the economy of computation, and the lack of statistical and/or physical justification for the introduction into the equations of variables beyond the cut-off point.

Example 1.13. A typical example of x_1 of graph (3) in Fig. 1.3 is river runoff related to various groups of causative factors, which are interdependent variables, with x_2 symbolizing the group of precipitation variables, x_3 of evaporation and other climatic variables, x_4 of antecedent moisture conditions, x_5 of river basin geometry, x_6 of soils and geology, x_7 of vegetation, and similar groups of interdependent variables. It is possible to list a large number of these variables affecting the runoff-rainfall relation. However, practical equations can often predict runoff, with sufficient accuracy, by using 5 to 10 variables similar to the above. All others, being of smaller individual effects, may usually be neglected in practice, together with the errors in the selected variables.

1.5. Stochastic Nature of Water Demand.

Randomness in water requirements. One is easily convinced, because of the multitude of presently available observations, that hydrologic phenomena are of random nature. It is less obvious to a practicing hydrologist or engineer to consider water demands, water deliveries, and/or water consumptions, as random variables. The engineer that has predicted the water demand of a city, power requirements of a region, or any water demand for other purposes, notices substantial differences between past estimates and water consumption at the predicted times. Future water demand should not be computed deterministically by unique curves as a function of time, but should be estimated as random variables. The predicted mean demand as a function of time should be assessed and supplemented by expected variations, measured by the variance or other parameters of dispersion.

Water consumption of a region or in a river basin has both similar and different characteristics in comparison to the water resources available in nature. In both cases, random variations are superposed on seasonal or periodic fluctuations. The differences are in frequencies, amplitudes, and phases of these periodic movements, as well as in the relative importance and character of random components. Differences are also in other deterministic components of time series. On the one side, mean water requirements per unit time interval usually increase, but may also decrease with time by simple or complex trends. These trends are functions of numerous social factors, such as population increase or decrease, changes in standard of living, various economic and social developments, technological innovations, water prices, etc. On the other side, the continuous development of water resources and various changes in nature create non-homogeneity in time series of the available water resources. As the water demand increases with time and the available water is approximately a stationary variable or time-invariant periodic-stochastic process, sooner or later the demand exceeds the supply. In matching demand and supply, the stochastic nature of both has significant effects on planning the development and use of water resources.

Example 1.14. Figure 1.4 presents a seven-day water deliveries in the city of Denver for a period of ten years. Three factors are included in this delivery: the consumed water by evaporation and evapotranspiration, the return flow through the sewage system, and the return flow through the ground water. The upward trend in this water delivery is weak because it represents only an increased water demand for an approximately fixed central part of the large Denver metropolitan area. The main feature of the time series of Fig. 1.4 is the superposition of a random variation on the basic cyclicity of the year. Figure 1.5 presents the monthly series of water deliveries in the city of Dallas, Texas, with an upward trend, cyclicity of the year, and random variations.

Example 1.15. Water consumption of an irrigated area is composed of within-the--year cyclicity and a random component superposed on this periodic movement. The chance aspects of irrigation water consumption result from the randomness in precipitation, evaporation, heat supply, and similar climatic random factors of irrigated areas, which are usually superposed on the seasonal variations. This random component in irrigation water consumption may or may not be related to random components in the water supply of rivers or ground-water acquifers.

Planning of water resource developments. The water resource specialist is required to plan structures and measures in water resource

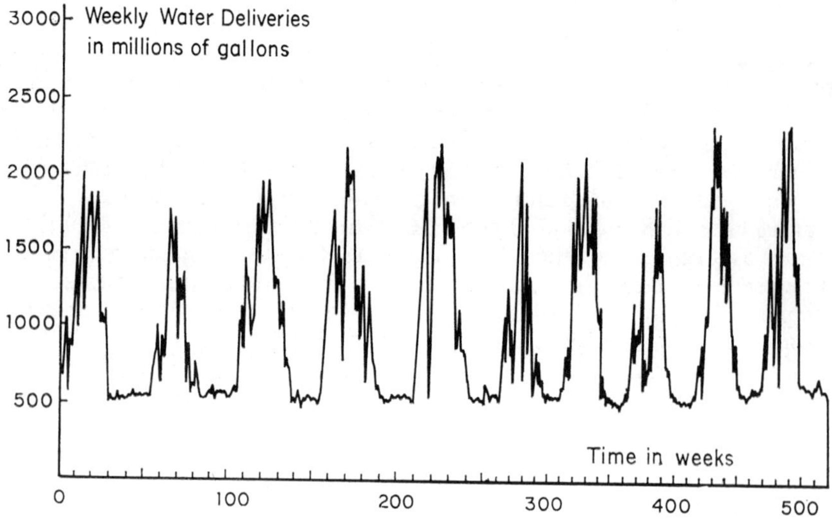

Figure 1.4. Weekly water deliveries in the city of Denver, Colorado for the period of ten years, as a periodic-stochastic process with a mild upward trend.

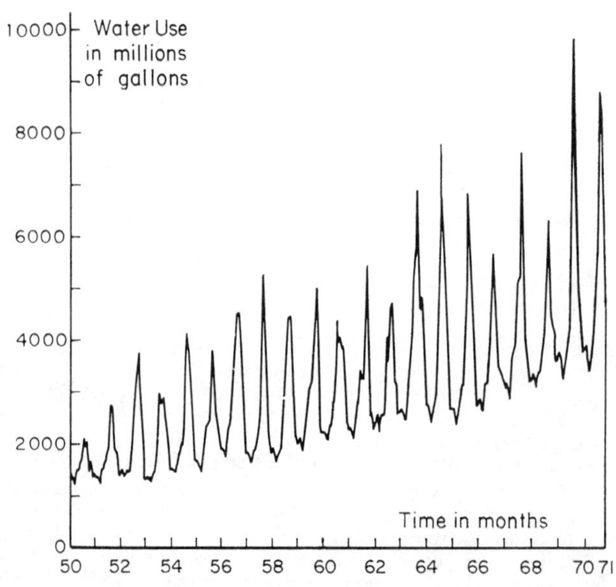

Figure 1.5. Monthly water supply to the city of Dallas, Texas for the period 1950–1971, as a periodic-stochastic process with a strong upward trend.

conservation, development and control by facing both the stochastic character of hydrologic natural and man-influenced processes and the stochastic nature of water requirements. Therefore, the modern planner has the task of optimizing economic, social, and engineering values between the two stochastic processes, which may not be independent, one continuously changing around the trend of water demand and the other approximately stationary in the average water yield. The application of probability theory, mathematical statistics, and stochastic processes in hydrology, in the analysis of water requirements, and in the planning of water resource developments, provides the necessary techniques for the analysis and description of random variables involved.

Example 1.16. A typical example of random components in input (water supply) and output (water deliveries) is the design of surface water storage facilities. The input is the periodic-stochastic process of river inflows and the similar process of sediment inflows; the output includes the periodic-stochastic process of evaporation from water surface, and a similar process of seepage losses, but the main output is the water outflow for various users and forced spillover or releases; outflows are often composed of periodic water requirements superposed by many chance variations in this demand. Each of these five processes, water inflow, sediment inflow, evaporation, seepage, and water deliveries, has a stochastic (chance) component which may or may not be mutually indepedent.

1.6. **Problems for Discussion and Solutions**

1. Give three typical random phenomena and for each of them give three descriptive variables from atmospheric hydrology, surface-water hydrology, ground-water hydrology, and water-quality hydrology.

2. Give three examples of sampling (observing, recording, measuring, surveying) hydrologic random variables in various dimensions: in time, along a line, over an area, in space (random vectors), in both time and space, and in more than four dimensions.

3. Give three typical examples of inconsistency (systematic errors) and three examples of non-homogeneity in hydrologic data.

4. List several hydrologic processes that are approximately deterministic processes.

5. List several hydrologic stochastic processes that have small or negligible deterministic components.

6. Discuss river basin response to rainfall input as a predominantly deterministic process with non-negligible stochastic components. What are the sources of stochastic components in river basin response? Why is the unit hydrograph of a river basin not a constant with time?

7. Give several examples where the joint application of deterministic and stochastic methods may best solve hydrologic problems.

8. Discuss the way the stochastic component in water demand is solved in practice by increasing or slowing the pace of constructing water resources projects, or by transferring the randomness of the water demand into the randomness of the allocation of investments.

9. Represent schematically the water supply and water demand in a region as functions of time, as two deterministic-stochastic processes. How these two processes may be mutually dependent and through which factors?

CHAPTER 2

RANDOM VARIABLES AND THEIR DISTRIBUTIONS

Each science with random phenomena has some particular characteristics that require proper application of theorems and formulae of probability theory. A brief review of approaches to probability theory and their implications on the analyses of hydrologic random phenomena are followed by discussions of random events, random variables, and general definitions of probability distributions (univariate, bivariate, multivariate, and conditional). The discussion of functions of random variables concludes the chapter.

2.1. Probability and Random Phenomena of Hydrology.

Definitions. Probability theory deals with events and variables of random phenomena. Any measurement of a random variable, regardless of the variable value obtained, is called an observation, a realization or an outcome. These three terms are synonymous in this text. The random event is the variable value obtained in a measurement.

The effect of each chance factor of a combination of causative factors on a random variable changes from one observation to the next, so that the variable value of a particular outcome cannot be predicted uniquely. Though it appears that no regularity exists in realizations of a random variable, analysis of a large number of observations does reveal regularities which are probability governed characteristics. The collection of characteristics of random variables of a random phenomenon represents its description.

Example 2.1. Peak discharge of the maximum annual flood is a random variable of the random phenomenom of river runoff. This discharge is controlled by a large number of causative factors which combine in various ways each year during flood seasons to produce different peak discharges from year to year. Statistical properties of flood peaks are estimated from past observations with the accuracy of these peaks depending on available data.

Example 2.2. Instantaneous pressure and velocity vector at a point in a turbulent stream are random variables of the random phenomenon of turbulence. Numerous causative factors, which change constantly, affect the outcomes of pressure and velocity vector at given points and at given instants. It is impossible to predict the individual values of pressure and velocity that will occur at a point and at a given future time. All observations taken together, however, produce some regular stochastic properties of the phenomenon of turbulence.

Example 2.3. If a day is considered rainy when there is precipitation, then outcomes as rainy and nonrainy day represent the two events. In this case only two discrete random events exist.

Example 2.4. The level of a natural lake is a random variable that theoretically has an infinite number of possible random events. This variable is continuous. One cannot find an upper practical limit for the maximum level, while the lowest level is predetermined by the elevation of the lake at zero outflow. This is a random continuous variable bounded from below. Limitations in the accuracy of level observations, however, restrict the number of possible events to readings of say each centimeter at a simple gauge. Every number expressed in hundreds of a meter in the limits of the experienced lower and upper lake elevations is an event. An infinite set of possible events of a continuous variable is replaced, in practice, by a finite set of discrete random events.

Regularities in random phenomena. Although the exact outcome in examples 2.3 and 2.4 cannot be predicted accurately, a particular regularity exists in a series of observations. If the rainfall is observed for N days, and if m is the number of rainy days, with $m < N$, the fraction m/N is the relative frequency of occurrence of rainy days. The nonrainy days have the relative frequency $(N - m)/N$. If N increases, the ratio m/N converges to a constant that is equal to or greater than zero and equal to or smaller than unity, provided no basic change of precipitation phenomon occurs in nature.

Similar results are obtained for each event in the case of a large number of discrete random events. If the total time of continuous observations of lake level is T, and the total time for which the levels occurred in the range H_1 to H_2 is ΔT, with $\Delta T < T$, then the ratio $\Delta T/T$ is the relative frequency of the lake level for that range. The ratio $\Delta T/T$ converges to a constant with an increase of T, provided no basic change occurs either in the climate, the properties of inflows, or in the lake and outlet configurations during these observations.

If the causative factors and their individual impacts on precipitation or the lake level remain the same with time, a necessary condition in this case, and if N or T increase to infinity, then their ratios m/N and $\Delta T/T$ converge to constant values. These constants are conceived as probabilities of the corresponding random events.

It is difficult to test by limit theorems how rapidly the frequencies $f = m/N$ or $f = \Delta T/T$ converge to the corresponding probabilities with an increase of N or T because hydrologic sequences are usually neither independent series nor simple dependent processes. However, in some turbulent phenomena of fluid mechanics, in various field surveys, or in sediment investigations in laboratory flumes, with a rapid variation of their random variables, it is possible to accumulate relatively long records. Observing these variables for several hours or days may produce close estimates of the probabilities by the computation of the corresponding relative frequencies.

Examle 2.5. The upper graph of Fig. 2.1 gives the annual number of rainy days while the lower graph represents the relative frequency of rainy days at the Fort Collins, Colorado, precipitation station for the period of 76 years, 1887 through 1962, as the length of observations increases. This frequency converges approximately to 0.21 for rainy days, and to 0.79 for nonrainy days. Frequencies may be close to probabilities of rainy and nonrainy days, provided climate, measuring techniques, and the station location and surroundings do not change during the period of observation. These conditions of stationarity seem to be satisfied for annual values of rainy days of Fig. 2.1, except at the beginning. A hypothesis may be postulated and tested that the series of annual rainfall days has an upward trend. Dailly precipitation with less than 0.01 inches (0.25 mm) was considered initially only as a trace equated with a nonrainy day. For a more accurate measurement of rainfall during the second half of the observational period, the trend may have been produced by fewer recorded rainy days in the first half of the period. A trend of the above type, apart from systematic errors in observations, may also be a result of sampling variation.

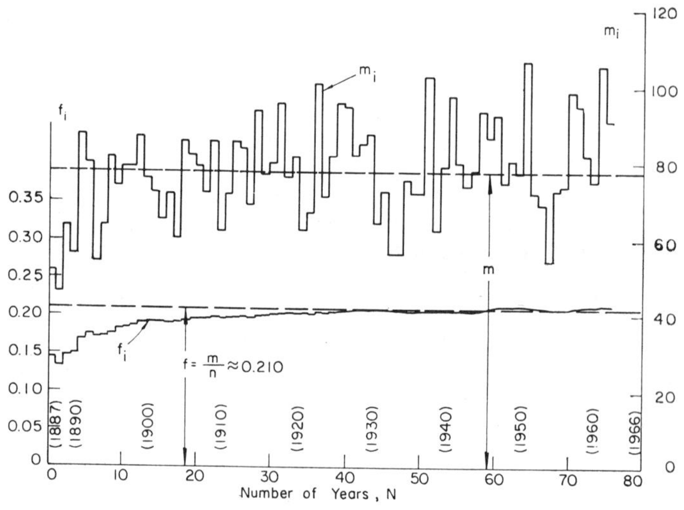

Figure 2.1. Number m_i of rainy days per year (upper graph) for the Fort Collins, Colorado precipitation gauging station (1887—1962), and the relative frequency f_i of rainy days (lower graph), and its change with the length of observation.

Determining probabilities of elementary random events. Probability theory is synonymous with probability calculus, which deals with random events in the sense that probabilities of new and usually more complex or composite random events are computed from probabilities of simple or elementary events. In examples 2.3 and 2.5 the rainy and nonrainy days are elementary events. A composite event is, for example, the occurrence of five consecutive years with the number of rainy days less than the average annual number. The probability of this event can be computed from probabilities of elementary events.

Current probability theory is based on the axiom that probabilities of elementary random events are known. How they are obtained is not usually the concern of probability theory. Probabilities of elementary random events are derived either by logical derivations or by experimental or observational methods.

The method of logical derivation is both the classical and the oldest approach of determining probabilities. Probability is defined as the ratio of favorable events to the number of total events, $p = m/N$, with $m \leq N$. If a die is perfectly cast and of a homogeneous substance, the probability of each face coming up in every trial is 1/6, where $m = 1$ and $N = 6$. In a perfect set of 52 cards, the probability of drawing each particular card is 1/52, with $m = 1$ and $N = 52$. For a perfect thin coin, the probability of tossing a head or tail in each throw is 1/2. These are logical derivations that can be easily supported by a large number of trials. Relative frequencies of these random events converge under the most ideal physical conditions of chance games to the above probabilities logically derived. This approach, however, cannot be usually used in hydrology because these ideal conditions do not exist.

The second approach in obtaining probabilities is by experimentation or observation. The exact probabilities are never known, but are approximated by the computed relative frequencies. Nearly all probabilities of hydrologic random events are determined in this way.

Most mathematicians accept the concept of probability based on set theory. A set of elementary random events is defined, and to each of them a number between zero and unity is attached as the probability of occurrence. It is not essential how these numbers are obtained as long as they specify probabilities of elementary random events in a set, and they satisfy some well defined basic conditions.

Example 2.6. The equiprobable (equally likely) events best applied in hydrology are some geometric probabilities. If the probability of hail or intensive rain is the same at any place over a large area A, then any smaller area A_0 to be hit in this region by a storm has the probability $p = A_0/A$.

Example 2.7. If a die is biased (not a perfect cube or not of a homogeneous substance), probabilities of the six faces coming up must be determined also by an experimental method, just as approximate probabilities in hydrology must be determined. In an experimental method, it is important that conditions do not change from one trial to the next, or from observation to observation.

2.2. Approaches to Probability, and Hydrology.

Three approaches to probability theory. Probability theory to be applied to an earth science like hydrology is based on axioms that must be satisfied, at least approximately. There are three, basic, historical approaches to probability: the classical, Laplacian approach of equally likely or equiprobable events; the approach based on set and measure theories; and the relative frequency or statistical approach.

Classical approach to probability. The classical approach is based on the concept of equiprobable or equally likely elementary random events. It is difficult to find hydrologic variables for which this approach can be meaningfully applied, even if attempts are made to idealize a natural hydrologic environment, like the concept of the perfect die, coin, or deck of cards. This Laplacian approach is good for the probability calculus of chance games with elementary events being of equal probabilities. In hydrology this approach, however, may be applied only to exceptional cases.

Example 2.8. The events of a yes/no type, with 50% chance each, are the most typical examples of these exceptions. The values of a variable above or below the median have equal probabilities of 0.50. If a continuous variable is divided into m class intervals, each with the same probability $p = 1/m$, the classical Laplacian approach is still applicable.

The set and measure theories approach to probability. The set approach to probability, with a probability attached to each random event, is the prevailing axiomatic approach in science at present, and it well satisfies mathematical logic. It is basically an approach to the probability of discrete random events. It assumes known probabilities of elementary events. This is rarely the case in hydrology for which only relative frequencies with various degrees of inaccuracy can be obtained. Nevertheless, this approach to probability theory is widely used. The fundamental formulation was proposed by Kolmogorov in 1933 [1,2] by assigning probabilities or numbers in the closed interval (0,1) to an initial class of elementary events, as the subset of the total set of random events. Methods of computing probabilities of composite random events from probabilities of elementary random events is the subject of probability theory. This approach, then, has an important property, namely the mathematical theory starts after probabilities are assigned to elementary random events.

Relative frequency approach to probability. An attempt was made by von Mises in 1931 [3,4] to introduce relative frequency or a statistical approach to probability theory. If a random event occurs m times in N trials, in experiments or in observations, the ratio $f_i = m/N$ is its relative frequency. For $N \to \infty$, only then does $f_i \to p_i$, or the relative frequency tends to the probability of the event. The error $\delta_i = p_i - f_i$ is always present in practical cases. This approach implicitly assumes that the experiments are made on a stationary (time-invariant) process, or on a homogeneous (line-invariant) process. The probabilities are $(f_i + \delta_i)$, and the errors δ_i are carried into probability calculus. This approach is attractive from the point of view of physical reality because only frequencies of various events are known in hydrology. Because there is a substantial mathematical difficulty in carrying out computations of probabilites of new random events in this approach, it has not gained wide acceptance. The difference δ_i and its treatment is the factor that differentiates the last two approaches to probability.

Comparison of the last two approaches. The set and measure theories approach starts with probabilities replaced by relative frequencies. When probabilities of new events are derived, the statistical inference by testing hypotheses or assigning confidence or tolerance limits takes care of these error differences. This is an a posteriori approach for treatment of errors in assigned probabilities; probability theory avoids dealing with these errors, leaving them to mathematical statistics.

The relative frequency or statistical approach to probability theory is an a priori approach for treating errors in relative frequencies by carrying these errors along into probability calculus. Because of cumbersome mathematical difficulties in this approach, most specialists do not find attractive to introduce it to probability theory. Attempts by some mathematicians to reject the relative frequency approach on grounds other than the cumbersome mathematical problems involved do not change the basic problem, namely of an a posteriori and an a priori treatment of differences between the unknown probabilities and the known relative frequencies.

Mathematical logic and technique, and the practical needs of an applied science, may be two different things. The set approach to probability theory has divorced the mathematical statistics from the probability theory, and made them two different but connected parts of mathematics, even though the field of statistics extensively uses the probability theory. The relative frequency approach to probability theory was an attempt to merge probability theory and mathematical statistics into one part of mathematics.

Effects on hydrology. Any development in probability theory may affect hydrology, and new concepts may further improve methods of treating hydrologic random variables and their stochastic processes. In conclusion, hydrologists should use the current approach to probability based on set and measure theories, and apply its many useful theorems, but they should be aware of new developments, particularly on the conceptual level.

The set approach to probability theory usually assumes discrete events. Most hydrologic variables are continuous. The discrete event approach, as somewhat artificial, is forced on continuous hydrologic variables. Again mathematical expediency is in conflict with the character of natural phenomena. It can be argued that any continuous variable and any continuous sequence may and often are well approximated by discrete events and sequences, if small intervals are used between discrete events or between observations. However, this is always only an approximation. A better way is to replace point events by interval events, as it is used in the concept of probability density which changes over that interval.

Discrete and continuous variables. A discrete variable can have only discrete random events. A continuous variable can take any event in a finite or infinite interval. The basic hydrologic processes are continuous for continuous variables, and in practice are usually approximated by discrete series of discrete variables.

Any controversy between the concepts of discrete and continuous stochastic variables and series affects hydrology. The reasoning is often that variables and series in practice are always discrete, either because they are observed at discrete positions or times, or because values are rounded, say 3.21, 3.22, 3.23, ..., as the prescribed discrete random events, or because of the uncertainty principle which limits the accuracy of any measurement. Artificiality of discreteness, approximations in measurement, and the uncertainty principle do not prove that variables with their true values are not continuous. In observing a discharge between zero and infinity, any value can occur in nature, and not just those defined by the set of discrete events, such as sets of countable events. It is difficult to find basic hydrologic variables and series that are not continuous by nature, at least in the major range of their variations.

2.3. Random Events.

Elementary and composite random events. All possible random events of an observed variable may be reduced to a set of simple or elementary events. The combination of elementary events gives sets which are all other events, usually called the composite events. All elementary and composite events taken together represent the sample space, and all its sets are possible random events. Therefore, it must be determined in each particular sample space what constitutes the set of elementary events.

If n is the number of all elementary events, then the results of experiments have: $\binom{n}{1}$, one-element events; $\binom{n}{2}$, two-element events; ...; $\binom{n}{n-1}$, $(n-1)$ — element events; one sure or certain event (the total sample space), and one impossible event (all values outside the sample space). This gives 2^n random events in all. Therefore, a sample space has n elementary events and $(2^n - n)$ composite events. Any elementary or composite event is a subset of sample space, and every subset is a random event. If no two random events can happen simultaneously, they are incompatible or mutually exclusive. If one random event from a set must occur in any observation, the set is said to be exhaustive. If an event is the sum of other events, it is decomposeable into mutually exclusive events, usually elementary events.

The mathematical operations with sets are not reviewed here. They can be found in any elementary book on probability theory or mathematical statistics. It is assumed they are known, at least in their simplest aspects.

Example 2.9. Current practice in hydrology replaces a continuous variable by a number of discrete values or random events. A discharge is given in m^3/sec or cfs and is expressed with two significant figures beyond the decimal point, such as 3.52, 3.53, and so forth. The third figure beyond the point has an error of $\pm$ 0.005, so that the true value of discharge is somewhere between 3.515 and 3.525 for the event 3.52. There is no need to express a discharge in more than these two significant figures beyond the decimal point in this case because errors in measurement do not justify a more accurate division. Because any observation of a continuous hydrologic variable has inherent errors, a variable that can take any value from zero to infinity is thus transformed to a discrete but infinite set of random events.

Example 2.10. To compute the numbers of elementary and composite random events in the case discussed in Example 2.9, a discharge of 25 m^3/sec is assumed to be a practical upper limit, or the probability of values greater than 25 m^3/sec is negligible. The range is 0 – 25 m^3/sec. If the discharge measurement is accurate only to two figures beyond the decimal point, the number of all possible elementary events is 2501, zero and 25.00 included; this is a finite but large set of elementary events. If one is interested in the probability of flows between 5.005 and 6.005 m^3/sec, this composite random event, as a subset of the sample space, has 100 elementary events, 5.01, 5.02, ..., 5.99 and 6.00. The probability of this event depends on probabilities of 100 elementary events. If the set theory is applied, this example with $n = 2501$ elementary events has 2^{2501} random events, or $2^{2501} - 2501$ composite random events. This indicates the difficulty in using the concept of discrete random events in hydrology when the number of elementary random events is large.

Axioms of probabilities. Probability theory is founded on the basic axiom that every elementary random event, x_i, has a probability measure, p_i, with these properties:

(1) $0 \leq p_i \leq 1$, or p_i is a positive number between zero and unity, these two limits included; and

(2) $\Sigma p_i = 1$, $i = 1, 2, \ldots, n$, or the sum of probabilities of all elementary random events is always unity.

The first property may be restated for all random events, rather than just for elementary events of a sample space, so that with each random event of this space a non-negative number $0 \leq p \leq 1$, called the probability, is associated.

For the pairwise mutually exclusive events, the total probability of n of these random events is the sum of their individual probabilities. The probability of a certain event is unity, and the probability of an impossible event is zero. The probability of zero, however, does not imply that the random event is impossible. The probability of either or both of two arbitrary events (not necessarily mutually exclusive) is the sum of individual probabilities minus the probability of the joint occurrence of these two events.

The reader is advised to review the elements of the axiomatic construction of the probability theory beyond the very brief summary given above. For example see reference [5, pages 59—66].

Probabilities of random conditional events. The random conditional event is defined as the random event A to occur, provided random event B has occurred. This conditional event is currently denoted by $A|B$. The converse event, $B|A$, is different from the event $A|B$. The probabilities of these random conditional events, $0 \leq P(A|B) \leq 1$, or $0 \leq P(B|A) \leq 1$, are often called conditional probabilities. Since the expressions of these probabilities are identical to probabilities of random conditional variables, $x|y$ or $y|x$, they will be presented in the next subchapter on random variables. It is then sufficient to replace the conditional variables in these expressions by conditional events. Further discussion of probabilities of random conditional events is given in Chapter 3, under 3.3.

Objective and subjective definitions of the probability of random events in hydrology. The objective definition of probability, which is the only definition acceptable for hydrology, requires that each random event has a number for the probability that must be between zero and unity, zero and unity included. Some hydrologists use the term probability without referring to these numbers, $0 \leq P \leq 1$. This is not acceptable because it represents the subjective definition of probability, which is mainly rejected by natural sciences.

Also encountered in hydrologic literature are expressions like *reasonable probability*, *reasonable risk*, or *reasonable expectancy*, without the figure between zero and unity being specified to define this reasonable probability, risk, or expectancy. These concepts without numbers attached usually mean different things to different individuals. They often refer to composite random events, such as the extreme values. If the probability P of a composite random event of flood peak Q_i is defined as all possible annual flood peaks greater than Q_i, and if it is small, say $P(Q \geq \geq Q_i) \leq 0.001$, then it is tempting to call it an event of reasonable small probability, or expectancy. By avoiding associating composite events with these small probability numbers, however, various ambiguities may be introduced.

It is surprising that engineers use this undefined expression of reasonable probability, risk, or expectancy though they attach numbers to as many other statements as possible. Probability theory in hydrology should not be different from this general and sound engineering practice of quantifying every statement whenever possible and meaningful. Both of these qualities are fulfilled for probabilities of hydrologic random events.

Determinists of hydrology often advance an argument against the probabilistic approach to floods, that a small creek in the Rocky Mountains or elsewhere cannot have a flood as large as the Mississippi River at New Orleans. Floods greater than such a flood have a finite but small probability of occurrence on a creek. The fault in this reasoning is the same as the question, what is the probability that a chicken would lay an egg as large as itself? It is easy to show that the probability of floods occurring on a small creek greater than a Mississippi River flood is so small that even 1,000 times the age of the earth, which age is about $4.6 \cdot 10^9$ years, is not a sufficient return period (the reciprocal of this probability) for an average occurrence of such floods. Thus, it is attractive to neglect all random events in hydrology with the probability 10^{-6} or less (the return period of a million years or more). It is worthwile risking this neglect except in exceptional cases of unusually high stakes. However, it is a small step from this neglect of events of very small probabilities to embark on the concepts of the maximum precipitation intensity, or the maximum probable precipitation of the subjective probability type.

Example 2.11. A typical example of incorrectly using the term probability in hydrology is the well-known hydrometeorological concept of the probable maximum precipitation. It is often used in hydrology for estimating values of the similarly incorrect concept of probable maximum flood, without ever associating a number in the range $0 \leq P \leq 1$ to these probabilities of maximum precipitation or maximum flood.

2.4. Random Variables

Definition of random variables. For the practical purpose of applying probability theory and statistics to hydrology, a simple definition of random variables is given in this text, although it may not meet a rigorous definition of random variables in probability theory. Any quantity X is a random variable if for a given real number x, which the quantity X may or may not assume by observations, a probability P exists that X is equal to or less than x. This statement is denoted by $P(X \leq x)$. If the oxygen content in water is considered as the quantity X, and for any positive value x there is a probability P that $X \leq x$, then X is a random variable. In this case of oxygen content, it is a positive-valued random variable because it can assume only positive values, zero included. If the quantity X satisfies the condition

$$0 \leq P(X \leq x) \leq 1, \qquad 2.1$$

it is a random variable. This should imply, however, that the sequence of various observations of X is governed by the laws of chance.

Randomness of composed hydrologic processes. Some hydrologic time series are known to be a composition of deterministic components in the form of annual and/or daily periodic movements in some of its parameters, and a stochastic component that is a random variable. If the stochastic component is not too small in comparison with the periodic components, or, in statistical terms, if the variance of a stochastic component is not small in comparison with the mean sum of squares of deviations from the mean of periodic components, the quantity X of Eq. 2. 1 is still considered a random variable in colloquial probability language. Strictly speaking, the random variation of parameters of deterministic components from sample to sample, and the stochastic component of a hydrologic process, are the only random factors inside that series. If the stochastic component is so small that it could be neglected, and the random variation of parameters of deterministic components is also small, the remaining periodic components would not be considered as a random process, though Eq. 2.1 could still be applied to the quantity X. Therefore, the additional condition to Eq. 2.1 in the definition of random variables is that outcomes in observing the quantity X must be governed by laws of chance. If outcomes are exclusively a deterministic sequence of values, such as the case with a pure trend and other transient functions, with periodic movements, and similar, the variable is deterministic. The composition of random variables and periodic components is an important subject of hydrology, regardless whether they can or can not be treated separately as a superposition of a stochastic and a deterministic process.

Example 2.12. Series of monthly precipitation and monthly runoff, as random variables, are composed of periodic and stochastic components. Often, components are of the same importance in the explanation of the variance of these variables.The discharge of a response hydrograph of an impervious surface for a well defined precipitation input (say, a unit time, uniform unit amount of rainfall) is a deterministic variable. In both cases, Eq. 2.1 may be applied. The additional condition of outcomes being governed by laws of chance for a random variable, however, is satisfied only in the first case.

Simplified symbols for random variables. Symbols X and x in Eq. 2.1 are used in such a way that X denotes the random variable, and x denotes any real number that the variable can assume. However, the symbol x is also used in this text to denote random variables. If a probability is given in hydrology for continuous variables, say $F(x) = 0.75$, it is the probability of all values $X \leq x$ being 0.75. Using the symbol F(x) instead of $P(X \leq x)$ is permissible as long as it is understood they have the same meaning.

Multidimensional random variables. If a quantity has several dimensions X, Y, Z, ... , and there are such real numbers x, y, z, ... , so that

$$0 \leq P(X \leq x,\ Y \leq y,\ Z \leq z,\ \ldots) \leq 1, \tag{2.2}$$

this quantity is a random variable with several dimensions or components that are also random variables. For three dimensions, the quantity is a random vector. If only X is investigated, it is a univariate. For joint investigation of X and Y, they are bivariate. For X, Y, Z, ... , they are multivariate. The outcomes of these variables must be governed by laws of chance.

Example 2.13. Turbulent velocity at a point is a random vector. Permeability, which randomly fluctuates in space and with directions in a ground water porous medium, may be conceived as a random tensor. They are multivariate, in three or six dimensions.

2.5. Univariate Distributions.

Distributions of discrete variables. If X can have only discrete values of x, their probabilities are finite values or zeros, and they are called the probability mass. The repartition of probability mass over the discrete values of x is the probability mass function or graph (p. m. f.), designated in this text by p(x).

A distribution function (d. f.), written as F(x), of a random variable X is a unique, single-valued, real non-decreasing and nonnegative function $F(x) = P(-\infty < X \leq x)$. For every point in the sample space, it is defined by the properties: $F(x_2) - F(x_1) \geq 0$ for $x_2 > x_1$, $F(-\infty) = 0$, $F(+\infty) = 1$, and $F(x + 0) = F(x)$. For discrete values of x, this d. f. is

$$F(x_i) = P(X \leq x_i) = \sum_{j=-\infty}^{i} p(x_j). \tag{2.3}$$

Probability distributions are either empirical curves, or they can be expressed by theoretical functions. Distributions of discrete variables are non-decreasing step graphs from zero to unity.

Discrete cases of probability mass in hydrology are: all counting problems, with events which are integers, 0, 1, 2, ... , n, ... ; any yes-or-no, either-or problem; two events problems such as + 1, − 1, or problems usually given with probabilities p of yes and q of no, with p + q = 1; and similar. These cases occur whenever the probabilities of all values below and above a given value, x_0, are of interest. Often, this is used where probabilities of zero and non-zero values are studied. Various discrete distribution functions applicable in hydrology are discussed in Chapter 7.

Example 2.14. The number of storms per a given time interval, the number of rainy days per year, the number of flood peaks above a given discharge per year, and similar counting cases are discrete random variables. Two events of yes-no are rainy or nonrainy days, freezing or non-freezing temperatures, polluted or non-polluted river discharges for a given standard, and so forth.

Example 2.15. Figure 2.2 gives the example of the probability mass graph (estimated by the relative frequencies of discrete random events, 0, 1, 2, ,) and of the corresponding probability distribution graph, for the number of rainy days in the first ten days of May, at the precipitation station Austin, Texas for the period 1861 − 1967.

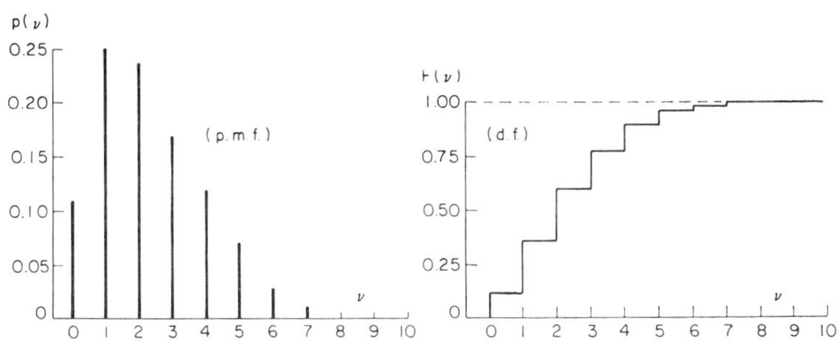

Figure 2.2. The relative frequency mass, or the probability mass p (υ), (left graph), and the distribution function F (υ), (right graph), of the number υ of precipitation days in the first 10 days of May, for the precipitation station Austin, Texas for the period 1861 − 1967 (107 years).

Distributions of continuous variables. If a random variable X is continuous, its value x has a probability density so that only a finite interval of x has a finite probability. The repartition of probability densities over x-values is the probability density function (p. d. f.), denoted in this text by f(x). The distribution function (d. f.) of continuous random variables is

$$F(x) = \int_{-\infty}^{x} f(x)\,dx, \quad \text{or} \quad \frac{dF(x)}{dx} = f(x). \qquad 2.4$$

Figure 2.3 represents schematically this continuous case, both for f(x) and F(x).

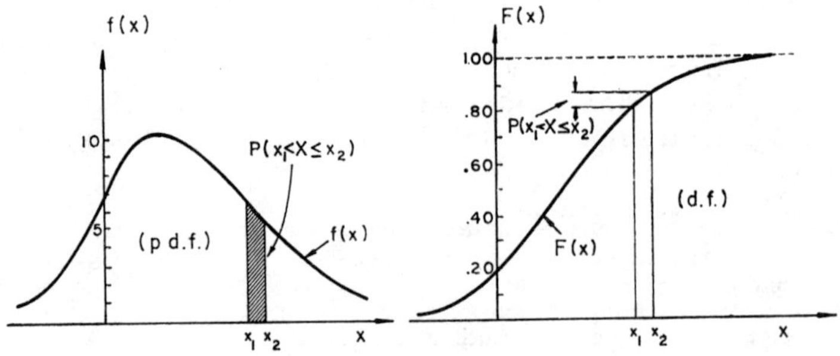

Figure 2.3. Schematic representation of the probability density function (p.d.f.), given by f (x), (left graph), and the probability distribution function (d.f.) given by F (x), (right graph), of a continuous variable, x.

Most hydrologic analyses involve continuous hydrologic variables, either treated as continuous or made discrete in an appropriate way. Continuous hydrologic variables are either positive-valued only, or they have both positive and negative values.

A tendency in practice is to replace a continuous density curve by a histogram which gives relative or absolute frequencies for selected class intervals. It is easy to imagine the class relative frequency f_i divided by the class interval length Δx, to obtain the frequency density at the center of the class interval. If this work is repeated for enough points, the frequency density curve is obtained as an estimate of the probability density curve. Probabilities and relative frequencies are without dimension, while frequency or probability densities have the inverse dimension of the dimension of x. In this latter case, the integral of f(x) between x_1 and x_2 gives the dimensionless probability value, as shown in Fig. 2.3. Various continuous probability distribution functions applicable in hydrology are discussed in Chapter 7.

Example 2.16. The river discharge of perennial streams, lake outflows, the ground water level or outflow, water oxygen content, relative humidity of the air, dissolved matter in the ground water, and so forth, are all continuous variables.

Example 2.17. The so-colled cumulative hydrologic variables (those for which the physical equivalent can be stored in a space) are all positive-valued variables, such as vapor content, soil moisture, heat, water, snow, ice, sediment, pollutants, dissolved solids in water, oxygen content in water, and other continuous variables. The total content in a given space, or measurements at a point in a given instant, shows only values that are either equal to or greater than zero, or $X \geq 0$. By definition and by their physical properties, these random variables cannot be negative.

Example 2.18. Whenever an arbitrary reference datum is selected, continuous random variables can have both positive and negative values. Such variables are temperature, lake level, river stage, ground water level, and similar. When the direction of water flow may be either downstream or upstream, both positive and negative values of discharge may result. Examples are the river discharge in an estuary where the water can flow in both directions because of tides, or a flash flood of a tributary, which may cause the water to temporarily flow upstream on the main stream of relatively small water surface slopes.

Distribution of mixed variables. Distributions may be combined of both discrete and continuous variable values, or of probability mass for some points and probability densities for some intervals. Typical examples are distributions of variables of intermittent phenomena in hydrology. For the value $x = 0$, there is a finite probability p_0, while for values $x \geq 0$, the probability densities $f(x)$ are used, as shown by the representation given in Fig. 2.4.

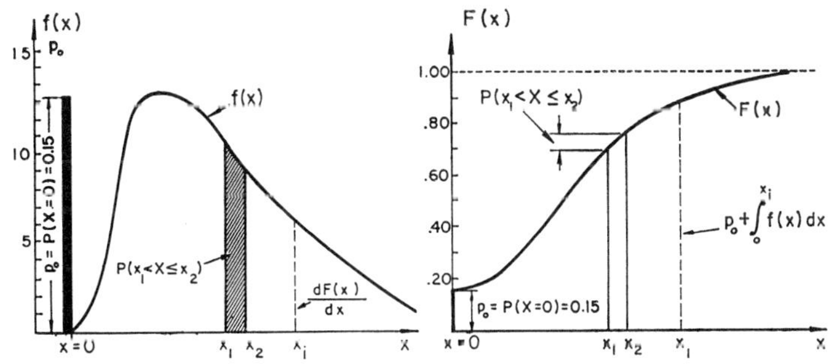

Figure 2. 4. Combination of probability mass p_0 for $x = 0$, and probability densities $f(x)$ for $x \geq 0$, (left graph), and their probability distribution, (right graph), of a mixed random variable (discrete-continuous).

This case is common in hydrology whenever a magnitude can have zero values or other boundaries besides all other continuous positive values. The probability of values below x_i may be expressed for the case of zero discrete value as

$$F(x_i) = p(x = 0) + \int_0^{x_i} f(x)\,dx. \qquad 2.5$$

It is common practice in hydrology to design populations of random variables in such a way that only random events either of discrete or of continuous variables are included in each population. This approach avoids mixed distributions. Every boundary in the occurrence of a well defined hydrologic variable is the source of discrete values with the probability mass.

Example 2.19. Daily rainfall has the probability mass $p(x = 0)$ of nonrainy days while $f(x)$ are probability densities of daily precipitation amounts of rainy days.

Example 2.20. Classical examples of discrete boundary values in hydrology are controlled reservoirs and ground water aquifers. The probability distribution of the water levels in an operational reservoir usually has a lowest level, either imposed by the operation of the reservoir or by the release structures, and a highest level, usually maintained by spillway control gates or valves, as the two boundary values with finite probabilities. The frequencies are relative times that these levels will be maintained during the empty or full storage capacity. Levels between these two boundaries have probability densities. Similarly, if a shallow ground water aquifer can be dried out, its impervious base level is the lower boundary. If high infiltration can cause the ground water table to emerge to the ground surface, it is then the upper boundary. Both should have finite probabilities, while all table levels in between, as random events, should have probability densities.

Example 2.21. Common lower boundaries of positive-valued variables in hydrology do not need to be zeros. If the constant small water flow Q_0 is diverted permanently, or percolates continuously, from a large perennial river to an intermittent small river, this Q_0 becomes for the small river a lower boundary with the probability mass, while random events $Q \geq Q_0$ have probability densities.

2.6. Distributions of Bivariates, Multivariates, and Conditional Variables.

Definitions of bivariate, multivariate, and conditional variables. Distinction is sometimes made between the terms *variate* and *variable* by using the first for a discrete and the latter for a continuous variable. This distinction is not made in this text, the two terms are considered synonymous and used interchangeably.

If two variables X and Y are investigated as they jointly occur so that there is a probability $P(X \leq x, Y \leq y)$, the case is bivariate. The corresponding distribution $F(x,y)$ is the bivariate distribution of X and Y, or the joint distribution in two dimensions.

If more than two variables are involved, such as $X_1, X_2, X_3, \ldots, X_n$, so that there is a probability $P(X_1 \leq x_1, X_2 \leq x_2, \ldots, X_n \leq x_n)$, the case is multivariate. The corresponding distribution $F(x_1, x_2, \ldots, x_n)$ is the multivariate distribution of n variables, or the joint distribution in n dimensions.

If two or more variables are considered, but under the condition that probabilities of one or more variables are required provided the outcomes of one or more other variables have occurred or are already known, it represents the case of conditional variables. For two variables X and Y there are two conditional variables: $X \mid Y$ (X given Y), and $Y \mid X$ (Y given X), for which the probabilities $P(X \leq x \mid Y)$ or $P(Y \leq y \mid X)$ are relevant. The above symbols of conditional variables $X \mid Y$ and $Y \mid X$ refer to discrete variables X and Y. For X and Y continuous variables these two probabilities are expressed as $P(X \leq x \mid Y \leq y)$, or $P(Y \leq y \mid X \leq x)$. The conditional part (or provided that) may have different expressions, say $P(Y \leq y) \mid x_1 < X \leq x_2)$ or similar conditions. In other words, the symbol $Y \leq y \mid X$ refers to the variable Y provided that the range of X to occur is specified. Distributions of these two conditional variables are $F_1(x \mid y)$ and $F_2(y \mid x)$. Similar conditional variables and their distiributions are related to more than two variables, like $X_1 \mid X_2, X_3, \ldots, X_n$, or X_1 occurs provided that $X_2, X_3, \ldots, X_n$ have occurred; or $X_1, X_2, \ldots, X_m \mid X_{m+1}, \ldots, X_n$, or distributions of one or more variables are sought, given that one or more other variables have occurred or are known.

Just as for the univariate case, the bivariate, multivariate, and conditional variable cases may be composed of only discrete values with the probability mass, of only continuous values with probability densities, and of mixed values with both the probability mass and probability densities. Bivariate and multivariate distributions and distributions of conditional variables are important to hydrology because various relations of two or more hydrologic variables are rather of a stochastic than of a deterministic nature. The use of multiple correlation analysis and other multivariate analyses are current methods of establishing statistical hydrologic relations, and they involve these types of distributions. For the advanced definitions of these variables and their distributions, see reference [6, pages 30—68].

Example 2.22. Bivariate cases are the joint occurrence of a rainfall variable at two stations, of a rainfall variable and the corresponding runoff variable in a river basin, of various properties of the same phenomenon like the joint occurrence of the storm duration and the total storm precipitation at a given station and for a given time of the year, and similar. In hydrology the usual transfer of information from one variable to another, or from one point to another for the same magnitude, involves bivariate cases. Other examples are: water runoff-sediment transport, lake inflow-lake level, river flow--dissolved solids content, ground water table and related evaporation, and so forth.

Example 2.23. Classical multivariate cases in hydrology are: runoff and many related precipitation and river basin variables; snow melt and related variables of accumulated snow, climate factors, expected precipitation during the melting season, and others; water quality and related variables of runoff, ground water contribution, river basin characteristics, and others.

Example 2.24. In designing flood spillways for a given reservoir, the three random variables: rainfall, snowmelt, and the concurrent reservoir level, determine the maximum reservoir level for the given spillway and reservoir characteristics. This may be treated as multivariate case and distribution.

Example 2.25. Cases of random conditional events are: river A exceeds a given flood discharge Q_a at the confluence with river B, while this latter river exceeds the flood discharge Q_b; a dry year occurs in the adjacent region A, provided that it occurred in the region B, and similar. Cases of conditional variables are: the flood occurs provided that the reservoir level is above a given position; a given total snow melt runoff in a flood occurs provided more than a given snow quantity is on the ground, and so forth.

Bivariate and multivariate distributions. The probability of X and Y being in an interval $(x_1, x_2; y_1, y_2)$ for a bivariate distribution F (x,y), denoted by $\Delta^2 F(x,y) \geq 0$, is

$$\Delta^2 F(x,y) = P(x_1 < X \leq x_2;\ y_1 < Y \leq y_2) = \sum_{X>x_1}^{X\leq x_2} \sum_{Y>y_1}^{Y\leq y_2} p(x,y), \qquad 2.6$$

which represents the sum of all values of the probability mass in that two-dimensional interval, with the bivariate being discrete. The following properties are valid for any bivariate, discrete or continuous: $F(-\infty, y) = F(x, -\infty) = 0$; $F(+\infty, +\infty) = 1$, and $F(x+0, y) = F(x, y+0) = F(x+0, y+0) = F(x,y)$. Distributions $F(x, +\infty) = F(x)$ and $F(+\infty, y) = F(y)$ are defined as the marginal distributions of x and y, respectively.

For the continuous random bivariates, the d.f. is

$$F(x,y) = \int_{-\infty}^{x} \int_{-\infty}^{y} f(x,y)\, dx\, dy, \tag{2.7}$$

and its p.d.f. is

$$f(x,y) = \frac{\partial^2 F(x,y)}{\partial x\, \partial y}, \tag{2.8}$$

so that the d.f., F (x,y), is uniquely determined by its p.d.f., f (x,y), and conversely. Marginal probability density functions (m.p.d.f.) are

$$f(x) = \int_{-\infty}^{+\infty} f(x,y)\, dy, \text{ and } f(y) = \int_{-\infty}^{+\infty} f(x,y)\, dx, \tag{2.9}$$

respectively for x and y.

Similarly, as for the bivariate case, Eqs. 2.6 through 2.9 are valid for multivariate cases with the corresponding addition of one or more variables. Equations 2.7 and 2.8 now become

$$F(x_1,x_2,\ldots,x_n) = \int_{-\infty}^{x_1} \cdots \int_{-\infty}^{x_n} f(x_1,x_2,\ldots,x_n)\, dx_1 dx_2 \ldots dx_n,$$

and

$$f(x_1,x_2,\ldots,x_n) = \frac{\partial^n F(x_1,x_2,\ldots,x_n)}{\partial x_1\, \partial x_2 \ldots \partial x_n}.$$

Bivariate and multivariate probability distribution functions applicable in hydrology are discussed in Chapter 7.

Distributions of conditional variables. For discrete variables, the probability mass function $p(y \mid x)$ of a random conditional variable $y \mid x$, for y to occur, given that x has occurred, is

$$p(y \mid x) = \frac{p(x,y)}{p(x)}, \tag{2.10}$$

In this equation, $y \mid x$ is the conditional variable y given x, p (x,y) is the bivarate p.m.f. of variables x and y, and p (x) is the p.m.f. for x to have occurred, or it is the marginal p.m.f. for x. In this case, there is nothing random about x, because the occurrence of x is a fixed and unquestioned condition. Equation 2.10 means the probability of y, provided that (or given that, or assuming that, or knowing that — whichever is relevant) x has occurred.

The distibution of a discrete conditional variable is

$$F(y \mid x) = \sum \frac{p(X \leq x, Y \leq y)}{p(X \leq x)}. \tag{2.11}$$

Therefore, the probability $p(x \mid y)$ is different from the probability $p(y \mid x)$ of Eq. 2.10. A basic difference exists between p (x,y) and $p(y \mid x)$. The first is the probability that both random events of x and y occur, while the second is the probability that a single event of y occurs provided that a given event of x must occur. In this latter case, x is not a random variable but a condition.

For continuous conditional variables, the p.d.f. is

$$f(y \mid x) = \frac{f(x,y)}{f(x)}, \qquad 2.12$$

in which f (x,y) is the joint p.d.f. of x and y, and f (x) is the marginal p.d.f. of x. This gives the d.f. of a conditional variable

$$F(y \mid x) = \int_{-\infty}^{x} \int_{-\infty}^{y} \frac{f(x,y)}{f(x)} dx\, dy. \qquad 2.13$$

Equation 2.12 may be expressed also in the form

$$f(x,y) = f(x) \cdot f(y \mid x) = f(y) \cdot f(x \mid y). \qquad 2.14$$

This last expression can be generalized for the multivariate case by

$$f(x_1,x_2,\ldots,x_n) = f_{1,\ldots,k}(x_1,\ldots,x_n) \cdot f(x_{k+1},\ldots,x_n \mid x_1,\ldots,x_k) \qquad 2.15$$

in which $k = 1, \ldots, n-1$

The concepts of conditional variables and their distributions are important subjects to hydrology; many theoretical and practical hydrologic problems may be reduced to the distribution of conditional variables. Hydrologic variables are usually dependent both in space and time. Many water resource problems often consist of and are related to the joint occurrence of two or more variables. In these cases, probabilities of conditional random events and distributions of conditional random variables are applicable. They deserve a much better treatment and more application in hydrology than they are presently receiving.

Example 2.26. Variables of a dry or a wet year to occur at a precipitation station, provided a dry or a wet year, or a wet or dry year, occurs at another precipitation station or stations, k miles apart, are conditional variables. Probabilities of these conditional variables are dealt with by S. Pinkayan [7] by dividing each of the pair of continuous variables x and y of annual precipitation at selected western United States stations into five discrete random events. For each station, the precipitation is defined as a very dry year if $X \leq x_1$, with $P(X < x_1) = 0.15$; a dry year if $x_1 < X \leq x_2$, with $P(x_1 < X \leq x_2) = 0.20$; an average year if $x_2 < X \leq x_3$, with $P(x_2 < X \leq x_3) = 0.30$; a wet year if $x_3 < X \leq x_4$, with $P(x_3 < X \leq x_4) = 0.20$; and a very wet year if $X > x_4$ with $P(X > x_4) = 0.15$. For two selected stations, the conditional variable $Y|X$ has $5 \cdot 5 = 25$ discrete random events. Each of them is studied by selecting a central station with its annual precipitation as the x-variable, with the y-variable being the annual precipitation of any station around the central station, k-miles apart. In this paper[7], mathematical models are developed of how the probabilities of these 25 conditional random events change with k, as $p(y|x) = f(k)$, by fitting various simple analytical functions to these empirical relations. Various tests of goodness of fit of these simple functions are performed, and the functions that best fit the data are selected.

Example 2.27. The probability of a storm occurring during a short time interval Δt, given that it has not rained for time t, is $\lambda_1(t) \cdot \Delta t$, where $\lambda_1(t)$ is a given function of t. It is usually a periodic function in most cases of precipitation stations, called the density of storms per unit time interval [8]. What is the probability that no storm will occur during time t? Let A be the random event of no storm occurring in (0, t), B the random event that storm occurs in $(t, t + \Delta t)$, and $\bar{B}$ the event of no storm during $(t, t + \Delta t)$. The probability P (A) is f (t), or it is a function of time. The probability of no storms in $(0, t + \Delta t)$ is $P(A, \bar{B}) = f(t + \Delta t)$, but it is also $P(A, \bar{B}) = P(\bar{B}|A) \cdot P(A) = [1 - P(B|A)] \cdot f(t)$. As $P(B|A) = \lambda_1(t) \cdot \Delta t$, because what happens in $(t, t + \Delta t)$ is independent of what happened in (0, t), then

$$\frac{f(t + \Delta t) - f(t)}{\Delta t} = - \lambda_1(t) \cdot f(t).$$

Because $\lambda_1(t)$ and f (t) are positive, the probability of no storm in the interval (0, t) decreases with an increase of t. For $\Delta t \to 0$, $\Delta f(t)/f(t) \to - \lambda_1(t)\, df$. If $f(0) = 1$, or no storm at $t = 0$, then

$$\int_0^t \frac{df(t)}{f(t)} = - \int_0^t \lambda_1(t)\, dt, \text{ so that } P(A) = f(t) = \exp\left[- \int_0^t \lambda_1(t)\, dt\right].$$

Mutually independent and dependent random variables. A bivariate (x,y), having the d.f. F (x,y), has a necessary and sufficient condition for x and y to be two independent variables

$$F(x,y) = F(x) \cdot F(y). \tag{2.16}$$

The product of the two marginal d.f. is equal to the bivariate distribution. Similarly, for continuous variables,

$$f(x,y) = f(x) \cdot f(y). \tag{2.17}$$

The p.d.f., f (x,y), is equal to the product of the two marginal p.d.f. The same condition is valid for the multivariate distributions, in which case Eqs. 2.16 and 2.17 are applied to three or more variables. If a bivariate distribution is given either as F (x,y) or as f (x,y), and if it can be proven that it can be factored into a product of a probability function of x times a probability function of y, then x and y are independent. If the converse is proven, they are dependent.

In the case of distribution functions of conditional variables, x and y are independent if, and only if,

$$f(y \mid x) = f(y); \text{ or } f(x \mid y) = f(x), \tag{2.18}$$

or if probability densities of conditional variables are equal to marginal probability densities of y or x, respectively. Similarly, for a multi-conditional variable, if

$$f(x_{k+1}, \ldots, x_n \mid x_1, \ldots, x_k) = f_{k+1, \ldots, n}(x_{k+1}, \ldots, x_n), \tag{2.19}$$

the variables $x_{k+1}, \ldots, x_n$ are independent of variables $x_1, \ldots, x_k$.

The case of Eq. 2.19 may be often encountered in hydrology when a multivariate case may be divided either into two groups of variables $x_1, \ldots, x_k$ and $x_{k+1}, \ldots, x_n$, or into more groups, and in such a way that the variables are independent between groups while the variables in each group are mutually dependent.

Example 2.28. The so-called independent variables in multiple regression equations may be such a case. The geometric variables of a river basin may be independent of its variables describing the precipitation, underground storage, antecedent moisture conditions, and similar.

Proving independence between random variables. Using Eqs. 2.16 through 2.19 to prove the independence between two or more hydrologic variables is not as simple as these equations may imply. It must be proven directly for any pair of x,y-values of a discrete bivariate, or any class-field of a continuous bivariate, that the discrete probabilities p (x,y) or the integrated f (x,y) over $\Delta x \Delta y$ class-fields are not significantly different, on a selected probability level, from the product of the corresponding frequencies of discrete univariate events or the product of the corresponding class frequencies of the continuous marginal x and y frequency distributions.

Another way to prove the independence of random variables by Eqs. 2.16 through 2.19 is in fitting the known bivariate or multivariate distributions functions, which can be factored. If the goodness of fit is accepted in a proper statistical test, i.e., by the chi-square test, it automatically infers the variables x, y, . . . , are mutually independent. The shortcoming of this method is that it is nearly impossible to exhaust all bivariate or multivariate distribution functions that can be factored; if all currently used functions are rejected the variables may still be independent.

The most current method of investigating whether hydrologic variables are independent is by postulating the type of dependence models, say linear or the non-linear models. Then statistical tests are used to prove that variables are not correlated in that type of dependence, with the prescribed criteria of non-significant correlation. Because it is difficult to implement the above simple probabilistic definitions of independence of Eqs. 2.16 through 2.19, approximate statistical methods are used to indicate dependence or independence among variables.

Example 2.29. Assume that it has been statistically proven that an empirical hydrologic bivariate distribution may be well fitted by the probability function

$$f(x, y) = \frac{1}{2\pi\sigma_x\sigma_y} e^{-[(x-\mu_x)^2/2\sigma_x^2 + (y-\mu_y)^2/2\sigma_y^2]},$$

with μ_x, μ_y, σ_x, and σ_y the means and standard deviations of x and y, individually. This function is then factored to

$$f(x, y) = \left[\frac{1}{\sigma_x\sqrt{2\pi}} e^{-(x-\mu_x)^2/2\sigma_x^2}\right]\left[\frac{1}{\sigma_y\sqrt{2\pi}} e^{-(y-\mu_y)^2/2\sigma_y^2}\right],$$

so that x and y are inferred as being two independent normal random variables.

2.7. Distributions of Functions of Random Variables.

Distributions of functions of one random variable. The computations of distributions of functions of a random variable are often required in hydrology. Let f (x) be the p.d.f., and F (x) the d.f. of a random variable x. The new variable y is given as a function, $y = g(x)$, of the random variable x, with $x = g^{-1}(y)$ representing the symbol for the function $y = g(x)$ solved for x. The new variable y has the p.d.f. h (y) and the d.f. H (y). The p.d.f., h (y), is related to the p.d.f., f (x), by

$$\frac{dH(y)}{dy} = h(y) = f(x) \left|\frac{dx}{dy}\right| . \tag{2.20}$$

Absolute values $\left|\frac{dx}{dy}\right|$ in Eq. 2.20 are a necessary condition because h (y) cannot be negative. Equation 2.20 can be rewritten as

$$h(y) = f[g^{-1}(y)] \cdot \left|\frac{dg^{-1}(y)}{dy}\right|, \text{ if } \alpha < y < \beta, \qquad h(y) = 0, \text{ otherwise,} \tag{2.21}$$

where $\alpha = \min(\alpha', \beta')$ and $\beta = \max(\alpha', \beta')$, with $\alpha' = \lim_{x \to -\infty} g(x)$ and $\beta' = \lim_{x \to +\infty} g(x)$. For h (y) given by Eq. 2.21,

$$f(x) \,|\, dx \,| = h(y) \,|\, dy \,| . \tag{2.22}$$

Example 2.30. By measuring water pressure at a point in a turbulent stream it may be found that its p.d.f. f (x) is well fitted by the normal p.d.f., designating the pressure p by x,

$$f(x) = \frac{1}{\sigma\sqrt{2\pi}} e^{-(x-\mu)^2/2\sigma^2}, \text{ for } -\infty < x < \infty,$$

where μ is the mean of x and σ is its standard deviation. Assuming the water velocity at the same point is proportional to $\sqrt{p}$, then $y = a\sqrt{p}$, where a is the proportionality constant, and y is the velocity. To find the p.d.f. of y from the p.d.f. of x, without transforming all observed values of x to y, but directly from Eq. 2.22, and by using Eqs. 2.20 and 2.21, a general scheme is first applied to the function $y = ax^n$. For $x = g^{-1}(y) = a^{-1/n} y^{1/n}$, and according to Eq. 2.21,

$$h(y) = f(a^{-1/n} y^{1/n}) \left| \frac{a^{-1/n}}{n} y^{(1-n)/n} \right|, \text{ for } \alpha < y < \beta.$$

For n = 1/2, this equation becomes

$$h(y) = \frac{2y}{a^2\sigma\sqrt{2\pi}} e^{-[(y/a)^2 - \mu]^2/2\sigma^2}, 0 < y < \infty,$$

or with $\alpha = 0$ and $\beta = \infty$.

It is sometimes feasible to determine the relations between the mean and standard deviation (μ and σ) of x, and the mean and standard deviation of y, and to substitute them in to the above equation for h (y). This example shows that a new variable which is a non-linear function of the observed variable, cannot have normal distribution if the observed variable has it. Turbulent pressure and turbulent velocity cannot have the same type of distribution function.

Example 2.31. An opposite procedure to that given in Example 2.30 is shown. It is known that logarithms of values of many hydrologic variables are close to being normally distributed. Designating y as logarithms of x, and x as the original values, say for the case of river discharges, then for y normally distributed and $y = \ln x$ (natural base)

$$f(x) = h(y)\left|\frac{dy}{dx}\right|, \text{ with } dy = \frac{dx}{x}, \quad -\infty < y < +\infty, \text{ or}$$

$$f(x) = \frac{1}{x\sigma_y\sqrt{2\pi}} \cdot e^{-(\ln x - \mu_y)^2/2\sigma_y^2}, 0 < x < \infty,$$

where μ_y and σ_y are the mean and standard deviation of $y = \ln x$, while those of x are μ_x and σ_x. This latter equation is the lognormal p.d.f. Relations between μ_x, σ_x, μ_y, and σ_y are given in Chapter 7.

Example 2.32. Recently, the U.S. Water Resources Council proposed that government agencies standardize the analysis of peak flood discharges by fitting the Pearson Type III distribution to logarithms of peak discharge. One of the expessions of Pearson Type III is [9, p. 152, Eq. 6.19]

$$f(y) = k\left(1 + \frac{y}{a}\right)^p e^{-py/a},$$

with $-a \leq y \leq \infty$, where -a is the lower boundary, the origin $y = 0$ is the position of the mode, k is the probability density at the mode, and p is the shape parameter. Because $y = \ln x$, where x is the flood discharge, $f(y)\,|\,dy\,| = \psi(x)\,|\,dx\,|$, and $dy = dx/x$, then

$$\psi(x) = \frac{k}{x}\left(1 + \frac{\ln x}{a}\right)^p e^{-p \ln x/a}$$

is the distribution used for flood peaks.

Distribution of a function of two random variables. For the function $z = g(x,y)$, where x and y are two random variables, z is also a random variable if $g(x,y)$ is real, single-valued, and defined in every point of the x,y-plane, except for a set of events with probabilities zero. All points $g(x,y) \leq z$ represent the set E_z. Then the d.f. of z is

$$H(z) = P[g(x,y) \leq z] = P[(x,y) \varepsilon E_z]. \qquad 2.23$$

In each particular case, the values (x,y), which belong to the set E_z, must be found in order to obtain $H(z)$.

Example 2.33. The variable x is the discharge of river A and the variable y is the discharge of river B. Let both rivers be diverted for some beneficial objective. Because the quality of their waters is different, they produce a total economic effect $z = ax + by$. The new variable z will have a distribution that depends on x and y, but is changed by the weights a and b. The probability $P[(ax + by) \leq z] = P(x, y)$, so that it is sufficient to determine for every pair x, y the value of z and assign to it the corresponding probability $P(x, y)$.

Distributions of a two-dimensional function of two random variables. The two-dimensional function (y_1,y_2) of the two random variables (x_1,x_2), expressed as $y_1 = g_1(x_1,x_2)$ and $y_2 = g_2(x_1,x_2)$, has the d.f. $H(y_1,y_2)$, provided that y_1 and y_2 belong to the set of points $E(y_1,y_2)$ for which $g_1(x_1,x_2) \leq y_1$ and $g_2(x_1,x_2) \leq y_2$ in the (x_1,x_2)-plane, or

$$H(y_1,y_2) = P[g_1(x_1,x_2) \leq y_1;\ g_2(x_1,x_2) \leq y_2] = P[(x_1,x_2) \varepsilon E(y_1,y_2)]. \qquad 2.24$$

The marginal d.f. $H_1(y_1)$ can be found directly without going through $H(y_1,y_2)$ simply by finding the set $E(y_1)$, so that

$$H_1(y_1) = P[g_1(x_1,x_2) \leq y_1] = P[(x_1,x_2) \varepsilon E(y_1)], \qquad 2.25$$

and analogously for the marginal d.f. $H_2(y_2)$, by finding the set $E(y_2)$.

A continuous function of two continuous random variables is $y_i = g_i(x_1,x_2)$, $i = 1,2$. The p.d.f. of (x_1,x_2) is $f(x_1,x_2)$, and $x_i = g_i^{-1}(y_1,y_2)$, $i = 1,2$. The p.d.f. of (y_1,y_2), in the form $h(y_1,y_2)$ at all points (y_1,y_2), is

$$h(y_1,y_2) = f(x_1,x_2) \cdot |J|. \qquad 2.26$$

In Eq. 2.26

$$|J| = \left|\frac{\partial x_i}{\partial y_i}\right| = \begin{vmatrix} \frac{\partial x_1}{\partial y_1} & \frac{\partial x_2}{\partial y_1} \\ \frac{\partial x_1}{\partial y_2} & \frac{\partial x_2}{\partial y_2} \end{vmatrix} \qquad 2.27$$

is the Jacobian determinant having non-zero values at all points (x_1,x_2), with these points replaced in Eq. 2.26 by $g_1^{-1}(y_1,y_2)$ and $g_2^{-1}(y_1,y_2)$, respectively. This gives

$$\int_{(x_1,x_2)} f(x_1,x_2)\, dx_1\, dx_2 = \int_{(y_1,y_2)} f[g_1^{-1}(y_1,y_2), g_2^{-1}(y_1,y_2)]\, |J|\, dy_1 dy_2. \qquad 2.28$$

Equation 2.28 can easily be extended to more than two-dimensional functions of more than two random variables.

Applications in hydrology. Distributions of functions of random variables have direct application in hydrology, enabling computations of distributions of functions from distributions of random variables without first computing function values from the data available on random variables. If variables x_1 and x_2 are given by their respective samples, the normal statistical procedure is first to select a general distribution function, either by experience, physical consideration, or by trial and error. Then, its parameters are estimated by the best estimation methods. Finally, the goodness of this fit is tested. If the distribution of functions $y_1 = g_1(x_1,x_2)$ and $y_2 = g_2(x_1,x_2)$ are needed, but y_1 and y_2 are directly computed from the sample data of x_1 and x_2, the above procedure for selecting a distribution function, of estimating its parameters, and testing the goodness of fit must be repeated. This approach is usually followed in hydrologic practice.

The distribution of y_1 and y_2, $F(y_1,y_2)$, may be determined directly without following the computational route, if Eqs. 2.20 through 2.28

are applied. It is assumed, however, that new tests of the distributions of y_1,y_2, and (y_1,y_2) variables will show a goodness of fit similar to that of the x_1,x_2, and (x_1,x_2) random variables. If the fitting procedure of distributions are made independently for random variables and for functions of random variables, the resulting distributions may not be related as required by Eqs. 2.23 through 2.28.

With the availability of digital computers, it may be much simpler to determine sample values of functions of random variables by computations, by applying programs available for the estimation of parameters, and by testing the goodness of fit of distributions selected, rather than to use Eqs. 2.20 through 2.28. Because the type of distributions for functions of random variables is automatically determined as soon as these functions and the distributions of basic random variables are known, and because they might not be used in the fitting procedure, speak against the above alternative of a pure computational approach.

Since the joint distribution of two variables is often already known in hydrology, the distributions of new functions, being sums, differences, products, or ratios of the two variables are often required. If (x,y) is a continuous bivariate with the p.d.f. $f(x,y)$ for $x \geq 0$, $y \geq 0$, and zero elsewhere, then [10, p. 58—62]

the p.d.f. of $u = x + y$ is

$$f(u) = \int_0^{\infty} f(x, u - x)\,dx, \tag{2.29}$$

the p.d.f. of $v = xy$ is

$$f(v) = \int_0^{+\infty} \frac{1}{|x|} f\left((x, \frac{v}{x}\right) dx, \tag{2.30}$$

and the p.d.f. of $w = y/x$ is

$$f(w) = \int_0^{\infty} |x|\, f(x,wx)\,dx. \tag{2.31}$$

Example 2.34. The bivariate distribution of flood peaks of two rivers is determined from the data of two gauging stations, one on each of them immediately upstream of their confluence. Often, the distribution of flood peaks on the main river below the confluence is of interest. Let assume that logarithms, $\ln x_1$ and $\ln x_2$, of peak discharges x_1 and x_2 are normally distributed with μ_1 and μ_2, σ_1 and σ_2, the means and the standard deviations of $\ln x_1$ and $\ln x_2$, respectively. The correlation coefficient between $\ln x_1$ and $\ln x_2$ is ρ. Because of the linear correlative dependence among flood peaks, the bivariate probability density $f(x_1, x_2)$ is

$$f(x_1, x_2) = \frac{1}{2\pi\sigma_1\sigma_2 x_1 x_2 \sqrt{1-\rho^2}}\, e^{-\frac{1}{2} Q(x_1, x_2)} \tag{2.32}$$

where

$$Q(x_1, x_2) = \frac{1}{1-\rho^2}\left[\frac{(\ln x_1 - \mu_1)^2}{\sigma_1^2} + \frac{(\ln x_2 - \mu_2)^2}{\sigma_2^2} - 2\rho\,\frac{(\ln x_1 - \mu_1)(\ln x_2 - \mu_2)}{\sigma_1\sigma_2}\right] \tag{2.33}$$

The distribution of the peak discharge of the main river, as the variable $u = x_1 + x_2$, is according to Eq. 2.29,

$$f(u) = \frac{1}{2\pi\sigma_1\sigma_2\sqrt{1-\rho^2}} \int_0^\infty \frac{1}{x_1(u-x_1)} e^{-\frac{1}{2}Q(x_1, u-x_1)} dx_1. \qquad 2.34$$

The integration of Eq. 2.34 can be carried out by a finite differences method. The problem is a classical hydrologic dilemma, as in the case of whether data on peak discharges of two joining rivers have more information about floods than data on flood peaks at the main river below the confluence.

Example 2.35. Let suppose that the probability distribution of river water discharge x_1 and its suspended sediment discharge x_2 is given as a bivariate distribution in the form of Eq. 2.32. The concentration is defined as $w = x_2/x_1$, and its probability distribution is, according to Eq. 2.31,

$$f(w) = \frac{1}{2\pi\sigma_1\sigma_2\sqrt{1-\rho^2}} \int_0^\infty \frac{|x_1|}{w x_1^2} e^{-\frac{1}{2}Q(x_1, wx_1)} dx_1. \qquad 2.35$$

Similarly as for Eq. 2.34, Eq. 2.35 may be integrated by a finite differences method.

2.8. **Problems for Discussion and Solutions.**

2.1 What is the difference between a random event and the outcome (observation, realization) of an experiment or trial?

2.2. Demonstrate that the relative frequency of a random event, defined as the interval x_1 to x_2 of a continuous random variable, converges to its probability with an increase of the sample size. Select a hydrologic variable, such as daily river flow, bed elevation when it is a random variable, daily rainfall, velocity or pressure at a point in a turbulent stream, river water temperature, or a similar variable which is given either as continuous or discrete series. Show graphically how this relative frequency, $f_i = m/N$ or $f_i = \Delta T/T$, changes and approximates the probability p_i, with a stepwise increase of the sample size until all available series is used.

2.3 Consider m large independent storms of limited areal coverage, which hit with a probability 1/n any one of a total of n small and equal areas ($n > m$) in a region, A. Find the probability that (a) each of specified m areas receives a storm; and (b) m arbitrary areas each receives a storm. Use the classical definition of probability, and Boltzmann, Bose-Einstein, or Fermi-Dirac statistics, depending how the complete group of equally likely events is defined, [5, pages 39—42].

2.4 In Example 2.26, the wet and dry years of a continuous variable of annual precipitation are reduced to five discrete and elementary random events (very dry, dry, average, wet and very wet years). Find the total number of random events and the number of composite events.

2.5. Show that the pairwise mutually exclusive events, like rainy and nonrainy days at a precipitation gauging station, are dependent random events.

2.6. If daily precipitation amounts at gauging stations A and B are dependent variables, are rainy days at A, and nonrainy days at B, dependent or independent random events?

2.7 Select the largest observed flood discharge of a river of about 100,000 sq. mi. and determine what the rainfall intensity should be to produce this large flood over a river basin of only 100 sq. mi., for which a flood frequency curve already exists, by using the unit hydrograph approach. Determine also the approximate probability of floods greater than this flood on the small river basin by using the fitted double exponential function of largest extreme values to a flood frequency curve.

2.8. Give several hydrologic examples of univariate, bivariate, and multivariate random processes composed of periodic and stochastic components, and discuss the composition and the physical justification for these components.

2.9. Give several hydrologic examples of discrete, continuous, and mixed random variables.

2.10. Give several hydrologic examples, beyond those given in the previous text, of positive-valued variables, and of variables with both positive and negative values.

2.11. Represent schematically the probability distribution of reservoir levels, with two discrete boundaries, as discussed in Example 2.20.

2.12. Give several hydrologic examples of each of the three types of conditional events, $(A \mid B)$, $(A \mid B_1, B_2, \ldots, B_n)$, and $(A_1, A_2, \ldots, A_m \mid B_1, B_2, \ldots, B_n)$, and of conditional variables $(X \mid Y)$, $(X \mid Y_1, Y_2, \ldots, Y_n)$, and $(X_1, X_2, \ldots, X_m \mid Y_1, Y_2, \ldots, Y_n)$.

2.13. A bivariate p.d.f. of x_1 and x_2 is given by Eqs. 2.32 and 2.33. Write the p.d.f. of the conditional variable $x_2 \mid x_1$ for this case.

2.14. A non-soluble floating pollutant is dumped in the middle of an elongated lake, L_1-distance from its upper end and $L - L_1$ from its outlet. Winds blow randomly in two opposite directions of the narrow valley, with p = probability of winds blowing upstream of the valley, and $q = 1 - p$ blowing downstream, with a displacement of each wind blowing equal a-distance units, with L_1/a and L/a integers. What is the probability p_1 that the upstream end is reached by the pollutant before it reaches the outlet? What is the probability of the opposite event, p_2? These probabilities of conditional events may be solved by the famous ruin-of-a-gambler problem. Find solutions by the following equations, given in reference [5, p. 76–80],

$$p_1 = \frac{1 - (q/p)^{(L-L_1)/a}}{1 - (q/p)^{L/a}}; \quad p_2 = \frac{1 - (p/q)^{L_1/a}}{1 - (p/q)^{L/a}},$$

and the particular values: $L_1 = 4.5$ km, $L - L_1 = 5.5$ km, $p = 0.60$, $q = 0.40$, and $a = 0.25$ km. How realistic is this approach to a practical case?

2.15. The water supply of an important consumer is provided from three sources of water, so that the total demand q_t is covered as $q_t = q_1 + q_2 + q_3$. The joint probability that in a year the sources will supply the requested quantities is $P(Q_1 \geq q_1, Q_2 \geq q_2, Q_3 \geq q_3) = 0.93$. The reservoirs and conveyance structures limit individual supplies to q_1, q_2, and q_3, respectively. Find the probability $P(Q_t < q_t)$ of the supply being less than q_t. (Use the concept that at least one of the complementary events to Q_1, Q_2, and Q_3 occurs).

2.16. A variable x is normally distributed, with the p.d.f. given in Example 2.30. Find the p.d.f. of the functions: $y = e^{-ax}$, $y = 1/x$, and $y = a + bx + cx^2$.

2.17. Take two confluent rivers, with the series of daily discharge available and determine their flood frequency curves; then fit them by the double exponential functions,

$$F(Q) = \exp\{-\exp[-(a_i x + b_i)]\}$$

with $a_i = a_1$ and $b_i = b_1$ parameters for the first river, and $a_i = a_2$ and $b_i = b_2$ for the second river. Find the frequency distribution of floods by using these two equations and ρ computed, as one method, and by using Eq. 2.23 and the set E_z, as the other method, and compare the two results. As an alternative, find the frequency distribution of floods of the main river by using the above two distributions and the concepts involved in Eq. 2.23. For simplicity, assume that the floods from one river are independent of floods from the other.

2.18. The variable x is normally distributed as given in Example 2.30, and the variable y is lognormally distributed as given in Example 2.31. Write distributions for the following functions of these two variables: $z_1 = x + y$, $z_2 = x - y$, $z_3 = x/y$, $z_4 = y/x$, and $z_5 = xy$. Discuss how the integrals can be best solved.

CHAPTER 3

VARIOUS PROBABILITY TOPICS APPLIED TO HYDROLOGY

All theorems and topics of probability theory may be applied to hydrology provided the hypotheses underlying these theorems are satisfied, at least approximately, for hydrologic random events or variables. This chapter treats combinatorial calculus in its most condensed form, some applications of geometric probabilities, Bayes's theorem, limit theorems, and laws of large numbers. To better discuss these two last topics in a summary form concepts of Markov simple and homogeneous chains, mathematical expectation, and variance are introduced.

3.1. Combinatorial Analysis

Definitions of permutations and combinations. The selection of r objects from n distinct objects, with their order arrangement taken into account, constitutes a permutation and all these different arrangements represent the number of permutations. Not only the identity of each object but also their order of arrangement is important. If two of r objects are interchanged in their respective positions, a new permutation is obtained.

The selection of r objects from n distinct objects, without reference to their order of arrangement in a group of r objects, constitutes a combination. All these groups represent the number of combinations. It is evident that number of combinations is smaller than the number of permutations because several permutations of r objects have the same number of objects differently arranged. It follows that the total number of permutations is obtained by taking every possible combination and arranging each in all possible ways.

Example 3.1. The sampling scheme is important in assessing the potential results of weather modification by cloud seeding in a mountain region, in which n small river basins are involved. A number r of n basins, with $r < n$, should be instrumented for various observations in connection with the weather modification undertaking. The total number of groups of r basins, out of n basins, regardless of their order or sequence in each group, is the number of combinations. If the 7th, 15th, 23rd, and 29th basin out of 32 basins constitute a group, the order is not important in this case, and it is irrelevant whether 7, 15, 23, 29, or 23, 15, 29, 7 or any other arrangement is obtained.

The sequence of basins in the group may be important, however, because of communication, economy, or some other constraint in solving these problems. In this case, the total number of groups of r basins, out of n basins with their order or sequence taken into account, is the number of permutations. Sequence, or arrangement, or order, of r objects must be important in order to use permutations.

Combinatorial formulae and probability. Combinatorial formulae for the numbers of combinations and permutations are often used in probability theory, especially when combinations and permutations are all possible, favorable or unfavorable composite random events out of a given number of elementary random events. The most condensed exposition and application of these formulae is used in this text. For more detailed reading, abundant literature is available on this subject [1, 2, 3, and others].

Example 3.2. The case discussed in Example 3.1 should produce the number of composite events that are r basins out of n distinct river basins. The n basins may be conceived as elementary random events (say an occurrence of a seeded storm over each of them), while combinations or permutations of these basins may be considered as composite random events.

Number of permutations. The total number of permutations is obtained by considering r positions in a sequence as fixed, and determining the number of ways in which objects can be placed in these r positions. For the initial position, any one of n objects can be chosen to fill this position. For the second position, any one of n − 1 remaining objects can be selected to fill this next position. The r-th position is filled by choosing any one of the remaining (n − r + 1) — objects. Hence, the total number of permutations of n objects taken r at a time, denoted by ${}_nP_r$, is

$$ {}_nP_r = n(n-1)\dots(n-r+1) = \frac{n!}{(n-r)!}. \qquad 3.1 $$

Equation 3.1 also gives the number of different samples of size r that can be obtained from a finite population of size n. For r = n, Eq. 3.1 becomes:

$$ {}_nP_n = n!. \qquad 3.2 $$

In many cases, n objects may not be all distinct or they may not be considered as distinct. The case of k distinct kinds of objects out of n objects are often relevant, such that there are r_1 objects of the first kind, r_2 of the second kind, . . . , and r_k of the k-th kind, with $r_1 + r_2 + \dots + r_k = n$. If r_i-like objects are still identifiable or different among themselves, then they can be arranged in $r_i!$ ways. This applies to each of k distinct kinds of objects. The total number of these permutations is the product $r_1!\, r_2! \dots r_k!$. If all n objects were different, and n objects taken all at a time, the total number of permutations of n objects is n!. Then

$$ P = \frac{n!}{r_1!\, r_2! \dots r_k!} = \frac{n!}{\prod_{i=1}^{k} r_i!} \qquad 3.3 $$

is the total number of permutations of n objects in which there are r_1 — alike, r_2 — alike, . . . , r_k — alike.

Example 3.3. For the case of Example 3.1, let n = 32 and r = 4, or four small river basins are observed out of 32 basins subject to cloud seeding. If sequences of selected basins are important, Eq. 3.1 gives ${}_{32}P_4 = 32!/28! = 32.31.30.29 = 863{,}040$, which represents a large number of composite events.If a constraint is made, however, that 32 basins are divided in four regional groups, each composed of eight adjacent and distinct river basins, and from each group one basin is selected for observations, then one basin from each group may be combined in $8^4 = 4096$ ways. However, the sequences of groups may be combined in $4! = 24$ ways. Therefore, if the sequence of four selected basins is relevant, the number of permutation in this later case is $P = 4!\ 8^4 = 98{,}304$.

Number of combinations. Let ${}_nC_r$ denote the total number of combinations of n objects taken r at a time. The number of arrangements of r objects in r places is r!. Hence, the total number of permutations is given by multiplying the number of combinations by r!, so that ${}_nP_r = {}_nC_r\, r!$. By using Eq. 3.1,

$${}_nC_r = \frac{n(n-1)\dots(n-r+1)}{r!} = \frac{n!}{r!(n-r)!} = \binom{n}{r}. \qquad 3.4$$

It is clear that ${}_nC_r = {}_nC_{n-r}$, so that this identity can serve to save time in comuputations.

It is important to stress that ${}_nC_r$ are coefficients of the general binomial expression $(a + b)^n$, or

$$(a+b)^n = \sum_{r=o}^{n} {}_nC_r\, a^{n-r} b^r, \qquad 3.5$$

where ${}_nC_r$ is given by Eq. 3.4. It can be proven that

$${}_nC_r + {}_nC_{r-1} = {}_{n+1}C_r, \qquad 3.6$$

and

$${}_{n+2}C_{r+1} = {}_nC_{r+1} + 2\,{}_nC_r + {}_nC_{r-1}. \qquad 3.7$$

Example 3.4. Examples 3.1 and 3.3 assume n = 32 and r = 4 and the value r! = 24. Then Eq. 3.4 gives ${}_{32}C_4 = \binom{n}{r} = \binom{32}{4} = {}_{32}P_4/4! = 35{,}960$. It is 1/24 of the number of permutations. However, if the four small river basins should not fall too closely, but each one is in the selected four regional groups of eight basins then each group has ${}_8C_1 = \binom{8}{1} = 8$, so that the total number of combinations is reduced from 98,304 to $8^4 = 4096$, which is 1/24 times the number of permutations for the same constraint.

Applications in hydrology. Combinatorial analysis is used in hydrology either for discrete variables, or for continuous variables divided into intervals that are *states*, but it is used mainly for computing the number of favorable, the number of unfavorable, and the total number of random events. The *state* is defined as an interval of a continuous random variable, with the outcome of a variable in any part of that interval being of equal importance to the solution of a hydrologic or water resource problem. Usually, there are n distinct items, each having r different states.

Example 3.5. A water resource development scheme of a region contains n = 5 important reservoirs. They are interdependent because their sequences of impoundments and releases affect the optimal operation. In optimization, the five continuous variables of reservoir levels are replaced by discrete and distinct states in the form of layers, including the boundary states of a full and an empty storage. Each storage capacity is divided into m = 6 states, the two boundary states plus four intermediate layers of equal volume. Sequences of impoundments and releases of water of various layers of n reservoirs may or may not be important. In the former case, the number of permutations, and in the latter case, the number of combinations are relevant for the computations of the total number of events, or the number of general states is the number of combinations. For each reservoir $C_i = \binom{m}{1} = m$, so that the total number of combinations is $C = m^n = 6^5 = 7776$. In sequences of impoundments and releases of n reservoirs being important, the number of combinations of reservoirs is n!, and the number of permutations is $P = n!\, m^n = 5!\, 6^5 = 933{,}120$. It is clear that no optimization procedure would be feasible to inverstigate such a large number of cases. Therefore, sampling schemes usually combined with the constraints that reduce the number of permutations must be used in practical optimizations.

Example 3.6. A design for sampling the water quality in a lake requires 10 points to be sampled at each of 10 elevations, for a total of 100 points. At each individual survey, the number of sampling points is 50. How many combinations are possible for any survey? Equation 3.4 gives $C = \binom{100}{50}$ 100!/50! 50!. By using the tables found in [4], $C = 9.3326 \cdot 10^{157}/2 \cdot 3.0414 \cdot 10^{128} \approx 1.53 \cdot 10^{39}$. It is evident why observations in practice are made at fixed points in every survey. Any random sampling would involve only a small fraction of all possible combinations, when the number of sampling points in each survey is much smaller than the total number of points.

3.2. Geometric Probabilities

Geometric interpretation of probability. A fixed large area of size A has another small area $A_0 \leq A$, which is singled-out in some way. If a particle is thrown and randomly hits the area A, and only A, the question arises what is the probability of hitting the singled-out area A_0. If the same reasoning is applied to a line L, and only L, what is the probability that a particle falling only on the line L would be within L_0, a small segment of L. In both cases the probability is determined by the approach of equally likely events, if any point of the area or the line has the same probability density of being hit,

$$p = \frac{A_0}{A} \quad \text{or} \quad p = \frac{L_0}{L}. \tag{3.8}$$

This is a basic postulate of geometric probability, with the assumption that all events in the area A, or along the line L, are equiprobable random events. In other words, if area A is divided into small equal elementary areas ΔA, with or without overlapping, or if line L is divided into the small elementary intervals ΔL, also with or without overlapping, all of them have the same probability of the particle falling on them. In this way, elementary points as random events are replaced by elementary areas or elementary intervals.

If the random hit of a particle is replaced by a very small area $\Delta A < A_0$ falling within A, in the first case, or a very small interval $\Delta L < L_0$ to occur within L, the question arises as to the probability that area ΔA falls within the area A_0, if $\Delta A < A_0$? Similarly, what is the probability that ΔL falls within L_0 if $\Delta L < L_0$? This case depends on the sizes and shapes of areas A, A_0, and ΔA, or the relative lengths of L, L_0, and ΔL. These are more complex random events than for the case of a particle falling within an area or on a line. The famous Buffon's problem of a needle L long, randomly thrown and intersecting any parallel line in a system of lines that are $d > L$ apart, is somewhere between the above two problems. In Buffon's problem the probability of the event is $P = 2L/\pi d$ [5].

Problems of an area partly or fully covering another area inside a large region, or similarly for an interval partly or fully covering another interval on a large segment of the line, are relevant to hydrologic applications. There are many examples. The transposition of an observed storm in a region to various river basins, the probability of an area being hit by hail, the probability of a river section or lake area being frozen during a particular cold wave, the probability of a fish population being poisoned along a river section, and similar problems may be treated by these geometric probabilities. Geometric probabilities enable a visual explanation of various random events, and their probabilities are understood by visualizing Venn--diagrams, often used in set theory.

Possible random events with zero probability. For a continuous variable x, any individual event x_i has the probability density $f(x_i)$, and only its intervals have finite probabilities. As an individual event x_i of a continuous random variable has a zero interval, its probability is zero or $P(X = x_i) = 0$. However, this event is possible. Often it is defined as $P(X \neq x_i) = 1 - P(X = x_i) = 1$. In this case, expressions *almost never*, or *almost always* events are used, where *almost* may be replaced by the term *except for events with zero probability*.

This case of possible random events with zero probability is illustrated by way of geometric probabilities. Throwing a disk of a given diameter d to fall exactly over a circular mark of the same diameter has the probability zero, though this event is possible. The event has a probability density different from zero. However, the event that more than 50 percent of the disk area falls within the mark has a finite probability.

Example 3.6. Transposition of storms is a classical hydrologic example of geometric probabilities. Assume that many storms were recorded in a region, with their areal distribution of isolines of the total rainfall. Storms can be described by probability distributions of their various variables of significance to flood events. If the largest flood peak of all storms is required for a small river basin, the greatest observed storm in a region transposed to the most critical position over that basin will produce the largest flood peak of all observed storms, and of all their possible positions over the basin. This flood peak is a possible event, but the probability that it will be produced by such an event and from such a position is zero. As greater storms may occur in the future, however, their transpositions in highest flood peak producing positions also have probabilities of occurrence equal to zero, though they are possible events. Difficulties with the current concept of transposition of storms in the highest flood producing position is that the resulting floods cannot be assigned meaningful probabilities without further investigations.

Geometric probabilities as applied to storm transpositions. It is possible to transpose storms and at the same time to assign probabilities to resulting floods. With the digital computer, the seemingly complex approach, as outlined here, may eventually be trivial. This same approach can be used to assess the probability of hail of various intensities and durations over a given area, provided many hail events are well recorded for some time over a wide region.

Let us assume that a region of relatively homogeneous storm characteristics has been well observed for a period of time by a dense network of rainfall gauging stations, and that a number of severe storms has been recorded. If storms are sorted in the order of magnitude according to their potential of producing flood peaks in a particular river basin of that region, then each storm may be assigned a probability of occurrence. Simultaneously, the preceding rainfall for each storm provides information for the computation of antecedent moisture conditions over the area to which the storm is transposed. Let us also assume that a transfer mechanism is available for determining the flood hydrograph out of a transposed storm, with its antecedent precipitation and all other river basin factors given. As hyetographs at several points of storms may be available, the time history of the storm may also be known. By dividing the total storm rainfall at various points into increments of equal time units, and by using either a lumped or a distributed system (sub-basins) and the proper integration procedure, each position of a storm over a river basin would result in a flood hydrograph. The purpose of this is to obtain the probability distribution of large flood peaks, say for a sufficient number of Q_{max} values that satisfy the condition $Q_{max} \geq Q$, with Q = large flood peaks.

If N largest storms for N years of observations are selected from the region, one for each year, any of these storms has the probability of occurrence over the region $P_i = 1/N$. Transposition of all storms that can produce large floods with $N_i > N$, rather than only the largest annual storms, gives a partial series of floods. This is a problem similar to taking all floods greater than a given flood value at a station in the partial series approach to flood analysis. Regardless of whether the annual or partial series of storms is used, probabilities can be attached to storms that have occurred over that region.

Storms may have elongated isolines. Therefore, during a transposition the storm isolines may also be rotated. It may or may not be assumed that all directions have the same probability densities. If all directions are equally likely, for a given position of the storm center, m equally likely directions of storms may be selected so that the probability of one of N storms in one of m directions, for a selected center position, is $P_j = 1/mN$.

The center of a storm may then be transposed at a set of selected positions of x,y-coordinates over the area of a river basin. A grid of points (x,y) should be selected with the center C of each storm located at each point, and for each position the storm is rotated in m positions. Figure

3.1 shows, schematically, the location of storm center C at the coordinate origin.

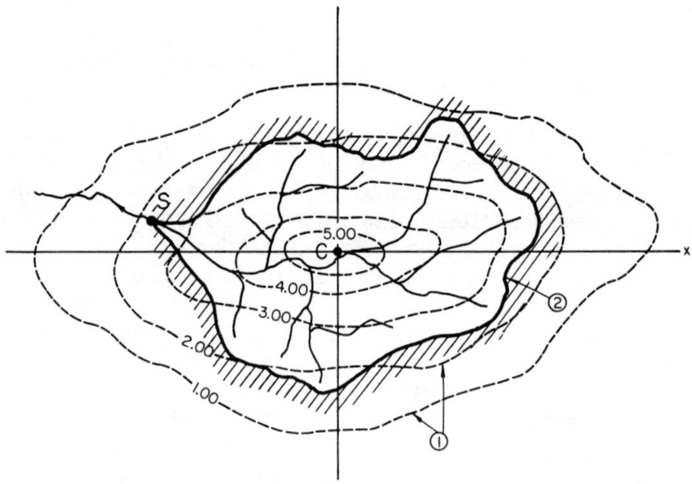

Figure 3.1. A transposed storm with the isolines in inches of equal rainfall, (1), over a river basin, (2), for the estimation of probability of flood peaks at the river gauging station, S. The storm center C is at the coordinate origin.

Each position and direction of a given storm produces a flood hydrograph by using the proper rainfall to runoff transfer technique. The flood descriptor (flood peak, flood volume, etc.) is obtained for that position and that direction. For a sufficient number of positions and a given direction, descriptors that permit the drawing of their isolines are obtained.

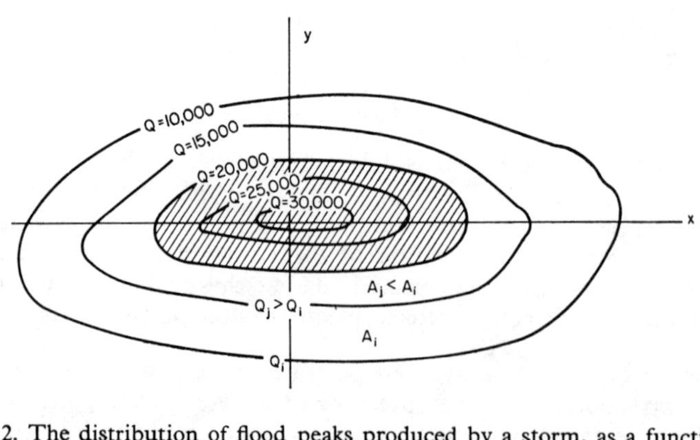

Figure 3.2. The distribution of flood peaks produced by a storm, as a function of the position of storm center set in a given direction in the coordinate system (x, y) of Fig. 3.1.

Figure 3.2 shows such isolines for the flood peak. In this case, the areas of the positions of centers, which produce $Q_{max} \geq Q$, can be obtained as A_i. For a given Q = 20,000 cfs of Fig. 3.2, the probabilities of floods greater than Q = 20,000 cfs are

$$P(Q_{max} \geq 20{,}000) = \frac{A_i}{A}. \qquad 3.9$$

Because this flood is produced from a given storm of probability 1/N, and a direction of probability 1/m, its probability is

$$P_i = P(Q_{max} \geq 20{,}000) = \frac{1}{m}\frac{1}{N}\frac{A_i}{A}. \qquad 3.10$$

If this procedure is repeated for all N storms and all m directions of transposed storms, then the final probability is approximately

$$F(Q_i) = P(Q_{max} \geq Q_i) = \frac{1}{mNA}\sum_{j=1}^{mN} A_j, \qquad 3.11$$

in which A_j are all areas enclosed by isolines of the flood descriptor Q_i for all N storms, and all m transposed storm directions. Probabilities of floods for various values of Q_i enable the computation of the flood frequency curve, $F(Q_i)$.

The time factor can be also introduced with rainfall intensities as a function of time at several points of recorded storm. Also, probability of the directions of isolines of all storms may be studied, and if they come out to be significantly different from 1/m, these new probabilities may be used.

The derivation of Eq. 3.11 is based on selecting the largest storm for each year from the climatologically homogeneous region and the approximate validity of computed probabilities for each storm, and the position and direction of each storm. Equation 3.11 for a sufficiently small Q_i gives $F(Q_i) = 1$, because each storm may give $A_j = A$ in this case. Therefore, the approximation of the flood probability function of Eq. 3.11 may be more reliable for larger values of Q_i, with $A_j \ll A$, than for its small values.

Another approach is to select all storms (partial series) that have the maximum point rainfall of the storm greater than a given rainfall critical value. The number υ of storms per year is Poissonian distributed as $P(\upsilon) = e^{-\lambda}\lambda^{\upsilon}/\upsilon!$, with $\lambda = n/N$ the average number of these storms per year. For any storm transposed over a basin the probability of $Q_{max} > Q_i$ becomes

$$P(Q_{max} \geq Q_i) = \frac{1}{n}\sum_{j=1}^{n}\sum_{i=1}^{s}\frac{A_j}{A}\cdot F_i(D), \qquad 3.12$$

in which A_j as above in Eq. 3.11, and $F_i(D)$ represents a nonuniform distribution of storm directions for s-segments of storm directions, D.

By using the random number $\lambda = n/N$ of selected class of storms (with the maximum point rainfall of the storm greater than a prescribed rainfall critical value), it can be shown that

$$F(Q_i) = 1 - e^{-\lambda P(Q_{max} \geq Q_i)}$$

which in the final form becomes

$$F(Q_i) = 1 - \exp\left[-\frac{1}{NA}\sum_{j=1}^{n}\sum_{i=1}^{s} A_j F_i(D)\right]. \qquad 3.13$$

Equation 3.13 shows that even for a very large number n of selected storms, for a relatively small value of maximum point critical rainfall in the storm, as the criterion for selecting the storm, Eq. 3.13 always gives $F(Q_i) < 1$.

This approach requires a reliable storm-flood hydrograph transfer technique; a sufficient number of points for the position of storm center; digital computer oriented computations; relative homogeneity of storm characteristics over the region; sufficient amounts of data on flood storms, and similar conditions. If storm patterns differ over the region, and they can be properly described, storms transposed over a river basin can be adjusted to the conditions over it.

3.3. Applications of the Theorem of Total Probability and the Bayes's Theorem.

Theorem of total probability. Suppose an event A can occur only jointly with one of several mutually exclusive random events, E_1, E_2, ..., E_n, so that $\Sigma P(E_i) = 1$, $i = 1, 2, \ldots n$, or it is the probability of a certain event. In that case,

$$P(A) = \sum_{i=1}^{n} P(A, E_i). \qquad 3.14$$

As each joint probability $P(A, E_i)$ can be expressed (see Eq. 2.10) by

$$P(A, E_i) = P(E_i) \cdot P(A \mid E_i), \qquad 3.15$$

then Eq. 3.14 becomes

$$P(A) = \sum_{i=1}^{n} P(E_i) \cdot P(A \mid E_i). \qquad 3.16$$

Equation 3.16 is known as the theorem or the formula of the total probability. The condition for applications of Eq. 3.16 is only that events E_i are mutually exclusive or incompatible. These events may have different probabilities, or they are not equiprobable events. The set E_i does not need to be exhaustive because A may not depend on some of them (or A event and some E_i events are mutually independent).

Figure 3.3 shows the Venn-diagram for this case. Equation 3.14 requires that the probability P (A) is the sum of joint probabilities P (A, E_i), or the sum of probabilities of eight joint events (A, E_i) of Fig. 3.3. Equation

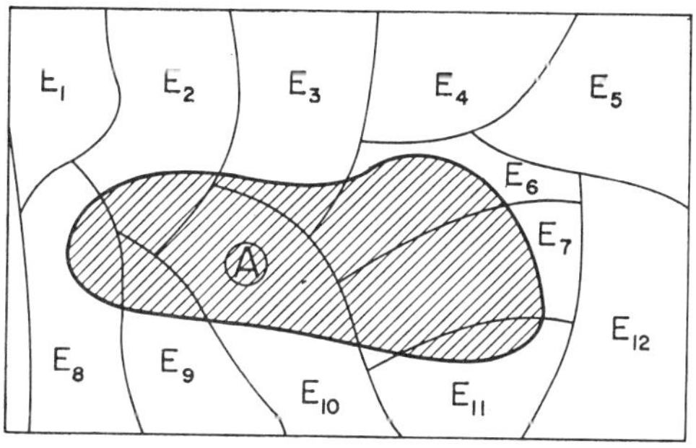

Figure 3.3. Venn diagram for the representation of the theorem of total probability in the form of geometric probabilities.

3.16 means that P (A) is the sum of probabilities of each event E_i, multiplied by the corresponding probabilities of conditional events $A \mid E_i$.

Example 3.7. A plan is being developed for flood protection in the valley of a large river by building a flood control reservoir on the largest tributary. This tributary contributes different water volumes during large floods in the valley. The event A is the exceedence of a large flood volume in the valley, and $E_1, \ldots, E_n$ are the events of flood volumes being in the given ranges of volume on the major tributary. Because the basin is large and because the storm events rarely cover all the basin, the contribution of this major tributary varies in a large range of flood volumes during a large flood in the main valley.

The conditional event $A \mid E_i$ means that the main river exceeds its critical flood volume, provided a tributary event, E_i, occurs. Equation 3.16 tells how to compute the probability P (A) of floods in the valley in terms of probabilities of flood volumes on the major tributary. When the flood control reservoir is built, and new probabilities $P(A \mid E_i)$ and $P(E_i)$ are estimated on the basis of flood routing through the new reservoir, the new probability of event A may be obtained.

Bayes's theorem. In many applications, the value $P(A \mid E_k)$ is given and the value $P(E_k \mid A)$ must be determined, or *what must have happened* should be found. In Example 3.7, the probabilities $P(A \mid E_i)$ are known, and it may be asked if event A has happened, what are the probabilities of various conditional events $E_k \mid A$, or probabilities of various contributions by the tributary to the flood in the main valley.

From the basic equation of conditional probabilities, given by Eq. 2.10, it follows

$$P(A, E) = P(A) \cdot P(E \mid A) = P(E) \cdot P(A \mid E). \qquad 3.17$$

Then this equation gives, for P (A) replaced by Eq. 3.16,

$$P(E_k \mid A) = \frac{P(E_k) \cdot P(A \mid E_k)}{\sum_{i=1}^{n} P(E_i) \cdot P(A \mid E_i)}; \quad k = 1, 2, \ldots, n. \qquad 3.18$$

Equation 3.18 is known either as Bayes's theorem, Bayes's formula, or the formula of hypotheses. Problems can be solved for probabilities of events $E_k \mid A$ for the known probabilities of conditional events A, relative to E_i ($i = 1, 2, \ldots, n$), and the known a priori probabilities of E_i (or suggested by intuition) before the trial that produces an event, A, occurs. In that case, the a posteriori probabilities of Eq. 3.18, which are equal to probabilities of any conditional event of $E_i, \ldots, E_n$ relative to A, can be determined or furnished by the analysis.

Applications in hydrology. The application of Bayes's formula in hydrology is a controversial subject, as it is in many other disciplines. Anderson [6, p. 13] states that »Bayes's formula is practically of infrequent technical value in hydrology although certainly conditional probabilities themselves are involved in many problems.« One of the reasons for this limited application is the requirement for a sufficiently accurate knowledge of probabilities of occurrence of various hypotheses in problems involving Bayes's formula. An instructive example is given by Cavadias [6, p. 29] in a written discussion of Anderson's paper [6]. This example provides insight for the analysis of economic benefits of various hydrologic forecasts of different accuracies, in comparison with the problems solved only by the use of simple probabilistic statements of various states, based on previous hydrologic observations.

Example 3.8. Three factories F_1, F_2, and F_3 in an industrial area occasionally release lethal waste into a river. Probabilities of fish killing ingredients (as event A) being in the waste releases of the three factories are: $P(A \mid F_1) = 0.75$, $P(A \mid F_2) = 0.85$, and $P(A \mid F_3) = 0.05$. The a priori probabilities of factories releasing this particular waste are equal, $P(F_1) = P(F_2) = P(F_3) = 1/3$. The factories are not related in their operations. When a waste release has occurred, what is the probability of killing the fish and what are probabilities, of each factory being responsible. Difficulties in solving the problem is in the determination of the above probabilities; first, the probabilities of critical concentration in each waste release of three factories, and second, the probabilities of waste release. They can be obtained only by observation. Using the formula of total probability of Eq. 3.16, the probability is $P(A) = 0.75/3 + 0.85/3 + 0.05/3 = 0.55$. However, the individual a prosteriori probabilities of each factory being responsible for the effect are obtained from Bayes's formula as $P(F_1 \mid A) = (0.75/0.55)/3 = 0.45$; $P(F_2 \mid A) = (0.85/0.55)/3 = 0.52$, and $P(F_3 \mid A) = (0.05/0.55)/3 = 0.03$. Therefore, the investigation should start at F_2, proceed to F_1 if F_2 is not found responsible and finish with F_3 if neither F_2 nor F_1 are responsible. This problem can be analyzed without involving Bayes's formula of Eq. 3.18.

Example 3.9. Two nearby river gauging stations A and B regularly report the ice coverage of a navigable river. Two events are of interest for navigation: S_1 ice-covered and S_2 free-of-ice events of this river. During the period of icing, or from the earliest and the latest observed ice-cover, long-term observations show that the river is covered by ice 30 percent of the time, or $P(S_1) = 0.30$, while 70 percent of time it is free, or $P(S_2) = 0.70$. From experience, it was known that reports of station A were incorrect 2 percent of the time, while those of station B were incorrect 8 percent of the time.

At a given moment, station A reports ice cover (S_1), while the station B reports the river to be free of ice (S_2). Which report is valid? The probability of A reporting correctly while B reporting incorrectly, as the event X, is $P(X \mid S_1) = 0.98 \cdot 0.08 = 0.0784$, and of the event Y, of A incorrect and B correct, is $P(Y \mid S_2) = 0.92 \cdot 0.02 = 0.0184$. By using Eq. 3.18, the probability of the report of station A being correct is $P(S_1 \mid X) = 0.3 \cdot 0.0784/(0.3 \cdot 0{,}0784 + 0.7 \cdot 0.0184) = 0.645$, while the probability of the converse is $P(S_2 \mid Y) = 0.354$. Either the stations should be asked to repeat the information or station A should be trusted for its original report.

Example 3.10. This example closely follows the discussion by Cavadias [6, p. 29]. A hydrologic service inquires about the usefulness of its forecasts for the operation of a set of storage reservoirs. Particularly, the operator is interested in how to use the stored water in reservoirs before the spring snow melt occurs to make space for the incoming inflow. If the reservoirs are not sufficiently emptied, the wasted spillover is not used beneficially. If the reservoirs are emptied too much, they stay unfilled during the summer when important water uses are needed, with resulting loss of benefit. It is assumed that Bayes's theorem can solve this problem, provided that the a priori probabilities and various costs can be computed with sufficient accuracy.

Flood occurrence as a continuous variable is divided in three states: E_1, E_2, and E_3, the respective occurrences of exclusive events of large, medium, and small floods. The previous observations give the following probabilities $P(E_1) = 0.20$, $P(E_2) = 0.30$, and $P(E_3) = 0.50$. The proper actions, for large, medium, and small floods, taken after the flood forecast is made, are: B_1, B_2, and B_3. The average losses of the potential benefit due to the action taken, assuming the forecast is perfect, is given in dollars, as a_{ij}-coefficients, and are:

	B_1	B_2	B_3
E_1	2,000	30,000	100,000
E_2	35,000	3,000	55,000
E_3	150,000	80,000	2,000

The forecast is given also as three states: A_1, A_2, and A_3, or large, medium, and small floods, respectively. Differences between E, B, and A are: E's are real occurrences of the three types of floods, B's are action taken by the operator, and A's are flood forecasts. Extensive experience with flood forecasts produces the following probabilities $P(A_j \mid E_i)$, i, j = 1, 2, 3:

	A_1	A_2	A_3
E_1	0.60	0.25	0.15
E_2	0.15	0.70	0.15
E_3	0.15	0.30	0.55

so that $P(E_i \mid A_1) + P(E_i \mid A_2) + P(E_i \mid A_3) = 1$, for i = 1, 2, 3. The annual cost of forecasting these floods is $1,200. The question is whether the forecast is warranted or not. The perfect forecast is $P(A_1 \mid E_1) = P(A_2 \mid E_2) = P(A_3 \mid E_3) = 1$, with all other probabilities of conditional events being zeros. The real forecast is then in between this perfect forecast and the simple statements of $P(E_1)$, $P(E_2)$, and $P(E_3)$ obtained from past observations. The three cases are compared.

The a posteriori probabilities $P(E_i \mid A_j)$ of E_i occurring given that A_j is forecast are computed by Eq. 3.18. As an example, $P(E_1 \mid A_1) = 0.20 \cdot 0.60/(0.20 \cdot 0.60 + 0.30 \cdot 0.15 + 0.50 \cdot 0.15) = 0.12/0.240 = 0.500$, with $P(E_2 \mid A_1) + P(E_3 \mid A_1) =$

= 1.00 − 0.50 = 0.50. The total probability of A_1 is given by Eq. 3.16, or $P(A_1) = 0.240$. In summary

	$P(E_i)$	$P(E_i \mid A_1)$	$P(E_i \mid A_2)$	$P(E_i \mid A_3)$
E_1	0.20	0.500	0.122	0.086
E_2	0.30	0.187	0.512	0.128
E_3	0.50	0.313	0.366	0.786

The probabilities of A_1, A_2, and A_3 are 0.240, 0.410, and 0.350, respectively. The expected cost of action B_1, if the forecast is A_1, is $C(B_1, A_1) = P(E_1 \mid A_1)\, a_{11} + P(E_2 \mid A_1)\, a_{21} + P(E_3 \mid A_1)\, a_{31} = 0.500 \cdot 2{,}000 + 0.187 \cdot 35{,}000 + 0.313 \cdot 150{,}000 = 54{,}400$. Similarly, the cost of other combinations of B_i and A_i are determined

	A_1	A_2	A_3
B_1	54,400	73,000	122,600
B_2	34,300	27,100	50,200
B_3	60,900	41,000	17,200

with the optimal strategy of forecast underlined. Multiplying each optimal strategy loss by the above probabilities of A_1, A_2, and A_3, the expected annual loss, in the case of performing reservoir operations according to forecast, is $C = 0.240 \cdot 34{,}300 + 0.410 \cdot 27{,}100 + 0.350 \cdot 17{,}200 = 25{,}400$. By using probabilities of E_1, E_2, and E_3 from historic data, $C(B_j, E_i) = a_{ij}\, P(E_i)$, $i = 1, 2, 3$; $j = 1, 2, 3$, then annual losses are

	$P(E_i)$	B_1	B_2	B_3
E_1	0.20	400	6,000	20,000
E_2	0.30	1,050	900	16,500
E_3	0.50	75,000	30,000	1,000
		76,450	36,900	37,500

with the best outcome of $B_2 = 36{,}900$ by always assuming medium floods. The minimum expected cost in case of a perfect forecast is $C = 0.20 \cdot 2{,}000 + 0.30 \cdot 3{,}000 + 0.50 \cdot 2{,}000 = 2{,}900$ Therefore, the cost range is 2,900 to 36,900, with the alternative in case of using results of a forecasting service being 25,400. The forecast is beneficial. Any improvement in forecast, both in the time lag and in the accuracy of discharges of floods, has a substantial margin for a further reduction of losses. Basic conditions are the good estimates of all a priori probabilities, as well as of the financial effects of various actions.

3.4. **Limit Theorems.**

Sequences of independent trials. If k random events can occur as independent events, $A_1, A_2, \ldots, A_k$, and if they have the probabilities $P(A_1) = p_1$, $P(A_2) = p_2$, ..., $P(A_k) = p_k$, then the probability of the joint event $(A_1, A_2, \ldots, A_k)$ is

$$p = P(A_1, A_2, \ldots, A_k) = P(A_1)\, P(A_2) \ldots P(A_k) = p_1 p_2 \ldots p_k, \qquad 3.19$$

as the condition for a sequence of independent trials.

The probability that in n trials there are m_1 of events A_1, m_2 of events A_2, and m_k of event A_k, is

$$P_n(m_1, m_2, \ldots, m_k) = \frac{n!}{m_1!\, m_2!\ldots m_k!}\; p_1^{m_1}\cdot p_2^{m_2}\ldots p_k^{m_k}, \qquad 3.20$$

which is the multinomial distribution law. For $k = 2$, $p_1 = p$, $p_2 = q = 1 - p$, $m_1 = m$, $m_2 = n - m$, Eq. 3.20 becomes

$$P_n(m) = \binom{n}{m} p^m q^{n-m}, \qquad 3.21$$

which is known as the binomial distribution law. The above independent trials leading to Eqs. 3.20 and 3.21 are Bernoulli trials. If n given trials are independent, then any number s of them are also independent. The properties of discrete distributions of Eqs. 3.20 and 3.21 are discussed in detail in Chapter 7, with examples.

The local limit theorem. For n, m_1, m_2, ..., m_k large, it is difficult to compute probabilities of Eqs. 3.20 and 3.21; there is a need for their approximations by asymptotic formulas. The resulting asymptotic relation for probabilities of Eq. 3.20 for $0 < p_i < 1$ ($i = 1, 2, \ldots, k$), that in n independent trials the events A_i occur m_i times ($m_1 + m_2 + \ldots + m_k = n$) is

$$\lim_{n\to\infty} P_n(m_1, m_2, \ldots, m_k) = \frac{\exp\left[-\frac{1}{2}\sum_{i=1}^{k} q_i x_i^2\right]}{(2\pi n)^{(k-1)/2}\sqrt{p_1 p_2 \ldots p_k}}, \qquad 3.22$$

and it converges in all m_i's for which the corresponding x_i are in the arbitrary finite intervals $a_i \leq x_i \leq b_i$, and where $q_i = 1 - p_i$, and $x_i = (m_i - np_i)/\sqrt{np_i q_i}$ depends on p_i, n, and m_i. Equation 3.22 is called the local theorem [7, p. 95].

The local De Moivre-Laplace theorem. Substituting into Eq. 3.22 the case of $k = 2$, $q_1 = q$, $q_2 = 1 - q$, $p_1 = p$, $p_2 = 1\text{-}p = q$, $m_1 = m$, $m_2 = n\text{-}m$, $x_1 = (m - np)/\sqrt{npq} = x$, $x_2 = -(m - np)/\sqrt{npq} = -x$, it becomes

$$\lim_{n\to\infty} P_n(m) = \frac{e^{-x^2/2}}{\sqrt{2\pi npq}}, \qquad 3.23$$

for all n for which x are contained in a finite interval, or for all practical purposes and for a large n. This is known as the local De Moivre-Laplace theorem [7, p. 98].

Example 3.11. For $n = 100$, $m = 20$, $p = 1/5$, $q = 4/5$, the exact value of $P_n(m)$ is 0.0993. By Eq. 3.23, for $x = 0$ and $\sqrt{npq} = 4.0$, $P_n(m) = 0.3989/4 = 0.0997$. Similarly, for $n = 25$ and $m = 6$, the exact value $P_n(m)$ is 0.1633, while Eq. 3.23 for $x = 0.5$ and $\sqrt{npq} = 2.00$ gives $P_n(m) = 0.3521/2.00 = 0.1760$.

The integral limit theorem. For m the number of occurrences of an event of the probability in each of n independent trials, $0 < p < 1$,

$$\lim_{n\to\infty} P\left\{ a \leq \frac{m - np}{\sqrt{npq}} < b \right\} = \frac{1}{\sqrt{2\pi}} \int_a^b e^{-x^2/2}\, dx, \qquad 3.24$$

or this probability converges in a and b, $(-\infty \leq a \leq b \leq \infty)$ as $n \to \infty$. Equation 3.24 is also known as the integral (De Moivre-Laplace) limit theorem [7, p. 104—109]. The theorem of Eq. 3.24 can be extended to several random events, $m_1, m_2, \ldots, m_k$, [7, p. 109—117].

Example 3.12. The probability that the relative frequency $f = m/n$ of a random event A deviates from its probability p for less than a smallquantity ε may be determined by the integral limit theorem. This absolute difference must satisfy the inequality $|f - p| \leq \varepsilon$, or $|m - np| \leq n\varepsilon$. Then

$$-\upsilon = -\frac{n\varepsilon}{\sqrt{npq}} \leq \frac{m - np}{\sqrt{npq}} \leq \frac{n\varepsilon}{\sqrt{npq}} = \upsilon, \text{ so that}$$

$$P[|f - p| \leq \varepsilon] = P\left[-\upsilon \leq \frac{m - np}{\sqrt{npq}} \leq \upsilon\right] = \frac{1}{\sqrt{2\pi}} \int_{-\upsilon}^{\upsilon} e^{-x^2/2}\, dx,$$

with $x = (m - np)/\sqrt{npq}$. If $n \to \infty$, then $\upsilon \to +\infty$ and $-\upsilon \to -\infty$, so in this case, $P[|f - p| \leq \varepsilon] = 1$. Applied to a hydrologic example, let us determine the probability that the relative frequency $f_i = m/n$ of the random event of annual maximum discharge differs from its probability $P = F(Q) = P(Q_{max} \leq Q) = 0.90$, by at most $\varepsilon = 0.05$ for the period of observation of $n = 50$. In this case, $\upsilon = 50 \cdot 0.05/\sqrt{50 \cdot 0.90 \cdot 0.10} = 1.178$. The above normal integral from the tables gives the value 0.761, so that $P[|f_i - p| \leq 0.05] = 0.761$.

Example 3.13. The problem of Example 3.12 may be reversed in looking for the number of years for which observations must be performed in order to have the absolute error-difference, between the relative frequency and the probability, at most $\varepsilon = 0.05$, with the probability at least $P = 0.95$. In this case, the normal integral for $P = 0.95$ gives $\upsilon = 1.96$, so that the answer is $n = \upsilon^2 pq/\varepsilon^2 = (1.96)^2 .0.90 \cdot 0.10/0.05^2 = 138$ years. This and the preceding example give an insight into the accuracy, under given probabilities of errors, of the relative frequencies empirically determined. Similarly, ε can be determined for given n and P. In summary, of the three factors: n, ε, and P, two must be given or assumed to determine the third value, provided p and q are known, or assumed in advance.

Bernoulli's theorem. Selecting ε as a positive constant and looking for the probability of $\left|\frac{m}{n} - p\right| < \varepsilon$, then

$$P\left\{\left|\frac{m}{n} - p\right| \leq \varepsilon\right\} = P\left\{-\varepsilon\sqrt{\frac{n}{pq}} \leq \frac{m - np}{\sqrt{npq}} \leq \varepsilon\sqrt{\frac{n}{pq}}\right\} \qquad 3.25$$

implies by virtue of the integral limit theorem of Eq. 3.24 [7, p. 117]

$$\lim_{n\to\infty} P\left\{\left|\frac{m}{n} - p\right| \leq \varepsilon\right\} = \frac{1}{\sqrt{2\pi}} \int_{-\infty}^{+\infty} e^{-x^2/2}\, dx = 1, \qquad 3.26$$

in which $x = (m - np)/\sqrt{npq}$, so that Eq. 3.26 states that the probability of the inequality $|m/n - p| \leq \varepsilon$ approaches 1 as the number of trials or observations increases. Equation 3.26 is known as the law of large numbers or Bernoulli's theorem. It has many applications in the theory of probability. This theorem can easily be extended to more events, with m_1, m_2, ..., m_k [7, p. 122].

Poisson's theorem. Equation 3.23 is a poor approximation for $P_n(m)$ of Eq. 3.21 when p and q approximate 0 and 1, or 1 or 0, respectively, or practically as p differs more and more from 1/2. It is poor approximation for $p = 0$ and $q = 1$, or $p = 1$ and $q = 0$. For very small p, it is necessary in the application of the De Moivre-Laplace theorem for n to be very large.

If $p \rightarrow 0$ as $n \rightarrow \infty$, then the Poisson limit theorem is [7, p. 122—128]

$$P_n(m) = \frac{\lambda^m e^{-\lambda}}{m!} \qquad 3.27$$

for $n \rightarrow \infty$, and $pn \rightarrow \lambda$ = a constant. Equation 3.27 is also called the Poisson distribution law. It is discussed in detail in Chapter 7, under discrete distribution functions.

Example 3.14. A hydrologic event has a very small probability $p = 0.001$. What is the probability of event occurring two or more times in 5000 observations. In this case, $\lambda = pn = 5$. The probability

$$P_n(m \geq 2) = \sum_{m=2}^{\infty} P_n(m) = 1 - P_n(0) - P_n(1).$$

From Eq. 3.27, $P_n(0) = e^{-5}$; $P_n(1) = 5 \cdot e^{-5}$, so that the required probability $P_n(m \geq 2) = 1 - 6 \cdot e^{-5} \approx 0.9596$. The probability $P(m \geq 2)$ obtained by the exact formula of Eq. 3.21 is 0.9597, so that the error by the Poisson's approximation is about 0.02 percent. The difference R_n of results between Eq. 3.21 and Eq. 3.27 for $P_n(0)$ may be approximately determined by $0 < |R_n| < \lambda^2/2(n - \lambda)$. For the above case of $\lambda = 5$, $n = 5000$, $|R_n|$ is less than 1/400.

3.5. Markov Chains

Definition of Markov chains. Let k be mutually exclusive but dependent random events, A_1, A_2, ..., A_k, and let s be the number of trials already performed. A simple Markov chain is defined as the sequence of the above random events, if the probability of conditional events of the occurrence of an A_j ($j = 1, 2, \ldots, k$) of the $(s + 1)$ th trial, provided that the event A_i ($i = 1, 2, \ldots, k$) has occurred at the s-th trial can be expressed only as a function of the events that occurred at the s-th trial, and not on the previous trials. Less simple Markov chains are those sequences when the outcome of an event A_j may be expressed on the two preceding events but not on the events previous to them, on the three preceding events but not on the events previous to them, etc. Usually and particularly in hydrology, the random events A_1, A_2, ..., A_k are states or the selected intervals of continuous variables, and trials are made at discrete times, or relate to time intervals. In this case and for the simple

Markov chains, the probability of transition of one state to another, or from the state A_i at time τ $(t_{s-1} < \tau < t_s)$ to the state A_j at time t $(t_s < < t < t_{s+1})$, can be expressed only in terms of the state A_i, and not on the previous states. These times may be also random quantities; they may be observations at equal intervals, t_0, $t_0 + \Delta t$, $t_0 + 2\Delta t$, ...; or they may refer to equal intervals. If the probabilities of conditional events do not depend on the number of trials or observations already performed, the Markov chains are homogeneous. These probabilities of conditional events are called probabilities of transition (transitional probabilities), p_{ij}, where the i-th event precedes the j-th event.

Example 3.15. The outflow of a ground-water aquifer depends on its water table. The recharge is assumed to be an independent infiltration process. The outflow during a month can be expressed as a function of the outflow of the previous month plus the effect of the random recharge in the month considered. The successive outflows may be simple Markov chains. In this case, the continuous variable of outflows is divided in k states, as discrete random events.

Example 3.16. Assume a lake has independent or dependent annual inflows. In the case of dependent inflows, they are approximately of the first-order linear autoregressive dependence. The free annual outfows may be close to be the simple and homogeneous Markov chains for the lake storage divided in various discrete states, or layers of equal height or water volume. When new reservoirs or lakes are built, the initial storage capacity and the initial operation make the outflows simple and nonhomogeneous Markov chains, if later the outflows follow the simple and homogeneous Markov chains.

Transition matrix. The simple Markov chains are completely described if all probabilities of transition are known, p_{ij} and p_{ji}, for i, j = 1, 2, ..., k. The square matrix

$$\pi_1 = \begin{vmatrix} p_{11} & p_{12} & \dots & p_{1k} \\ p_{21} & p_{22} & \dots & p_{2k} \\ \cdot & \cdot & \cdot & \cdot \\ p_{k1} & p_{k2} & \dots & p_{kk} \end{vmatrix} \qquad 3.28$$

constructed of probabilities of transition and called the matrix of transition probabilities describes uniquely the Markov chains.

The conditions that the simple and homogeneous Markov chains and their transition matrix must satisfy are: (1) the random events must be discrete, and a random variable is either discrete or the continuous variable is reduced to discrete states; (2) $0 \leq p_{ij} \leq 1$; (3) $\sum_{j=1}^{k} p_{ij} = 1$, since all probabilities in the passage from A_i to A_j must have a sum of unity, and (4) probabilities $p_{ij}(n)$ of transition from the state A_i at the s-th trial to the state A_j at the (s + n)-th trial can be derived from the above basic matrix π_1 of Eq. 3.28.

In this latter case and for an intermediate trial $s < s + m < s + n$,

$$p_{ij}(n) = \sum_{r=1}^{k} p_{ir}(m) \cdot p_{rj}(n - m), \qquad 3.29$$

which is known as Markov's equation. It can be shown that the square matrix π_n of $p_{ij}(n)$ transition probabilities is $\pi_n = \pi_1^n$. For $n = 2$, $\pi_2 = \pi_1 \cdot \pi_1 = \pi_1^2$. In this case, $p_{ij}(2) = \sum_{r=1}^{k} p_{ir} p_{rj}$, or it is a multiplication of the corresponding rows i and columns j, for all $r = 1, 2, \ldots, k$, as it is done in squaring a square matrix.

Limiting probabilities. It can be proven [7, p. 142—145] that for simple and homogenous Markov chains

$$\lim_{n \to \infty} p_{ij}(n) = p_j, \tag{3.30}$$

or that the probability of an event A, after a large number of transitional states, becomes independent of the occurrence of a previous but remote event A_i, for any $i = 1, 2, \ldots, k$, if n is sufficiently large. How rapid this convergence is depends on the transition probabilities. The chains, which satisfy Eq. 3.30, satisfy a theorem called ergodic.

Example 3.17. The wet and dry years, as defined in Example 2.26, of many rivers are approximately Markov chains, with five states present (very dry, dry, average, wet, and very wet) because of time dependence between flows of successive years. If the correlation coefficient ρ of succesive values of states, as the measure of dependence, is positive and not too large, the probability $p_{ij}(n)$ — where i = a previous state, n = number of intermediate years, and j = a new state — is approximately p_j if n is of the order 5 to 10. The smaller ρ the shorter may be n in order to approximate $p_{ij}(n)$ by p_j, with a given error ε.

The integral limit [De Moivre-Laplace] theorem of simple Markov chains. If only two events can occur, A and B, the transition probabilities are $P(A_{i+1} \mid A_i) = \alpha$, $P(B_{i+1} \mid A_i) = 1 - \alpha$, $P(A_{i+1} \mid B_i) = \beta$, and $P(B_{i+1} \mid B_i) = 1 - \beta$, with $0 < \alpha < 1$ and $0 < \beta < 1$, or the transition matrix is

$$\pi_1 = \begin{vmatrix} \alpha & 1 - \alpha \\ \beta & 1 - \beta \end{vmatrix}. \tag{3.31}$$

Since either A or B must occur in a trial, it can be proven that $P(A) = p = \beta/(1 - \alpha + \beta)$ is the limiting probability, and $P(B) = q = 1 - p$. It also can be proven [7, p. 146 – 153] that

$$\lim_{n \to \infty} P\left\{ a \le \frac{m - np}{\sqrt{npq(1 + \alpha - \beta)/(1 - \alpha + \beta)}} < b \right\} = \frac{1}{\sqrt{2\pi}} \int_a^b e^{-x^2/2}(1 + \omega_n)\,dx, \tag{3.32}$$

with a and b the two particular values of x, x_1 and x_2, obtained for given m_1, m_2, and n, p, q, α, and β. Equation 3.32 is the extension of the integral limit theorem to Markov chains, and where the small quantity ω_n approaches zero as $n \to \infty$, and $x = (m - np) / \sqrt{npq(1 + \alpha - \beta)/(1 - \alpha + \beta)}$.

If $\alpha = \beta$, this theorem gives the same results as in the case of the Bernoulli trials.

3.6. Mathematical Expectation, and Variance of Random Variables.

Two basic numerical characteristics of random variables. Of all characteristics of random variables, only two are discussed in this chapter, the mathematical expectation and the variance. These are necessary for understanding the law of large numbers found in the subsequent text. These two and the other numerical characteristics, and their applications in hydrology, are discussed in detail in Chapter 6, with examples and problems.

Definition of mathematical expectation. A discrete variable x has the probability mass function, p. m. f., p (x), and for any random event x_i, it gives $p(x_i)$. If the sum of series of products, $x_i \cdot p(x_i)$ with $i = 1, 2, \ldots, n$, converges absolutely to a constant, only if $n \to \infty$, then

$$\lim_{n\to\infty} \sum_{i=1}^{n} x_i\, p(x_i) = Ex; \qquad 3.33$$

this sum is called the mathematical expectation, the expectation, or the expected value. This last term and the symbols Ex and μ will be used exclusively, in this text. In this case x_i denotes both the random variables and the values these variables can take.

For a continuous variable and its p.d.f., f (x), and its distribution function, d. f., F (x), the definition of expected value is

$$Ex = \int_{-\infty}^{+\infty} x\, f(x)\, dx = \int_{-\infty}^{+\infty} x\, dF(x). \qquad 3.34$$

It will be shown later that the expected value is the mean of the population of a random variable.

Similarly, as in Eqs. 3.33 and 3.34, expected values are defined for bivariates, multivariates, for conditional variables, and functions of random variables. Not all random variables have expected values. The typical example are variables of Cauchy's distribution law.

Theorems on expected values. The expected value of a constant C is that constant, or EC = C. The expected value of a sum of two random variables is equal to the sum of their expected values, or

$$E(x + y) = E\,x + E\,y. \qquad 3.35$$

Similarly, the expected value of the sum of a finite number of random variables is the sum of their expected values. The expected value of a product of two independent random variables is the product of their expected values, or

$$E\,xy = E\,x \cdot E\,y. \qquad 3.36$$

For a product of a constant C and a random variable,

$$ECx = C\,Ex. \tag{3.37}$$

Definition of variance. Variance is defined as the expected value of the square of deviations of the variable from its expected value, or

$$Dx = E\,(x - Ex)^2 = \sum_{i=1}^{\infty} (x_i - Ex)^2\, p\,(x_i), \tag{3.38}$$

for the discrete random variables, and

$$Dx = E\,(x - Ex)^2 = \int_{-\infty}^{+\infty} (x - Ex)^2\, f\,(x)\, dx = \int_{-\infty}^{+\infty} (x - Ex)^2\, dF\,(x), \tag{3.39}$$

for the continuous random variables. Variance is designated in this text either by Dx or var x or σ_x^2, and the three symbols refer to definitions of either Eq. 3.38 or Eq. 3.39.

Theorems on variance. The variance of a constant C is zero. If C is a constant, then

$$DCx = C^2\, Dx. \tag{3.40}$$

The variance of the sum of two independent random variables is the sum of their variances, or

$$D\,(x + y) = Dx + Dy. \tag{3.41}$$

Similarly, for n random variables pairwise independent, the variance of their sum is the sum of their variances. The variance of the difference of two independent random variables is the sum of their variances, or

$$D\,(x - y) = Dx + Dy. \tag{3.42}$$

3.7. Law of Large Numbers.

Definition. Probabilities of random events, which are extremely close to zero and one, are of special interest in the application of probability theory to natural sciences. Particularly, it is the case with probabilities of events, which are very close to zero. These events usually look *nearly impossible, or almost impossible*. However, if the number of trials or observations, which involve a random event of extremely small probability, increases without a limit (or, in practice, if there is a very large number n of observations or trials), the probability of occurrence of that event will be as close to one as desired. This *almost impossible event* becomes *almost certain to occur*. The theorems that refer to this, both theoretical and experimental, are encompassed under the concept of the law of large numbers.

If $x_1, x_2, \ldots, x_n$ is a sequence of random variables (say, observations at times or points 1, 2, ..., n, of a random phenomenon), then a function of the first n variables is

$$y_n = f\,(x_1, x_2, \ldots, x_n). \tag{3.43}$$

If there is a sequence of constants $a_1, a_2, \ldots, a_n$, so that for any small quantity $\varepsilon > 0$

$$\lim_{n\to\infty} P\{|y_n - a_n| < \varepsilon\} = 1, \qquad 3.44$$

then the sequence of variables $x_1, x_2, \ldots, x_n$ follows the law of large numbers for the function of Eq. 3.43.

If the function of Eq. 3.43 is the arithmetic mean of n independent and identically distributed random variables, $y_n = \sum_{i=1}^{n} x_i/n$, and the sequence of constants is the same value a, then

$$\lim_{n\to\infty} |y_n - a| = 0. \qquad 3.45$$

Example 3.18. Pressure is measured in a turbulent stream in many points of a grid Δx, Δy, Δz apart in small volume V, and at all these points at many small increments Δt during the short time interval t. Their average value is determined. By increasing the volume V, and time t, and correspondingly, the number of points and instants of measurements, the difference between the computed arithmetic mean of these measurements and the expected value of the pressure at the center of the volume V becomes smaller and smaller. This example is an analogy to the phenomenon of Brownian motion of molecules of a gas. Brownian motion is a classical example of the theoretical and experimental application of the law of large numbers in physics.

Chebyshev inequality. One of the oldest form of the law of large numbers is the Chebyshev inequality,

$$P[|x - Ex| \geq \varepsilon] \leq \frac{Dx}{\varepsilon^2}, \qquad 3.46$$

where x is any random variable, Ex is its expected value, Dx is its finite variance, and $\varepsilon > 0$ is any real positive number.

The case of the random variable x in Eq. 3.46 being the arithmetic mean of n independent random variables indentically distributed, $\bar{x} = \sum x_i/n$, with $i = 1, 2, \ldots, n$, is often of practical interest. According to Eqs. 3.40 and 3.41, $D\bar{x} = \sum Dx_i/n^2 = nDx/n^2 = Dx/n$, with $i = 1, 2, \ldots, n$. In this case, Eq. 3.46 becomes

$$P\left\{\left|\frac{1}{n}\sum_{i=1}^{n} x_i - Ex\right| \geq \varepsilon\right\} \leq \frac{Dx}{n\varepsilon^2}, \qquad 3.47$$

for every n and every $\varepsilon > 0$. This equation shows an inequality basically involving a probability P, a small positive real number ε, and a number n of random variables. In any practical problem two of them must be given, or assumed, in order to find for the third a quantity from which it is either greater or smaller.

Example 3.19. A variable x, with Ex and Dx finite, represents the series of independent annual river flows. The expected value Ex is always replaced by the mean of the total series, as its best estimation. The sum of deviations

$$S_i = \sum_{i=1}^{i} (x_i - Ex) \qquad 3.48$$

is a dependent random variable even if x_i are independent variables, with $ES_i = 0$, and the variance $DS_i = iDx$, or the variance increases linearly with i. For a finite subseries of $i = n$, where n is a parameter, the sum S_i of Eq. 3.48 has a maximum $S_{max} \geq 0$ and a minimum $S_{min} \leq 0$. They are random variables as functions of x. The difference $R_n = S_{max} - S_{min}$ is also a random variable, and is called the range. This range parameter is relevant to the analytical investigation of water storage capacities. For an independent normal variable x with $Ex = 0$ and $Dx = 1$, Anis and Lloyd [8] give the exact expected value

$$ER_n = \sqrt{\frac{2}{\pi}} \sum_{i=1}^{n} i^{-1/2}, \tag{3.49}$$

and Feller [9] gives the asymptotic value of the variance of R_n as

$$DR_n = 4n\left(\ln 2 - \frac{2}{\pi}\right) \approx 0.218\ n. \tag{3.50}$$

To verify Eq. 3.49 experimentally, a very large number of independent, standard normal random numbers is generated on a computer by the Monte Carlo (data generation) method. Figure 3.4 shows the application of Eq. 3.47 to the mean range in the case $n = 4$ in Eq. 3.49. In Eq. 3.47, x_i is now replaced by individual values R_n (in this case by R_4), n by the number m of subsamples of size $n = 4$, Ex by ER_n of Eq. 3.49 which gives for $n = 4$ the value $ER_4 = 2.2218$, Dx by DR_n of Eq. 3.50 for $n = m$, in Eq. 3.47, and $\varepsilon = 0.01$. Then Eq. 3.47 becomes, for this example, for $n = 4$ in Eq. 3.50,

$$P\left\{\left|\frac{1}{m}\sum_{i=1}^{m} R_n - 2.2218\right| \geq 0.01\right\} \leq \frac{8720}{m}. \tag{3.51}$$

Curve (1) of Fig. 3.4 gives the limiting values of P computed from Eq. 3.51 for various values of m. Curve (2) shows the relative frequencies of the difference between the

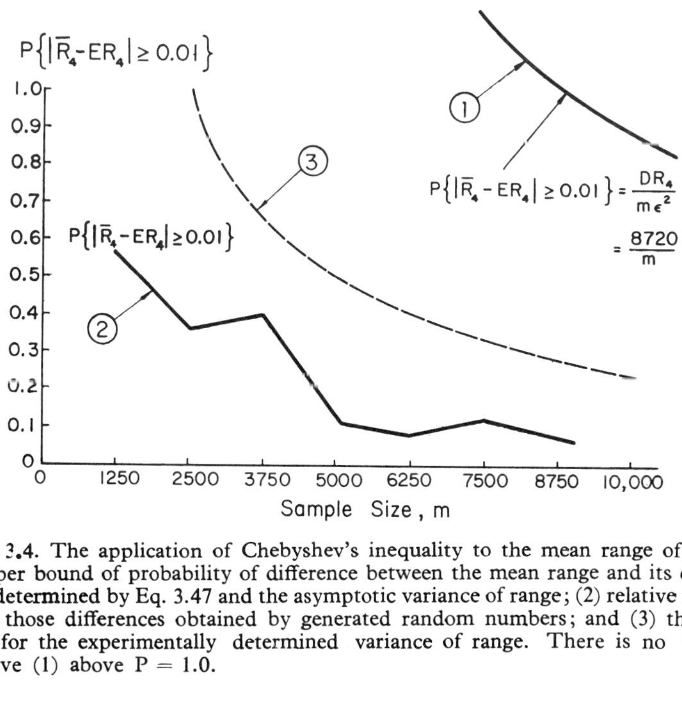

Figure 3.4. The application of Chebyshev's inequality to the mean range of R_4 : (1) the upper bound of probability of difference between the mean range and its expected value, determined by Eq. 3.47 and the asymptotic variance of range; (2) relative frequencies of those differences obtained by generated random numbers; and (3) the upper bound for the experimentally determined variance of range. There is no meaning for curve (1) above $P = 1.0$.

mean range $\bar{R}_4$ and the expected value $ER_4 = 2.2218$, with the mean computed from m range values, and each range in turn from a subsample of four values of x. For each sample of m ranges, only one value $\bar{R}_4$ is obtained and one difference $\Delta \bar{R}_4 = \bar{R}_4 - ER_4$. To determine the relative frequency of $|\Delta \bar{R}_4| \geq \varepsilon = 0.01$, 25 samples of ranges, each sample with the size m, are generated from random numbers of x. For each sample, $|\Delta \bar{R}_4|$ is determined, and finally the relative frequency of $|\Delta \bar{R}_4| \geq 0.01$ computed. Therefore, this relative frequency of curve (2) of Fig. 3.4 has an accuracy of about $1/25 = 0.04$. Curve (3) represents the upper bound obtained by Eq. 3.51 where DR_n of Eq. 3.50 is replaced by the variances of R_n, experimentally determined from a very large amount of random numbers of x. The asymptotic variance of Eq. 3.50 gives a much larger estimate of the upper bound than the exact variance would produce.

Application of Chebyshev inequality in hydrology. Example 3.19 shows the application of the law of large numbers as expressed in the form of Chebyshev inequality. As the number of observations increases, the probability of a difference tends to zero, bounded by probabilities determined from this inequality. The application may be extended to the use of experimental statistical method in hydrology. It means this inequality may be even used to prove experimentally that some equations, which are derived from hypotheses or assumed, fit well the curves obtained by the Monte Carlo method. This approach does not usually meet rigorous requirements of classical mathematical proofs, but in absence of these analytical methods, the law of large numbers, as a theoretical concept, justifies the use of experimental method for various proofs.

Example 3.20. In Example 3.19 the expected value and the asymptotic variance of the range were known as analytical expressions; this asymptotic variance is only an approximation. In the case of the normal independent variable x_i, with $Ex = 0$ and $Dx = 1$, and a new variable y of the first-order autoregressive dependence function of x_i,

$$y_i = \rho\, y_{i-1} + x_i, \qquad 3.52$$

where ρ is a selected constant $0 < \rho < 1$ (the first autocorrelation coefficient), and the exact values of ER_n of the range of y are not known, nor are the exact or the asymptotic values of its DR_n. Replacing i in Eq. 3.49 by DS_i, this equation becomes

$$ER_n = \sqrt{\frac{2}{\pi}} \sum_{i=1}^{n} (DS_i)^{-1/2}. \qquad 3.53$$

The hypothesis is advanced that Eq. 3.53 is not only valid for ER_n of independent variable x, but also for all linear dependence models of $y = f(x)$, provided DS_i is the variance of S_i of the variable y. Under this assumption the expected value of range of Eq. 3.52 is [10]

$$ER_n = \sqrt{\frac{2}{\pi}}\, (1 - \rho^2)^{-1/2} \sum_{i=1}^{n} i^{-1/2} \left[\frac{1+\rho}{1-\rho} - \frac{2\,(1-\rho^i)}{i\,(1-\rho)^2}\right]^{1/2}. \qquad 3.54$$

To verify Eq. 3.54 experimentally, the Chebyshev inequality is used. An example is $n = 2$ and $\rho = 0.2$ in Eq. 3.54, so that $ER_2 = 1.4449$. From the generated large samples, $DR_2 = 0.7476$, which replaces the unknown exact variance of R_2, in this example. For $\varepsilon = 0.01$, $\rho = 0.2$, Eq. 3.47 gives

$$P\,[\,|\bar{R}_2 - 1.4449| \geq 0.01] \geq \frac{0.7476}{m\,\varepsilon^2} = \frac{7476}{m}.$$

Figure 3.5 shows this case, with curve (1) giving the upper bound of Eq. 3.55, and curve (2) the relative frequencies computed for $|\Delta \bar{R}_2| = |\bar{R}_2 - ER_2|$ from 25 samples of ranges R_2, each with m values of R_2. As there is an empirical, rapid convergence of the relative frequency of $|\Delta \bar{R}_2|$ to zero inside the bound, it can be concluded that Eq. 3.54 satisfies the Chebyshev theorem, at least in the range of experimental values of m. Therefore, the hypothesis based on Eq. 3.53 seems acceptable, on the basis of the Monte Carlo or statistical experimental method.

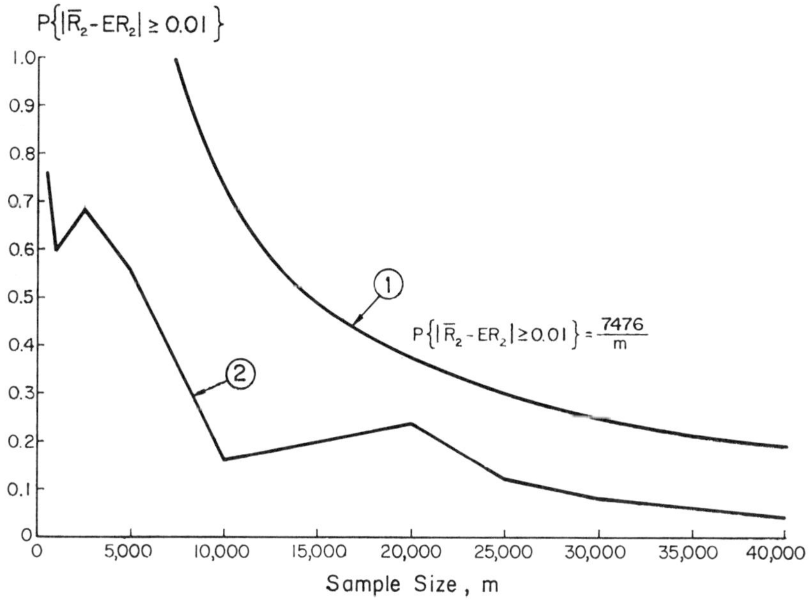

Figure 3.5. The use of Chebyshev inequality to experimentally verify the validity of Eq. 3.54 for the expected range of first-order linear autoregressive dependence of normal variables, based on a hypothesis: (1) the upper bound of Eq. 3.54 for $\bar{R}_2$; (2) relative frequencies of $|\Delta \bar{R}_2|$ obtained by the Monte Carlo method.

Chebyshev-Markov theorem. If a sequence of random variables $x_1, x_2, \ldots, x_n$ satisfies the condition that their variances are finite, and if

$$\lim_{n\to\infty} \frac{1}{n^2} D\left(\sum_{i=1}^{n} x_i\right) = 0, \qquad 3.55$$

then for every $\varepsilon > 0$ [7, p. 240]

$$\lim_{n\to\infty} P\left\{\left|\frac{1}{n}\sum_{i=1}^{n} x_i - \frac{1}{n}\sum_{i=1}^{n} Ex_i\right| < \varepsilon\right\} = 1. \qquad 3.56$$

This is called the Chebyshev, Markov or Chebyshev-Markov theorem. The above two Examples, 3.19 and 3.20, may also be interpreted in the form of the law of large numbers.

Strong law of large numbers. Since the law of large numbers is the fundamental proposition of the theory of probability, it is not surprising that many theorems have been derived regarding it (Bernoulli's, Poisson's, Khinchin's, and other variations of this law, which are not reviewed in this text). Many attempts also have been made for establishing general conditions for random variables, under which the law holds. Borel theorem states that the relative frequency m/n of an event A in n independent trials, with A occurring in each trial with the same probability p, converges to p with the probability one as $n \to \infty$. This is often considered as the strong law of large numbers. For this strong law, Kolmogorov condition, that the random variables $x_1, x_2, \ldots$ must satisfy [7, p. 257] is

$$\sum_{i=1}^{\infty} \frac{Dx_i}{i^2} < +\infty. \qquad 3.57$$

Also, the existence of the expected value is a necessary and sufficient condition for the strong law of large number to be applicable to a $x_1, x_2, \ldots$ sequence.

Central limit theorem. In practice, there are numerous cases when the measured effect is the sum of a large number of effects of independent random causative factors. To find the distribution of this over-all effect from the properties of causative factors, logical derivations from the limit theorems and the law of large numbers lead to several forms of theorems usually known under the name of the central limit theorem. One of the forms relates to the distribution of the sum of random variables. For

$$x = x_1 + x_2 + \ldots + x_n = \sum_{i=1}^{n} x_i, \qquad 3.58$$

the sum of n mutually independent random variables, with finite expected values $\mu_1, \mu_2, \ldots, \mu_n$ and finite variances $\sigma_1^2, \sigma_2^2, \ldots, \sigma_n^2$, the distribution of the variable $(x - \mu)/\sigma$ — where the expected value μ is the sum of n values of μ_i and the variance σ^2 is the sum of n values of σ_i^2 — tends to the normal distribution. The following Lindeberg condition must be satisfied for every $\tau > 0$,

$$\lim_{n\to\infty} \frac{1}{\sigma^2} \sum_{i=1}^{n} \int_{|x_i - \mu_i| > \tau\sigma} (x_i - \mu_i)^2 \, dF(x_i) = 0, \qquad 3.59$$

in which $F(x_i)$ is the distribution of the variable x_i. The central limit theorem in this case may be expressed in the form

$$\lim_{n\to\infty} P\left\{\frac{1}{\sigma} \sum_{i=1}^{n} (x_i - \mu_i) \leq z\right\} = \frac{1}{\sqrt{2\pi}} \int_{-\infty}^{z} e^{-z^2/2} \, dz, \qquad 3.60$$

which is also known as the Liapounov theorem [7, p. 310]. Instead of looking for the distribution of the variable $(x - \mu)/\sigma$, the distribution of x of Eq. 3.58 converges to the normal distribution, with the expected value of μ and the variance of σ^2, or for $n \to \infty$,

$$f(x) = \frac{1}{\sigma\sqrt{2\pi}} e^{-(x-\mu)^2/2\sigma^2}, \qquad 3.61$$

for all practical purposes.

The special case is when all independent random variables have the same distribution, $F(x_i)$, for which $\mu = n\mu_i$ and $\sigma^2 = n\sigma_i^2$. The convergence to the normal distribution is also valid in practical cases where the random variables of Eq. 3.58 are mutually dependent. The only difference is that convergence to normal distribution with $n \to \infty$ is slower for mutually dependent variables than for independent variables.

Three properties of the central limit theorem are of practical hydrologic interest: (a) for what types of dependence among variables this limit theorem is still valid; (b) how fast the distribution of a sum of independent and dependent random variables, of the same distribution $F(x_i)$, converges to the normal distribution, and (c) how fast is this convergence in the case of x_i variables, independent or dependent, having different distributions, $F(x_i)$.

The skewness coefficient C_s, and the kurtosis k, as parameters of a distribution $F(x)$, are defined in Chapter 6, with $C_s = \mu_3/\mu_2^{3/2}$, and $k = \mu_4/\mu_2^2$, where μ_2, μ_3 and μ_4 are the second, third, and fourth central moments of $F(x)$. Let all variables x_i in Eq. 3.58 be of the same distribution $F(x_i)$, with the same parameters $C_s(x_i)$ and $k(x_i)$. In practice, the parameter $C_s(x)$ of the sum x of x_i as it changes with n, starting at $n = 1$ and $C_s(x_i)$, measures the rapidity of convergence in central limit theorem to a symmetrical distribution for which $C_s(x) = 0$. Similarly, the convergence of $k(x)$ to $k = 3$ of the normal distribution with an increase of n, starting at $n = 1$ and $k(x_i)$, measures the rapidity of convergence of a symmetrical case of $F(x_i)$ to the normal distribution. Both $C_s(x) - f(n)$ and $k(x) = \psi(n)$ simultaneously measure the convergence to normal law with n, for given $F(x_i)$ with its parameters $C_s(x_i)$ and $k(x_i)$.

Figure 3.6 gives a schematic representation of the decrease of $C_s(x)$ with n, for x being the sum of independent and dependent variables.

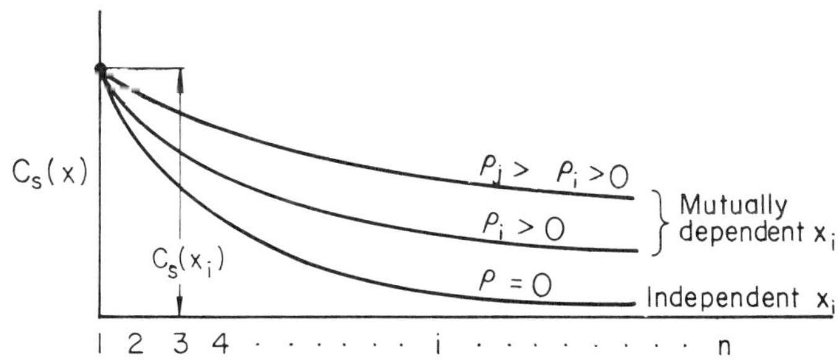

Figure 3.6. Convergence of a sum of n-random independent or dependent variables with the same distribution $F(x_i)$ to the normal distribution, measured by the decrease of skewness coefficient, $C_s(x_i)$, with n starting from $C_s(x_i) =$ constant for $n = 1$.

If x_i of Eq. 3.58 are independent variables, the skewness coefficient $C_s(x)$ converges quickly to zero as n increases. If x_i variables are dependent with the average correlation coefficients ρ between pairwise variables, then the convergence of $C_s(x)$ to zero is slower with n, and the greater ρ the slower this convergence. The central limit theorem may be well applied in hydrology, particularly in assessing when the distribution of a variable, as the sum of other variables, may be close to a normal distribution.

Example 3.21. The average skewness coefficient is greater for distributions of daily flows than for distributions of monthly flows of a runoff series, the latter in turn is greater than the skewness coefficient for distribution of annual flows, etc. The variable of monthly flow is the sum of 30 daily flows, and the variable of annual flow is the sum of 12 monthly flows. Similarly, the distribution of annual precipitation of a given station is usually more skewed than the distribution of the average or weighted annual precipitation in a region. In other words, the larger number of time interval values, or the larger number of points, for a given hydrologic variable, summed to form a new variable, the closer is its probability distribution to a symmetrical and a normal distribution.

3.8. **Problems for Discussion and Solutions.**

3.1. The annual precipitation x is observed N years. It is known from a regional study that this variable is approximately independent in sequence, and that the average characteristics of it do not change with time. Investigators of individual precipitation stations often find an upward or a downward trend, or even cyclicity in series of annual precipitation. Let the series be divided into m subseries, each n years long (or $m = N/n$), with m an integer, and for each subseries the mean computed, $\bar{x}_1, \bar{x}_2, \ldots, \bar{x}_m$. Find the probability of the random event of either $\bar{x}_1 < \bar{x}_2 < \ldots < \bar{x}_m$, or $\bar{x}_1 > \bar{x}_2 > \ldots > \bar{x}_m$ for an upward or a downward trend. Find it for $m = 4$, $m = 6$, and $m = 8$. For $m = 4$ find the probability of the event of an apparent cyclic movement among the means $\bar{x}_i$, such as $\bar{x}_1 > \bar{x}_2 < \bar{x}_3 > \bar{x}_4$, or $\bar{x}_1 < \bar{x}_2 > \bar{x}_3 < \bar{x}_4$. Find it also for $m = 8$. Use the combinatorial calculus. This example should show what the chances are for obtaining apparent (sampling) trends and cycles in an independent stochastic process, when N is relatively small (say $N = 20 - 60$ years).

3.2. A hydrologic system has n major subsystems, and with each subsystem a continuous variable is associated. Each of these continuous variables is represented by m discrete and distinct states. An optimization problem may require that all possible combinations of subsystems and states are investigated, regardless of which subsystem or state is chosen for investigation. Find the number of cases for investigation. If the sequence of investigation is important, find the number of permutations. Particularly, find them for $n = 5$ and $m = 4$.

3.3. A region of 6,000 square miles has experienced 72 hail storms in 10 years with each covering an average area of 8 square miles. An experimental agricultural area of four square miles did not experience a hail storm during these 10 years. What is the probability that at least one half of this experimental area would be hit by hail in the next five years, provided that the probabilities of each subarea in the region to experience the hail is the same?

3.4. River flows greater than q_1, q_2, and q_3 of three tributaries are mutually dependent events. The individual and joint probabilities are: $P(Q_1 \geq q_1) = 0.20$, $P(Q_2 \geq q_2) = 0.30$; $P(Q_3 \geq q_3) = 0.40$; $P(Q_1 \geq q_1;\ Q_2 \geq q_2) = 0.10$; $P(Q_1 \geq q_1;\ Q_3 \geq q_3) = 0.18$; $P(Q_2 \geq q_2;\ Q_3 \geq q_3) = 0.20$, and $P(Q_1 \geq q_1;\ Q_2 \geq q_2;\ Q_3 \geq q_3) = 0.05$. Find the probability $P(Q \geq q)$ of a flood on the main river if this can be caused by at least one of the events $Q_1 \geq q_1$, $Q_2 \geq q_2$, or $Q_3 \geq q_3$.

3.5. Three storage reservoirs have probabilities of water overflow at a given time 0.07, 0.05, and 0.09, and are independent events. The repair work is to be done below each dam during that time. If the repair site is selected at random, what is the probability that the water will spill over and interrupt the work?

3.6. Weather and river flow forecasting information is sent by routes R_1, R_2 and R_3, each with the same probability of being used. The probability of an error to be introduced in transmitting messages on these routes are 1/5, 1/4, and 1/3, respectively. If a message is received without errors, what is the probability that it was sent via route R_3?

3.7. The inverstigation of cloud seeding effects is often associated with the random selection of storms to be seeded. The selected ratio of storms to be seeded is p and nonseeded 1-p. The decisions of seeding and not seeding are independent Bernoulli trials. For the operation pursued for 10 years, each year there is an average of 18 storms, or 180 storms in 10 years, to decide on. If $p = q = 1/2$, find in this random selection of seeding, what is the probability of the seeded storms being between 45 − 55 percent of all storms. Find it another way by any approximate procedure.

3.8. The number of storms per year is also a random event. A total number of possible storms in a winter season is evaluated to be $n = 45$. Since the expected value of the number of storms per winter is $m = 18$, and since $Em = pn$, then $p = 0.40$. Find the solution of Problem 3.7 by considering the number of storms per winter as a random variable.

3.9. Find the probability that the relative frequency 0.21 of rainy days, as given in the discussion of Fig. 2.1, does not deviate from the probability of rainy days more than $\varepsilon = \pm 0.03$, assuming that the sequence of rainy and nonrainy days are independent Bernoulli trials. If the transition relative frequency from rainy to nonrainy day is $\alpha = 0.35$, the transition relative frequency of nonrainy to rainy days should be computed; and assuming the sequence of rainy and nonrainy days to be a simple homogeneous Markov chain, find again the probability of deviations $\varepsilon = \pm 0.03$.

3.10. Select a river with highly time dependent annual river flows, such as the St. Lawrence River at Ogdensburg, N.Y.*. Divide all discharges into five states of very dry (S_1), dry (S_2), average (S_3), wet (S_4), and very wet (S_5) states as described in Example 2.26. Find the matrix π_1 of transition probabilities p_{ij}, $i = 1, 2, 3, 4$, and 5, and $j = 1, 2, 3, 4$, and 5. The five probabilities p_i of the above five events are respectively: $p_1 = p_5 = 0.15$, $p_2 = p_4 = 0.20$, and $p_3 = 0.30$. Find n for various p_{ij} that the limiting probabilities of Eq. 3.31 become close to p_j values, by computing $\pi_1^2, \pi_1^3, \pi_1^4, \ldots, \pi_1^n$ matrices from the π_1 matrix, for $n = 1, 2, \ldots$.

Repeat the calculation for only three states, a dry year with $p_1 = 0.35$, an average year with $p_2 = 0.30$, and a wet year with $p_3 = 0.35$. How do the errors of estimates of p_{ij} change with the number n of states when the total number of years of observations is N?

3.11. Take a variable x with a long series of data and select at least three time intervals, such as a month, three months, or six months, and a year; or 1 month, 2 months, 3 months, 4 months, and 6 months, etc. Determine the new time series on the basis of the time intervals selected. For example, if monthly flow is selected to be the variables $x_1, x_2, x_3, \ldots$, then the first series may be the series of monthly flows with the first term $x_I = x_1$, the second series may be formed with the 3-month totals, that is $x_{II} = x_1 + x_2 + x_3$, and the third series may be made up of the yearly totals, so that $x_{III} = x_1 + x_2 + \ldots + x_{12}$, and similar. Select sufficient data so that at least 20 values are available in the series for the largest interval (for example, 20 annual values; 40, 6-month values; 240 montly values). For each of the three or more intervals, compute the skewness coefficinent, C_s. Plot C_s versus the interval, Δt. Discuss the case from the vantage point of the sum of independent or dependent random variables, and of the application of the central limit theorem in this case.

3. 12. Give several examples of the application of the central limit theorem in geophysical sciences, in general, and hydrology in particular.

3.13. What is the theoretical justification for using the normal distribution for random errors of a hydrologic quantity, assumed to be measured a large number of times?

---

* Data may be selected from Colorado State University Hydrology Paper No. 1, Fort Collins, Colorado.

CHAPTER 4

STATISTICS AND HYDROLOGY

Because hydrologic data are mostly given as limited amounts of all data in the form of samples, mathematical statistics is the main discipline enabling the extraction of full information from the data and the derivation of conclusions about the hydrologic population characteristics. Statistics are properties of samples necessary for description of mathematical statistical models, or for a direct use in hydrology. Concepts of basic risk, various uncertainties, and total risk are defined.

4.1. Relationship of Statistics to Probability Theory.

Infinite and finite hydrologic populations. Hydrologic random variables in most cases have infinite populations. Some variables may be either encountered or construed with finite populations. The infinite population is defined as possessing an infinite number of potential observations (outcomes, experiments, trials), and an infinite number of potential realizations of samples of any size. Any survey or any observational attempt cannot exhaust all values or all realizations of an infinite population. For large finite populations, the economy of information procurement forbids an exhaustive approach of obtaining data for the whole finite population, though it is a possible task.

Example 4.1. Assuming that a river basin and its climate stay unchanged forever, the entire populations of its annual precipitation, annual runoff, and many similar infinite populations cannot be observed. The population of all river basins of 10 square miles, properly defined and without an overlapping of areas in a region, is a finite population. It is not economically feasible to gauge all these basins, though it is a possible task. The number of streams of first, second, third, ..., order in a large river basin are finite populations.

Samples. Finite amounts of data collected for a random variable either from an infinite or a finite population constitute the sample drawn from the respective population. For finite populations, samples are usually much smaller than their populations. From practical point of view, samples are either small, medium, or large. However, there is no generally accepted convention that defines these three arbitrary divisions of samples of all possible sizes, though colloquial statistical language uses terms such as *small sample* and *large sample*. These words have relative meanings for various disciplines and problems.

Example 4.2. If 150 years of observations are available for a river or a precipitation gauging station, such as for the Rhine or the Göta rivers, or for many precipitation stations in the world, many hydrologists would consider them large samples. Whereas, if observations exist for only 10 to 20 years, the sample would be considered small. Even 150 years of data for certain hydrologic problems might be considered a small sample.

Descriptive and inferential statistics. A sample of a random variable can be analyzed in two basic ways. First, considerations and investigations of the sample data are without any reference to the properties of their population. This approach is usually referred to as the descriptive statistics. Secondly, inference is made from the sample data about the properties of their population. This approach is referred to as the inferential statistics, which is based on the use of probability theory. Once properties of their population are inferred, the next step is often the derivation of probability characteristics of functions of this random variable. The same approaches are used when the sample data refer to bivariates, multivariates, and conditional variables.

The first approach of descriptive statistics is a current hydrologic practice. Past observations of available samples are summarized in the form of empirical frequency distributions and in the form of computed parameters and other statistics. They are used as the expected distributions or parameters for evaluating the future performance of water resource projects or measures. No inference is derived about the population, and no limits are assigned to the computed hydrologic distributions or parameters. Often, the underlying expectations is that future samples — smaller than, equal to, or larger than the sample available — will have the same or similar properties as the available sample. This attitude may be misleading.

The best analysis of hydrologic statistical data is based on the application of inferential statistics. By providing future expected hydrologic characteristics in the form of estimated properties of the population of a variable, and by assigning errors, which are inevitably associated with these estimates of characteristics, pertinent hydrologic information for decision making may be drawn from a sample.

The present forms of a river basin, or of any other hydrologic environment, are limited when measured by geologic time scales. The climate also may change over long periods. However, when practical problems of water resource conservation, development and control are considered, and when the increasing human influence is either neglected, or taken into account, or corrected for, whichever is relevant to the problem, both the climate and hydrologic environments may be considered basically unchangable with time. This has the practical implication that the observed values of a hydrologic random variable as it changes in time can be considered as belonging to a unique infinite or finite population. The shorter the future period to which the properties of a hydrologic variable, derived from a sample, are applied, the more justified this basic concept of a unique population. This approach facilitates the meaningful use of inferential statistics in hydrology. If the climate and environments continuously change, the case may be understood as a super-population or a mixture of populations, which would be difficult to treat by classical inferential statistics.

Example 4.3. An earthquake has produced a large slide into a river and has created a new lake of significant dimensions. The river flow, the sediment transport, the water quality variable, and other observed quantities at the gauging station below the new lake have then undergone substantial changes. Their populations before and after the earthquake have now significantly different properties. If data of two populations, with a transient regime of many populations between the two, are mixed, incorrect conclusions may be drawn for the future performance of various projects. The hydrologic pratice is based at present in finding relations between properties of these two or more populations, and in reducing the total sample available to the sample of one population.

Bridge between samples and populations. The main problem in hydrology, as in other natural sciences with random phenomena, is bridging the gap between samples and their populations. It is indirectly a bridge between the sample available and all future samples, which may occur in the case of time series. The underlying concept is that characteristics of a sample provide estimates for the population characteristics of a random variable. These estimates serve as expected values of the future samples. This is characterized by Fig. 4.1 for hydrologic problems related to time

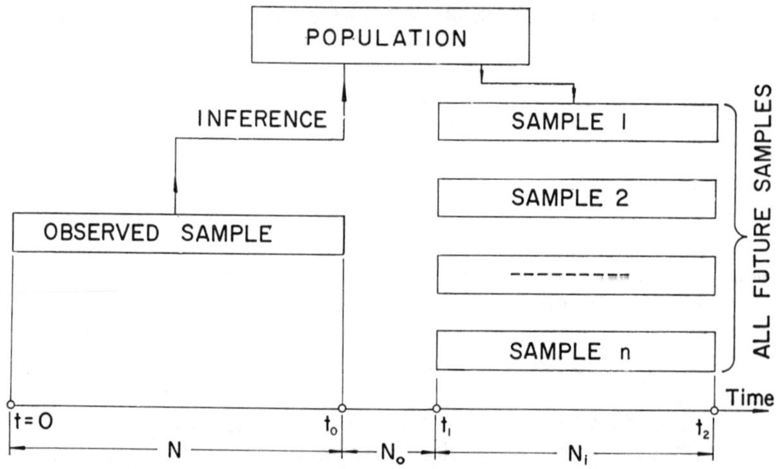

Figure 4.1. A schematic representation of statistical inference problems related to time series in hydrology: N = the sample available; N_0 = the period of data analysis, water resource planning, design and project construction; N_i = the sample size equivalent to the economic life of a project, for which a set of samples, each with a different probability density of occurrence, may be derived.

series. Several properties of time series during the economic life of a project are assumed to be derived from the characteristics of the population. Many samples of different characteristics, however, may occur and with such probabilities of exceedence that they should not be neglected in various analyses. At least, they may be considered as the limiting cases, and each with a given probability of exceedence, as the potential economic or safety boundaries of the future.

Inferential statistics is, therefore, concerned with the inference on population characteristics from the sample data, and with drawing the properties of various potential samples to be determined from the population characteristics. Techniques of mathematical statistics, which serve for making inferences about the population from the sample data, and about new samples from the population characteristics, are grouped under the subjects: estimation, sampling theory (including the drawing of confidence and tolerance intervals or limits), and testing of hypotheses. They are considered as parts of inferential statistics. Statistical inferences are mainly the inductive conclusions about distributions of random variables, and about distributions of their various sample statistics. They are a type of decision making based on probability theory. Chapters 8 through 10 examine problems relating to inferential statistics.

The literature on probability theory and statistics is abundant, particularly that referring to inferential statistics with the extensive use of probability theory. Although it is not feasible to review it here, some books the writer used as current sources of information are given in the references [1 through 6].

Deterministic versus statistical solutions in hydrology. It is common to find attitudes and even convictions among hydrologists that statistics is used only when it is not possible to exactly or completely solve a problem. In this interpretation, the exact solution is the deterministic solution of a problem. It is possible, however, to show that the deterministic solution is nothing else than a particular solution of the general probabilistic or statistical solutions, with the former being often much easier than the latter. In forcing a deterministic approach to a fundamentally probabilistic phenomenon, three errors may be committed: a part of the total information is lost, misleading results are obtained, and incorrect solutions are produced. Deterministic solutions of hydrologic problems most often use the arithmetic means of random variables as the main parameters. The use of means only is correct when variations about them are negligible. Probabilistic solutions are as exact as deterministic solutions. However, deterministic solutions do not often represent the full information available in the data of hydrologic variables.

Example 4.4. A unit hydrograph is supposed to be a deterministic solution of how a river basin responds to a unit-time and unit-amount of surface runoff, as the part of the total precipitation over that basin. Experience shows that the surface runoff of each storm that causes a flood produces a different unit hydrograph. The averaging of all unit hydrographs or any other approach in producing a representative unit hydrograph, means a loss of information about all river basin responses. The greater the variation from one unit hydrograph to another about the average or the representative unit hydrograph, the larger that loss of information.

Example 4.5. Some hydrologic variables occur in clusters of very low or very high values. A season may have very small monthly precipitation while another season may have very large values. If this is the case, the average monthly precipitation of each year may be a misleading parameter if these clusters are not identified and properly statistically described. The probability of monthly values being in an interval around this average monthly value may be smaller than probabilities for the same interval length in clusters of very low or very high values.

Example 4.6. A station in a desert region has the mean annual precipitation of 2.2 inches. However, the record shows that many years have no or negligible rainfall, while wet years often have 7 — 10 inches of precipitation. The mean 2.2 may be a misleading characteristic of this type of rainfall regime.

Example 4.7. The dimensional analysis is an already classical technique in searching for deterministic relations between dimensionless numbers in fluid mechanics and hydraulics, and is often also applied in hydrology. The same magnitudes appear sometimes in two or more dimensionless numbers. If such magnitudes are random variables, or if they are variables that are a combination of deterministic and stochastic components, and if either the observed values or their means, which are also random variables, are used in the dimensionless analysis, the spurious correlation may be introduced. It may show in turn a smaller spread of computed points about a deterministic analytical function or about an empirical curve of relations of dimensionless numbers than the true relation among the original magnitudes would exhibit.

4.2. **Statistics as Description Numbers.**

Statistics. Various characteristics of a random variable, whose properties are given by a sample, may be described by numbers. These numbers are either its particular values, or they are functions of some or of all sample values, or they are related to various frequencies. Values such as a specified x_i, functions of all x_i values, $f(x_i)$ as the frequency densities for given values x_i, $F(x_i)$ as the cumulative frequencies for given values x_i, and similar values all represent individual numbers obtained from a sample. Since all these numbers are functions of the sample of a random variable, they are also random variables. These sample numbers, which change from one sample to the other, are generally called statistics. The sample arithmetic mean, the largest sample value, the frequency of the 10-th smallest value, the largest frequency density of the sample, and similar numbers are all statistics derived from a sample.

An individual sample gives only a value of each statistic. For m equal samples observed from the same population, there are m values of a particular statistic. All of them usually differ. The sample value of a statistic is only one of an infinite number of random events that this statistic can take in a large number of samples from the same population. The population value of that statistic is a constant. Usually it is the arithmetic mean of all sample values of that statistic. The statistic of the observed sample and the population value usually do not coincide. Statistics are important numbers in hydrology. They usually refer to some property of hydrologic samples, important to various water resources problems, and are often called the descriptors of hydrologic variables or processes.

Example 4.8. The arithmetic mean, the median, the standard deviation, the quantiles of different frequencies, the mode , the autocorrelation coefficients, the storage capacity to obtain a guaranteed flow for the 3-year driest period of a sample, and similar statistics are currently needed for condensing hydrologic information for practical application.

Distributions of statistics. It is the task of probability theory and inferential statistics to derive the frequency or probability distributions of sample statistics. Figure 4.2 gives a schematic representation of these

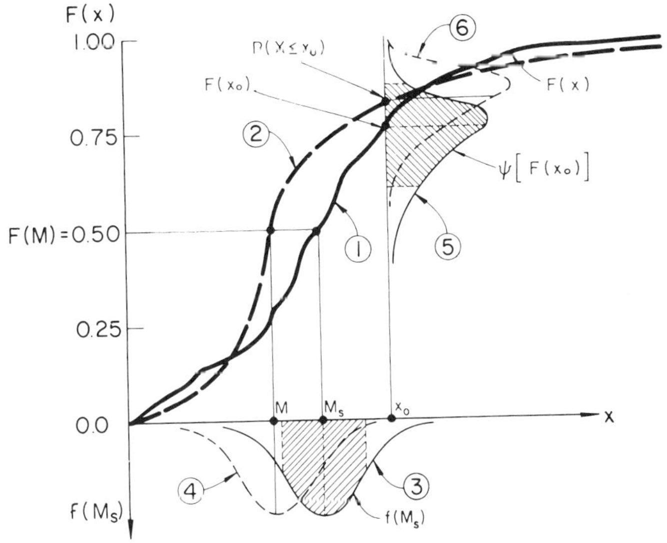

Figure 4.2. Schematic representation of distributions of sample statistics: (1) sample frequency distribution of random variable x; (2) assumed population distribution of x; (3) probability density curve of the sample median; (4) an equivalent to curve (3), centered about the assumed population median; (5) probability density curve of the sample relative frequency F (x_0) of a given x_0; and (6) an equivalent to curve (5), centered about the population probability P $(X \leq x_0)$ of x_0.

concepts, with sample distributions given as solid lines and the hypothetical population distributions as dashed lines. Two cases of distributions of statistics are shown: the distribution of sample median and the distribution of the cumulative frequency of a given constant value x_0 of the variable x.

The sample median M_s is determined for $F(x) = F(M_s) = 0.50$ because the population median M is not known. Its probability density curve is centered in such a way that M_s is the arithmetic mean of this density curve. The alternative case is when the population median M is known or assumed to be known, and the probability density curve of the sample median M_s is drawn around M in such a way that M is the arithmetic mean of this distribution. When M is unknown, one can state only the probability of M being in a given interval about M_s, if a large number of samples of the same size is assumed to be available. Assuming that for each sample M_s is computed, an interval drawn about M_s, and the number of cases determined for which the population value M would be inside this interval, to the total number of samples, is approximately that probability.

For a given value x_0, its cumulative frequency is $F(x_0)$; the probability density curve of $F(x_0)$, given as $\psi[F(x_0)]$, is also shown in Fig. 4.2. In this case, the frequency, $F(x_0)$, is a statistic and ψ is the symbol of its probability density curve; this symbol also gives the probability distribution of a relative frequency. Similarly as for the median, it is only possible to state what the probability is for the population value $P(X \leq x_0)$ to fall within an interval about the sample value of $F(x_0)$, if one assumes a large number of samples of the same size. Assuming the $F(x_0)$ value is computed for each sample, and the interval about it drawn similarly, as stated for the median, this probability can be computed Figure 4.2 shows that $\psi[F(x_0)]$ depends also on the population value $P(X \leq x_0)$, because the larger this P the more skewed $\psi[F(x_0)]$.

Inference by probability statements. The preceding discussion shows that statistical inference can be made only by probability statements. Once the statistic of a sample is estimated as to be the closest value to the

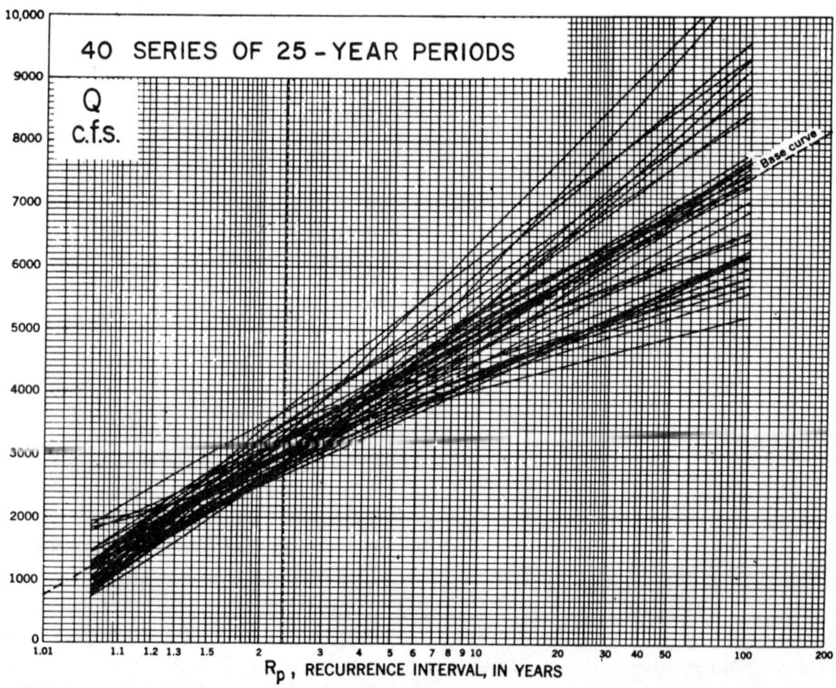

Figure 4.3. Frequency distributions of 40 samples, each of a 25-year size, of the maximum flood discharges on a Gumbel paper graph. Each sample is drawn from a finite population of 1,000 values without replacement. These values follow the double exponential distribution. [According to Benson, 7]

population value, and a probability density function of that statistic is either analytically derived or its frequency curve is computed by experimental methods, one is able only to find what the probability is that the population value falls in any given interval about the sample values for a large number of samples, each with its statistic and its interval about it. This is the limit of information about the population value that can be obtained from the sample data. Therefore, bridging the gap between the known sample and the unknown population, by using mathematical inferential statistics, can be made only by probability statements. The general pattern, derived from the law of large numbers, is that the larger a sample the smaller the variation of a sample statistic about its population value.

Example 4.9. Figure 4.3 shows an experiment performed by Benson [7]. A finite population of 1,000 flood discharges is obtained from the double exponential function, $F(x) = \exp\{-\exp[-\alpha(x-\beta)]\}$. From it, 40 samples, each simulating 25-year period , are randomly drawn as independent trials without replacement. The base curve plots as a straight line on Gumbel graph paper. The 40 frequency distributions of 25-year periods are also plotted against the recurrence interval or the return period, $R_p = 1/(1-F(x))$. They lead to several basic conclusions: (a) the sampling variation of these curves about the base curve represents an important error in estimating the population characteristics from a sample; (b) the sampling errors increase with an increase of the flood discharge, or with an increase of the return period, R_p; and (c) To require every frequency distribution curve of a relatively short sample to plot as the straight line, while its population distribution plots as such in selected transformed coordinates, is a current but erroneous attitude in practice. This example is instructive regardless that a bias is introduced by drawing samples from a finite population without replacement.

4.3. Mathematical Statistical Models.

Mathematical models as condensed information. The basic objectives in the use of stastistical methods is the analysis of outcomes of real or conceptual experiments or observations of random phenomena. However, one of the main tasks of this analysis is the construction of mathematical models that describe properties of random variables. The models facilitate the prediction, in probability terms, of properties of future outcomes of repeated experiments or observations. Mathematical models are considered as the most condensed form of information about random variables, especially in comparison with the tabular or the graphical representation of statistical information. Establishing good mathematical models in statistical information is often considered the best, final result of any statistical analysis.

Example 4.10. A series of annual runoff, Q_i, was found by tests to be a stationary (time-invariant) process, with Q_i lognormally distributed and following the first-order linear autoregressive model of dependence in time.

For $\bar{Q}$ and $\bar{Q}_n$, the arithmetic means of Q_i and $\ln Q_i$, and s and s_n the standard deviations of Q_i and $\ln Q_i$, respectively, the lognormal distribution of Q_i is

$$f(Q_i) = \frac{1}{Q_i s_n \sqrt{2\pi}} \exp[-(\ln Q_i - \bar{Q}_n)^2/2s_n^2], \tag{4.1}$$

and the time dependence model is

$$Q_i = \bar{Q} + r_1(Q_{i-1} - \bar{Q}) + s\sqrt{1 - r_1^2}\,\varepsilon_i, \tag{4.2}$$

in which r_1 is the first serial correlation coefficient of the Q_i-series (the correlation coefficient of successive two values), and ε_i is a standardized independent random variable (mean zero and variance unity).

With these two mathematical models, and the estimated parameters $\overline{Q}$, s, (or $\overline{Q}_n$, s_n), and r_1, the process of annual flow sequence is statistically described in the most condensed form.

Mathematical models as inductive inference. In general, drawing conclusions from sample data in the form of mathematical models represents inductive inference. In the case of random variables and their samples of data, this inference is referred to as the statistical inductive inference.

As statistical inference is a decision making process, based on probability, the mathematical models serve as a basis for the inference. Types of inferences are tests of hypotheses referring to distributions or statistics. Examples are: tests for probability distribution functions that best fit an observed frequency distribution; tests of significant diffferences of sample statistics from given or assumed values, such as the parameters of a probability distribution or any other population characteristic; tests about significant differences of various measures of sequential dependence in samples, from the assumed population values of time series, and so forth.

Statistical methods. Statistical methods deal with data obtained by repeated operations. If, in hydrology, these operations are not repeated, they can be conceived as being repetitive, because the observed processes behave as though they occur under essentially stable conditions. This stability, however, must be proven; conditions necessary for hydrologic stability are not the same conditions necessary for performing games of chance.

The derivation of mathematical models of applied statistics is usually connected with various prediction problems. In the process of solving practical statistical problems four steps are important: (a) selection of a mathematical model, (b) estimation of its parameters, (c) a check of the reasonableness of the model by various tests, and (d) derivation of conclusions from the model that affect the solution of the proposed problem.

4.4. Concepts of Risk and Uncertainty in Hydrology.

Definitions of risk and uncertainty. It is often convenient and instructive to make a distinction in hydrology between the concepts of risk and uncertainty. Risk is conceived, in this text, as a permanent population property of any random phenomenon. If the population distribution were known for floods, droughts, and average water yields, the risk would then be exactly known in any analysis and decision of problems related to floods, droughts, and water yields. The risk is then identified with probabilities of values greater than, or smaller than, a given value. This risk, specified further as the basic risk, is inherent in the phenomenon itself, and cannot be avoided without a change in population itself. By various flood control or drought control measures, or by flow regulations in the case of water yields, corresponding populations are changed. New types of risks are then associated with populations of these changed conditions because of new probabilities resulting for critical values.

Because inaccurate, deficient, or biased data are usually always available, and because population properties must be estimated from these data by some technique, various errors and information losses are represented as uncertainties. Differences between the population properties and properties derived from sample data, available for drawing inferences about the population, constitute the concept of uncertainties in this text. All uncertainties, added to the basic risk, are conceived as the total risk. This total risk is colloquially known as the hydrologic risk. Basic risk is usually decreased or minimized by various water resources developments and measures, while uncertainties are decreased or optimized by obtaining better data and by using better statistical techniques.

Sources of uncertainties. Sources of uncertainties are: random measurement and computation errors of observations; systematic errors in data; non-homogeneity; loss of information by approximating a continuous variable in nature by observations as a discrete variable; loss of information by approximating a continuous sequence in nature by observations in the form of a discrete series; sampling errors that result from a limited number of data; sampling errors that are the consequence of a limited number of observational points; use of inefficient statistical methods in extracting the information from a pool of data; errors produced by various ambiguities in defining the population characteristics of importance to a problem; errors introduced in filling in the missing data by the correlation and regression methods, and similar.

Random errors. If a magnitude Q is measured n times, the resulting values are Q_m, m = 1, 2, ..., n. The random errors $\varepsilon_m = Q_m - Q$ are symmetrically distributed about 0, or the arithmetic mean of ε_m is approximately zero. The standard deviation s_ε of ε_m is the measure of erorrs, and it is often proportional to Q, or

$$s_\varepsilon \approx \alpha Q . \tag{4.3}$$

Its square, var $\varepsilon_m = s_\varepsilon^2 \approx \alpha^2 Q^2$, is a measure of uncertainty in the form of error variance produced by these random measurement errors.

Example 4.11. Figure 4.4 is a schematic representation of random and sampling errors for the case of a distribution of flood discharges: (1) F (Q_t) is the unknown population distribution; (2) F (Q) is the unknown distribution of exact values of discharges of a sample; and (3) F (Q_m) is the known distribution of measured sample values. The probability density curve (5) is the distribution of Q_m about a given Q, for given F (Q_t) = F (Q) = F (Q_m), while curve (4) gives the probability density curve of sample values Q about the population value Q_t.

Systematical errors. Denoting Q_s as the measured discharge with both random and systematical errors, then the difference $\eta_s = Q_s - Q_m$ are systematical errors, or $Q_s = Q + \varepsilon_m + \eta_s$. The mean, $\bar{\eta}$, of n measurements Q_s of Q, is the measure of this error for any Q. Often these errors are approximately proportional to Q, as

$$\bar{\eta} \approx \beta Q, \tag{4.4}$$

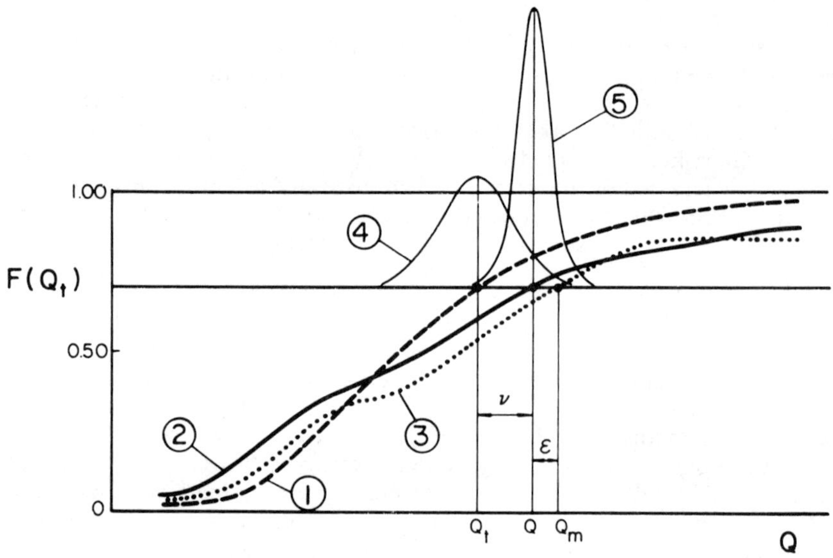

Figure 4.4. Effects of measurement and computation random errors: (1) population distribution, F (Q_t); (2) sample distribution, F (Q); (3) distribution with random errors, F (Q_m); (4) probability density curve of sample values, Q, for given Q_t , and (5) probability densities of Q_m, for given Q.

which is a measure of uncertainty produced by these undetected and remaining systematic errors

Non-homogeneity. Similarly, as for the systematic errors, non-homogeneity may be measured by the difference between the non-homogeneous and homogeneous values of Q, as $\zeta = Q_n - Q_h$. The mean $\bar{\zeta}$ of ζ for a given Q may be determined in several ways. It is a measure of uncertainty introduced by undetected and remaining non-homogeneity.

Example 4.12. It was shown in Example 1.11 by Fig. 1.2 that the series of maximum annual flood discharges of the Danube River at Orshova has non-homogeneity in the data. Figure 4.5 shows, in turn, two frequency distribution curves, (1) of the historical non-homogeneous sample of 110 years, and (3) of the homogeneous sample, approximately reduced on the flood producing conditions in river basin, which prevailed around the year 1950. For any F (Q), the difference ΔQ of the two empirical distributions gives a measure of this non-homogeneity.

Loss of information in replacing continuous variables and series by discrete variables and series. It can be shown by investigations that a replacement of a continuous variable x by discrete states or discrete values at intervals Δx only, a part of information is lost. Similarly,

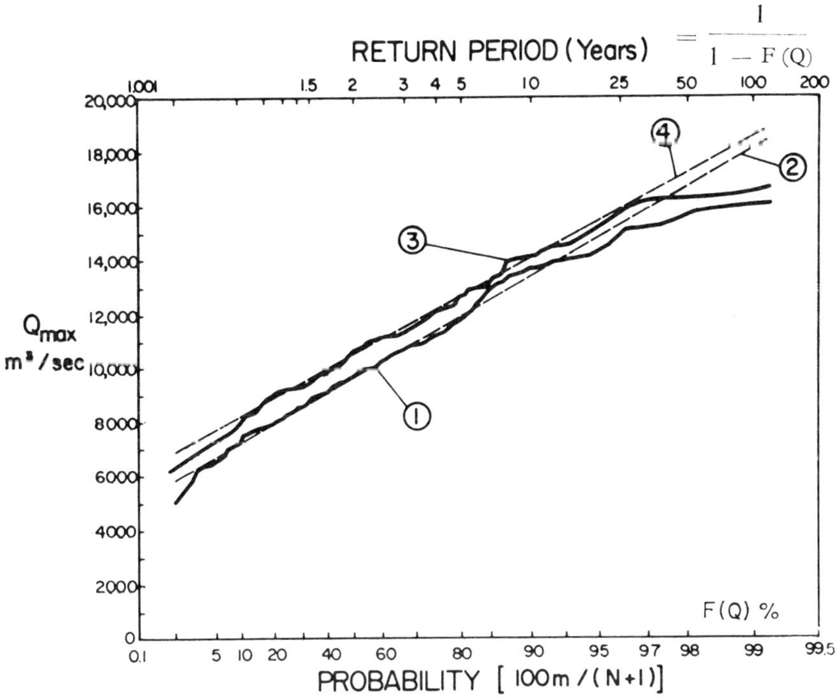

Figure 4.5. Frequency distributions of maximum annual flood discharge of the Danube River at Orshova (1840–1950), using Gumbel graph paper: (1) historical record; (2) straight-line fit to curve (1); (3) homogeneous sample on flood conditions prevailing in 1950, and (4) straight-line fit to curve (3).

in replacing a continuous series either by discrete values Δt apart, or by averages over Δt, some information is also lost [8]. The greater Δx and Δt, the larger this loss, but in turn less data must be observed and processed. This is a problem of optimization. Therefore, the procedures of making continuous variables and sequences discrete also introduce some uncertainty through the loss of information.

Sampling errors. Sampling errors, as differences between various statistics estimated from samples and their constant population values are the most important errors producing uncertainty in hydrologic information. For a given $F(Q_t)$, the differences $\upsilon_i = Q_t - Q_i$, where Q_t is the population constant and Q_i are values from various samples, are measured by their standard deviation, s_υ, and their variance s_υ^2 is a measure of uncertainty.

Example 4.13. Figure 4.2, Fig. 4.3 of Example 4.9, and Fig. 4.4, curve (4) of Example 4.11, all represent various aspects of sampling errors. The random errors and sampling errors in Fig. 4.4 may be combined to form the joint error. The method of combining these errors depends on whether these two types of errors are dependent or independent.

Inefficient statistical methods of data analysis. Estimation of a population property from a sample may be carried out by an efficient or an inefficient statistical method. An efficient method extracts all information contained in a sample. An inefficient method, as will be shown in the subsequent text, may be such as to extract only part of the information, equivalent to a sample of size N_i, which is smaller than the available sample of size N. In the age of digital computers it is easy to use the most efficient methods of estimation of population properties, if these methods are available. The differences in the extracted information represent a type of uncertainty.

Ambiguities in the selection of population characteristics. Often random variables or parameters are either incorrectly selected or some variables or parameters are neglected when analyzing a problem based on statistical analysis of hydrologic data. This may be equivalent to using a smaller portion of all information in a hydrologic sample, and therefore, may represent a type of uncertainty.

Example 4.14. Let us assume that the damage D by floods along a river is a linear function (though the linearization of this dependence may also represent a loss of information) of flood peak Q_m, flood volume above a critical discharge V, duration of flood T, and the sediment load in floods Q_s, as

$$D = a_0 + a_1 Q_m + a_2 V + a_3 T + a_4 Q_s. \qquad 4.5$$

If only the properties of populations of Q_m and V are estimated from a sample, with those of T and Q_s neglected, the problem involves uncertainty by the loss of information because of this neglect.

Filling in missing data by correlation and regression. Often in hydrology missing data in a sample can be filled in, or the sample can be extended by using samples of other stations or other variables. If this transfer of information is not properly measured, but assessed from the number of data filled in or extended, the amount of information may be smaller than originally considered. This also may be some kind of uncertainty in the analysis of hydrologic data.

Total risk. If uncertainties are properly measured as errors or losses of information, they can be added to the basic risk to produce the total risk resulting from a pool of hydrologic data.

Optimization of uncertainties. The use of probability theory, mathematical statistics, and stochastic processes enables the optimization of uncertainties in finding the corresponding optimal value of their sum for given means and time available. To decrease the overall risk in using hydrologic data in decision making, hydrologic samples should have non-homogeneity removed; accurate observations, measurements, and compilations of data; smaller sampling errors by longer observations in time and at more points over an area; the use of proper statistical methods for extraction of the major part of information in data; a limit to the use of the discrete approach to continuous variables and series to an acceptable loss of information, and similar. Decrease of uncertainties is the problem of optimizing the procurement of information.

4.5. Statistics in Hydrology.

Application of statistics in hydrology. Statistical analysis is used in hydrology where variables and processes are observed and stated in terms of numbers. Principles, possibilities, and limitations of statistical methods as applied to hydrology are more important than the elaborated techniques of their use. It is not sufficient to be just acquainted with techniques but one should be also acquainted with conditions under which they are valid. A mathematical statistical model in hydrology is no better than the basic axioms or hypotheses underlying it. Statistics can be meaningfully applied to hydrology anywhere complex natural processes are observed and stated in terms of numbers (or sometimes ranks). Statistics in hydrology is a selection and adjustment of statistical principles, methods, and techniques to a particular field of hydrologic phenomena, processes, environments, and variables.

Two approaches in the application. Two extreme positions may be discerned in the application of statistics in hydrology. First, some statistical methods are available for a given type of problem, if investigators adjust the problems and data, and accept the underlying hypotheses, in order to apply these methods or techniques. Second, the hydrologic physical reality is not changed, but the statistical methods are adjusted, developed, and supplemented to best suit the hydrologic reality. This second approach is best whenever feasible.

Objectives of applying statistics to hydrology. The main objectives of applying statistical principles and methods to hydrology follow.

(a) Interpretation of observations, because statistical methods are useful as long as they assist in the interpretation of observations.

(b) Search for hydrologic probabilistic regularities.

(c) Extraction of maximum information from hydrologic data, with methods available of knowing when that is approximately attained.

(d) Presentation of hydrologic information in condensed form, as graphs, tables of numbers, and mathematical equations, basically for decision making in water resources planning, conservation, development and control.

The basic objective of applying statistics in hydrology is, however, the derivation of information from observed hydrologic phenomena in the past, and the making of inferences about what is expected in the future. The advent of electronic computers, particularly of digital computers, and the accompanying rapid development of modern mathematical numerical methods, have given a particular impetus to the use of statistics in all natural sciences, particularly in hydrology.

Dangers in formalistic application of statistics. Misconceptions, misuses, and dangers in the application of statistics in hydrology usually are:

(a) expectations are often greater than the available data and the best statistical methods can produce,

(b) false expectations that statistical methods may in some way replace the scarcity or unreliability of data, which is impossible, and

(c) the tendency to replace the necessary judgment and interpretation of physical facts by statistical computations or statements.

4.6. **Problems for Discussion and Solutions.**

4.1. To have a finite number of possible outcomes what conditions should a finite hydrologic population satisfy?

4.2. What are the basic differences betwen descriptive and inferential statistics?

4.3. Give examples in hydrology, besides Examples 4.4 through 4.7, in which the classical deterministic approaches in solving problems mean: a partial loss of information, misleading, and incorrect solutions.

4.4 A statistic $\hat{\alpha}$ is obtained from a sample and its symmetrical distribution F $(\hat{\alpha})$ computed from the known type of distribution. Then an interval $\alpha_1 \leq \hat{\alpha} \leq \alpha_2$ is drawn around $\hat{\alpha}$, so that $\hat{\alpha}$ is assumed to be the expected value of $\hat{\alpha}$-sampling distribution, and $P(\alpha_1 \leq \hat{\alpha} \leq \alpha_2) = 2P(\alpha_1 \leq \alpha_i < \hat{\alpha})$ is a probability of that interval (α_1, α_2). Is it correct or incorrect to consider, on the basis of one sample, that the probability of the population value α to be in the interval (α_1, α_2) is the above $P(\alpha_1 \leq \hat{\alpha} \leq \alpha_2)$? Use Figs. 4.2 and 4.3 to support your answer.

4.5. Risk is a probability. Review various water resources problems and single out those that involve risk, based on probability statements.

4.6. Under what conditions should systematic errors (inconsistency) and non--homogeneity be considered independent or dependent on variable values, x_i?

4.7. The optimization of uncertainties are problems of design of hydrologic networks; selections of instrumentation; the training of technicians, who carry out observations; selection of statistical techniques for data analysis, and an understanding of types of problems that need hydrologic data. Discuss in principle how the various sources of uncertainties, outlined in the preceding text, should be optimized for given financial constraints.

4.8. Give examples of forcing the hydrologic reality to fit a given statistical technique and the converse, of adjusting and improving a statistical technique to well fit a hydrologic reality.

CHAPTER 5

EMPIRICAL DISTRIBUTIONS OF HYDROLOGIC VARIABLES

Most of the statistical information in hydrology comes from observed data of limited sample sizes. The empirical sample frequency distributions, as the estimates of the population probability distributions, are often the only available information. The practical methods of obtaining these empirical distributions are summarized in this chapter along with their representation and qualitative analysis.

5.1. Frequency and Cumulative Frequency Curves of Univariates.

Types of samples in hydrology. Observations of a variable are either in the form of a sample of discrete values or continuous recordings. Often, they are discrete or continuous series and observed in time or along a line. Three cases of samples are current: a sample of discrete observations of a discrete variable, a sample of discrete values of a continuous or of a mixed variable, and a recorded continuous series in time or along a line of a continuous or of a mixed variable. The frequency and cumulative frequency curves, which constitute empirical distributions, are obtained either from discrete sample data or from recorded continuous finite series. The total number N of discrete values in a sample, or the total length T of the recorded continuous series is the sample size.

Frequency graph and frequency distribution of discrete univariates. For a sample of N values x_i, of a discrete variable x, there are n_j values of an event x_j. These n_j numbers are absolute frequencies of x_j events. The ratios $f_j = n_j/N$ are their relative frequencies. Events x_j are represented by discrete points on the x-variable line. The frequency graph shows how the absolute or relative frequencies are distributed over these discrete events. It is conventionally represented by vertical barlines at these events. The sum of all absolute frequencies, n_j, is N, and the sum of all relative frequencies, f_j, is one. The graph on the left of Fig. 2.2 shows the relative frequency of the number of storms in the first 10 days of May, for the precipitation station at Austin, Texas, for

$N = 107$ years. The absolute cumulative frequency, N_j, of a discrete event is defined by

$$N_j = \sum_{i=1}^{j} n_i, \tag{5.1}$$

and the relative cumulative frequency, $F(x_j)$, is defined by

$$F(x_j) = P(X \leq x_j) = \frac{1}{N} \sum_{i=1}^{j} n_i. \tag{5.2}$$

These numbers N_j and $F(x_j)$ include all absolute or relative frequencies, respectively for Eq. 5.1 and Eq. 5.2, of all events $x < x_j$, and that of x_j. Cumulative relative frequencies of the number of large storms per year are also given in the graph on the right of Fig. 2.2, which is a non-decreasing step curve. In subsequent text, the two graphs of Fig. 2.2 will be called the frequency graph and the frequency distribution, either for the absolute or for the relative frequency case of discrete univariates.

Discrete series of continuous univariates. A continuous variable, x, has a sample of discrete observations, with the sample size N, in the following cases: if x is observed only at discrete times or discrete distances, at equal or random spacings; if the continuous series of x is integrated over equal successive and non-overlapping intervals, Δt for the time, or Δl for the length, and given as a sequence of discrete values; if a new sample is derived in selecting discrete values by some rule, such as the maximal value every year, the minimal value every season, and similar, and if the continuous series is digitized by reading out only ordinates at selected equal intervals, Δt, or Δl. The use of digital computers requires that a continuous series be replaced by a discrete series that sufficiently represents its fluctuation.

For a continuous variable, the elementary events are all (countable real) points on the line of x. The accuracy of the data limits the number of these events. For the example of the accuracy being $\pm$ 0.005 in every

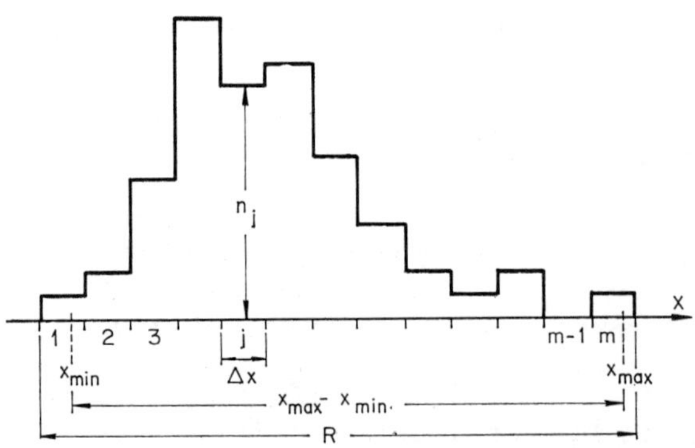

Figure 5.1. Graph for definitions of class intervals and related frequencies.

observation, and the data given within two significant figures beyond the decimal point, a difference of unity between x_1 and x_2 contains 100 events. For a range of 100 there are 10,000 possible elementary events. If a sample has only 250 values many possible elementary events of this accuracy will not show.

Class intervals, absolute and relative frequencies, and frequency densities. The range of a sample is the difference of the largest and the smallest observed value, $x_{max} - x_{min}$. It is often useful to sort all sample values x_i into subranges of the range to readily grasp their statistical properties. To this end, Fig. 5.1 serves for the definition of various concepts. The subranges are class intervals. The number of values falling within the j-th class interval is its absolute class frequency n_j. The class mark x_j, $j = 1, 2, \ldots, m$, is the representative value of this class interval. Usually, the class mark x_j is the midpoint of class interval, though it has a bias. All values in the class are treated as having the equal value x_j. The upper class limit and the first event after the lower class limit are the largest and the smallest values of x_i in a class interval. The class intervals are consecutive, non-overlapping, subranges of the total sample range. The range R between the upper limit of the last, m-th class, and the lower limit of the first class interval, should be somewhat greater than the sample range, so that the observed x_{max} and x_{min} fall within R. The sum of all n_j is the sample size N.

The absolute cumulative frequency of an upper class limit, x_j, is the sum of all absolute class frequencies for classes with $x_i \leq x_j$. The absolute cumulative frequency of a value, x_i, is defined for ungrouped sample data as the number N_i ($N_i < N$) of all values $x \leq x_i$. Figure 5.2 serves as an illustration of these definitions.

The relative frequency, f_j, of the j-th class interval is the dimensionless ratio of its absolute frequency, n_j, to the sample size, or $f_j = n_j/N$. The sum of all relative frequencies is one. The relative cumulative frequency for the value x_i is $F(x_i) = N_i/N$. The smallest and the largest values of $F(x_i)$ are zero and one.

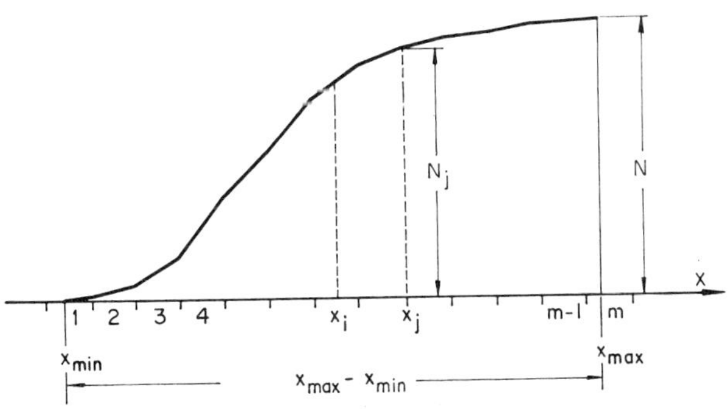

Figure 5.2. Graph for definitions of the cumulative frequency distribution.

The absolute and relative frequencies depend on the size of class intervals and on the position of class limits. The larger a class interval the greater these frequencies. To avoid this dependence on the class size and the class limit positions, and to replace the concept of frequency of discrete elementary events by frequencies related to finite intervals of continuous variables, the concept of frequency density must be introduced. The frequency density, d_j, at the class mark x_j of the j-th class interval, is the relative class frequency of that interval divided by the class length Δx, when Δx converges to zero around the mid-point x_j, or

$$d_j = \lim_{\Delta x \to 0} \frac{f_j}{\Delta x} \approx \frac{n_j}{N\Delta x} . \tag{5.3}$$

As Δ_x converges to zero this ratio, d_j, converges to a finite value greater than zero if $f_j > 0$. This value is defined as the frequency density at the point x_j. For a sufficient large number of j points in the range $x_{max} - x_{min}$, a curve of frequency densities may be defined. To obtain relative frequencies, the frequency density curve must be integrated in an interval. Therefore, the dimension of a frequency density is the inverse of the dimension of the variable x.

Frequency histogram, frequency polygon and frequency curve. The frequency curve or graph describes the manner the class frequencies are distributed over class intervals. To avoid confusion, it is necessary to always clearly indicate whether the absolute or relative frequencies, or frequency densities, are used. If frequencies for each interval are represented by horizontal bars, as in Fig. 5.1, the graph is called the frequency histogram. If the points with coordinates (x_j, n_j), or (x_j, f_j), or (x_j,d_j), are connected by straight lines, the broken line graph is the frequency polygon. If a curve is drawn through the points of frequency densities the resulting graph is the frequency density curve.

Frequency curve and frequency distribution of continuous variables. Frequency curves and frequency distributions of continuous variables are more or less smooth curves. However, they appear irregular because of the sampling errors involved. The smaller a sample, the less smooth are the curves, with more pronounced irregularities or wiggles. Experience shows the empirical frequency curves and distributions converge to smooth curves with an increase of sample size. A misrepresentation of these curves occurs when their smoothness is assessed visually and the smoothness is used to draw conclusions. Frequency distributions look to an observer much smoother than the corresponding frequency curves. Because the frequency density for a value x is the first derivative of the frequency distribution at x, the frequency curve looks much more irregular, with many more wiggles for small samples, than the distribution. This phenomenon may be the reason why practitioners and statisticians prefer to work with frequency distributions rather than with frequency curves.

For many hydrologic problems the question is how often a certain variable value is or is not exceeded. To answer this it is necessary to consider the frequency distribution rather than the frequency curve. As frequencies are assumed to be the closest values known for probabilities, the words *frequency distribution* and *probability distribution* refer to the sample and the population, respectively.

Example 5.1. Figure 5.3 gives the frequency polygon (1), and the cumulative frequency polygon, or the frequency distribution (2) for the relative frequencies of annual river flow, given in modular coefficients $K_i = Q_i/\bar{Q}$, of the Rhine River at Basle, Switzerland, for the period of observations 1808–1957 (150 years), with $\bar{Q} =$ = 36,253 cfs. This figure supports the statement that a frequency distribution looks much smoother than the corresponding frequency curve (in this case a polygon).

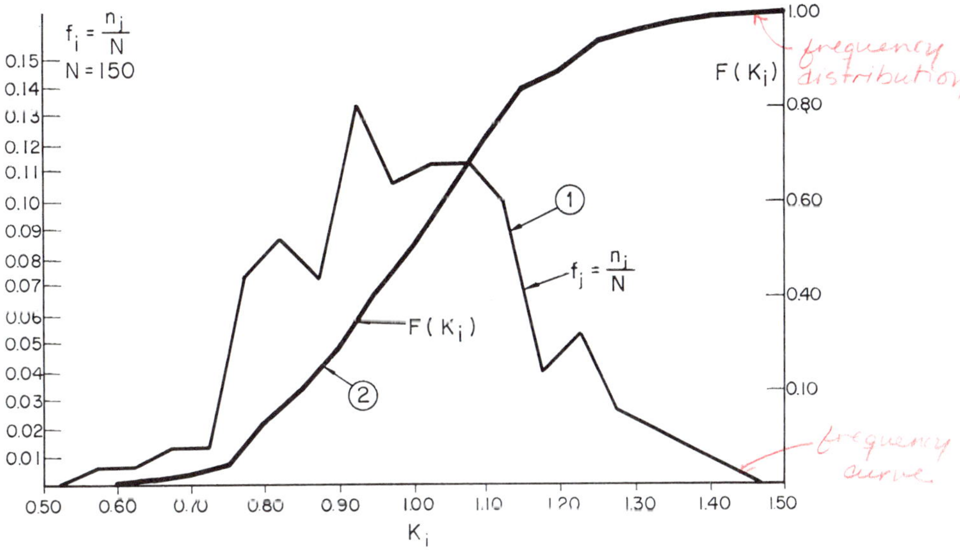

Figure 5.3. Polygons of relative frequency (1), and of relative frequency distribution (2), of the annual flow of the Rhine River at Basle, for 150 years (1808–1957), given in modular coefficients, $K_i = Q_i/\bar{Q}$, with $\bar{Q}$ the mean annual flow.

Frequency curves and distributions of a continuous series of continuous or mixed variables. The sample size is the total time of the series, T, or the total length, L, of recordings along a line. The frequency curve is obtained by dividing the x-variable range into class intervals, and by determining from the recorded graphs the total times, t_j, or lengths, l_j, the variable x stayed within each class interval, x_j. These are absolute frequencies. Divided by T or L, they give the relative class frequencies, $f_j = t_j/T$, or $f_j = l_j/L$. A more accurate method is to first determine the frequency distribution and from it the class frequencies as differences of cumulative frequencies for the upper and lower class limit of a class interval. For any value x_i of the variable it is easy to determine the total time, T_i, or the total length, L_i, for the values $x \leq x_i$ of the series. These are the absolute cumulative frequencies. Ratios $F(x_i) = T_i/T$, or $F(x_i) =$

$= L_i/L$, are relative cumulative frequencies. The more values, x_i, for which $F(x_i)$ are determined, the more accurate the distribution curve, F (x). For the time series, F (x) is the classical hydrologic duration curve.

Example 5.2. Examples of recording instantaneous values (intensities, rates) for the time T are all continuous time series. An example of a continuous recording along the line is the random roughness for a distance, L,along a longitudinal line of a conveyance structure.

Selection of the number of class intervals. The choice of the number and the width of class intervals depends on the sample size and the sample range of a variable. Class intervals should be sufficiently small so that the values falling within each class interval may be considered as approximately equal. It is desirable to make class intervals as large as possible, while still assuming that their values are approximately equal. Each class interval should contain several elementary events determined by the accuracy of the data. If data are given only as integers, and the class widths are $\Delta x = 10$, each class has 10 elementary events. As the first event greater than the lower class limit and the last event of the upper class limit determine the subrange of x in a class interval, the class mark should be shifted from the class mid-point to avoid a bias. As an example, for the class interval 350—360, and events 351, 352, ..., 360 in the interval, the class mid-point 355 should be shifted to 355.5. If the number of elementary events in the class interval is large, this bias can be neglected.

Sometimes rules or expressions are given for the number of class intervals, like Snedecor's suggestion that the class interval should not be larger than s/4, where s is the estimated standard deviation. An expression for the number of class intervals is [1]

$$m = 1 + 1.33 \ln N, \tag{5.4}$$

in which ln N is the natural logarithm of the sample size.

A sample of size 100 should have about 16 class intervals. Example 5.1 approximately satisfies Eq. 5.4. Regardless of these rules, the number of class intervals in hydrologic applications should be between 10 and 25, the choice depending on the sample size. If less than 10 class intervals are used, too much information may be lost. If more than 25 class intervals are selected, too many details are revealed that are not justified by sampling errors in class frequencies, the main properties of a distribution cannot readily be grasped, and the work becomes tedious.

Class limits are usually selected as common rounded numbers, with the assumption that class intervals include all values greater than the lower limit and smaller than and equal to the upper limit. Regardless of the method used for the selection of class intervals, it is a necessary condition that the first and the last class intervals be set up in such a way that they include the smallest and the largest observed value.

5.2. Plotting Empirical Univariate Distributions of Ungrouped Data.

Ungrouped data. The foregoing discussion was concerned mainly with the case of sample sizes large enough to require the data to be grouped into class intervals. If the sample is small, grouping of data is undesirable because many class intervals come out to have the absolute frequency of zero. Various problems in hydrology require the use of ungrouped data for their solutions. In order to facilitate the frequency analysis of ungrouped data, the data are arranged either in an increasing or in a decreasing order of magnitude. To represent graphically an empirical distribution, or to compare graphically a theoretical probability distribution function with an empirical frequency distribution, it is necessary to decide where to plot single observations x_m (m = 1,2, ..., N) of ungrouped data arranged in descending or ascending order of magnitude. To plot these data, the coordinates x_m and P (x_m) must be known for all values x_m, with P (x_m) being the plotting position in the coordinate system of x and F (x).

Example 5.3. Sampling a variable from the subseries of a series usually produces a short new sample. Such examples are: annual flood peak discharge, annual largest water volume for an interval Δt, annual smallest flow or level, annual smallest water volume for an interval Δt, storage capacity to regulate flows of a given number of years to an outflow regime, water yield for a period of n years, and so forth. In these and similar cases, the frequency distribution is determined from the ungrouped but ordered data.

General about plotting positions. The plotting position, P (x_m), of the m-th ordered value of x of the sample size N, and of the given distribution function F (x), is a function of m, N, and F (x). The m-th value has a distribution, which possesses the median, the mode, and the expected value. Theoretically, the probability of the expected value, P (E x_m), of this m-th value x_m, should be its plotting position. If F (x) is described by parameters a, b, c, ..., then P $(x_m) = \psi$ (m, N; a, b, c, ...). For every specific function F (x), with its parameters a, b, c, ..., there should be a particular equation for the plotting position P (x_m).

This means a probability function, F (x), should be assumed, P (x_m) computed, and the points of x_m and P (x_m) coordinates plotted in a graph. Then the parameters a, b, c, ..., are estimated and a test is performed to find how well F (x) fits the plotted points. If the fit is rejected, another function, F (x), is assumed, the computation of P (x_m) repeated, and the points replotted. Because P (x_m) is dependent on F (x), the plotting positions P (x_m), which are functions of F (x), are not practical though they may have a good theoretical justification.

If the function to be fitted is known, and a plotting position formula for that particular function is available, the method of P (x_m) being dependent on this F (x) may still be used efficiently. Differences between the two plotting positions P (x_m), one being the best position independent of F (x), and the other being related to it, are usually much smaller than the sampling errors in data. The m-th value x_m may be very much distant from E x_m. Even in the case of the known function F (x), the plotting positions independent of underlying probability functions are preferred, for several reasons, practicality being the most important.

Five conditions a plotting position should satisfy. Gumbel [2, p. 32—34] summarizes five conditions that a plotting position P (x_m) of the m-th ordered value should satisfy:

1. All observations can be plotted on graph paper between P (x) = 0 and P (x) = 1, without the impossible event of probability zero, and the sure event of probability one.

2. The plotting position of the m-th value is distribution-free, or it should be independent of the underlying distribution F (x), and it should lie between the observed frequencies m/N and (m — 1)/N, if the first condition is to be satisfied that P (x_m) is never either one or zero.

3. Probabilities of exceedence of the largest observed value or of non--exceedence of the smallest observed value should be close to the relative frequency 1/N, or the inverse of the sample size.

4. Observations plotted on the frequency scale should be equally spaced, or the difference between the plotting positions of the (m + 1)-th and the m-th ordered value should be a function of N only, independent of m.

5. The plotting position should be analytically simple, with a good intuitive meaning.

Plotting positions determined as probabilities of the expected value, the median and the mode of the m-th value x_m, satisfy only the latter part of the fifth condition and the first condition; they do not satisfy the remaining conditions.

Practical plotting positions. The most practical plotting position which fully satisfies all five of Gumbel's conditions is

$$P(x) = \frac{m}{N+1}, \qquad 5.5$$

in which m is the ordered sequence of x values, m = 1, 2, ..., N, and N is the sample size of the ungrouped data [3]. The property of Eq. 5.5 is that for m = 1 it gives P (x_1) = 1/(N + 1), and for m = N, it yields P (x_N) = N/(N + 1), the first value being very close to the frequency 1/N, and the last value being smaller than one. The use of Eq. 5.5 dominates at present the hydrologic applications.

The inverse of P (x) is called the return period or the recurrence interval which is

$$R(x) = \frac{1}{P(x)}. \qquad 5.6$$

If the sampling of variable values refer to a year, or a season, or a month, the return period then relates to these intervals, and is counted in years, or seasons, or months. For the plotting position of Eq. 5.5, the return period is

$$R(x) = \frac{N+1}{m} . \qquad 5.7$$

The return period for the plotting position of Eq. 5.5 gives, for m = 1 and m = N, the values $R(x_1) = N + 1$ and $R(x_N) = 1 + (1/N)$. For the sample size of N = 100, the return periods are 101 and 1.01, respectively. The first value should be 100, but the difference is negligible except for very small samples, and the second value is greater than one but close to it which is what it should be.

Another formula for the plotting position currently used is the Hazen compromised position [4]

$$P(x) = \frac{m - 0.5}{N} = \frac{2m - 1}{2N} , \qquad 5.8$$

which is between the two plotting positions

$$P(x) = \frac{m}{N} \qquad 5.9$$

and

$$P(x) = \frac{m - 1}{N} . \qquad 5.10$$

The disadvantage of Eqs. 5.9 and 5.10 is that the first gives P (x) = 1 for m = N, and the second gives P (x) = 0 for m = 1, which are the sure and impossible events, respectively. Neither should appear in a finite sample. Equation 5.8 satisfies all of the five conditions indicated except that for m = 1 it gives the return period R (1) = 2N, or a double value of the sample size. For m = N, then R (N) = 1 + 1/2N. On the above example of N = 100, the return period for the largest value m = 1 becomes R (1) = 200, and for m = 100, R (100) = 1.005. If one is interested only in the uppermost or the lowest part of the sample distribution, using Eq. 5.9 or Eq. 5.10 is appropriate for the two extremes.

Using Eq. 5.5 is recommended for plotting hydrologic distribution points regardless of the sample size, when the total frequency distribution curve is of a practical interest, provided there is no particular reason to do otherwise. When the investigator is interested in the highest values only, and the data are ordered from the greatest to the smallest values, Eq. 5.9 is most appropriate. For the same ordered series, and when only the smallest values are of interest, Eq. 5.10 gives best results.

Example 5.4. Figure 5.4 shows differences obtained for frequency distribution curves by the use of four plotting positions of Eqs. 5.5, 5.8, 5.9, and 5.10 for the annual flood peak discharge of the Colorado River, at Lees Ferry, Arizona, for the period 1923–

1937, or N = 15 years. Differences are significant for small samples. This example demonstrates the properties discussed for the four plotting positions given.

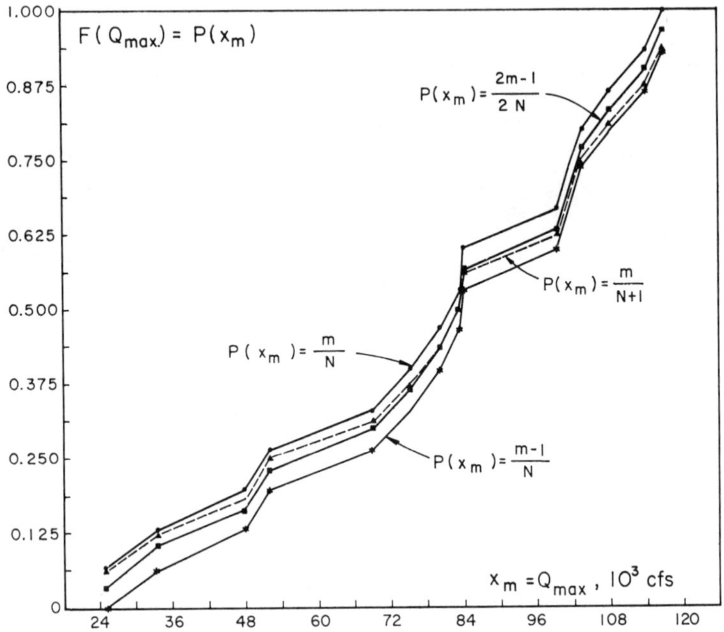

Figure 5.4. Distribution of flood peaks of the Colorado River at Lees Ferry, for the period 1923 – 1937 (15 years), obtained by four plotting positions, P (x_m).

Other plotting position formulas. The literature given in reference [5, p. 8.28 – 8.29] summarizes other expressions proposed for plotting positions. These expressions are not especially useful in hydrology because they do not have more advantages than Eqs. 5.5, 5.8, 5.9, and 5.10. Expressions for the plotting position, given in the above reference [5] and elsewhere, produce relatively small differences in comparison with the four plotting positions discussed, when these four plotting positions are properly applied.

5.3. **Multivariate Frequency Distributions.**

Empirical frequency distributions of bivariates. For the simultaneous occurrence of two variables, it is necessary to set up class intervals and class marks of the two-dimensional frequency distribution and for each variable, in the manner previously described. The variables x and y have now their corresponding ranges $x_{max} - x_{min}$ and $y_{max} - y_{min}$, divided into m_1 and m_2 class intervals, respectively. They form m_1m_2 joint class fields. These class fields are represented in such a way that

TABLE 5.1

Joint Frequencies of Class Fields of the Length of Two Axes x and y (the Largest and the Intermediate Axes) for 200 Coarse Gravel Particles

x, inches / y, inches	2.0 – 2.2	2.2 – 2.4	2.4 – 2.6	2.6 – 2.8	2.8 – 3.0	3.0 – 3.2	3.2 – 3.4	3.4 – 3.6	3.6 – 3.8	3.8 – 4.0	$_yf_a$	$_xf_i$
2.22 2.55	5 0.0250 0.0042	3 0.0150 0.0064	0.0070	0.0050	0.0038	0.0036	0.0052	0.0020	0.0018	0.0010	8	0.040
2.55 2.88	7 0.0350 0.0147	12 0.0600 0.0220	8 0.0400 0.0250	1 0.0050 0.0175	0.0132	0.0126	0.0182	0.0070	0.0063	0.0035	28	0.140
2.88 3.22	5 0.0250 0.0168	6 0.0300 0.0256	10 0.0500 0.0280	6 0.0300 0.0200	3 0.0150 0.0152	2 0.0100 0.0144	0.0208	0.0080	0.0072	0.0040	32	0.160
3.22 3.55	1 0.0050 0.0184	7 0.0350 0.0280	7 0.0350 0.0306	8 0.0400 0.0220	5 0.0250 0.0166	2 0.0100 0.0137	4 0.0200 0.0230	1 0.0050 0.0088	0.0079	0.0044	35	0.175
3.55 3.88	3 0.0150 0.0158	3 0.0150 0.0240	1 0.0050 0.0262	6 0.0300 0.0188	4 0.0200 0.0142	5 0.0250 0.0135	6 0.0300 0.0195	1 0.0050 0.0075	0.0068	1 0.0050 0.0038	30	0.150
3.88 4.21	0.0158	1 0.0050 0.0240	5 0.0250 0.0262	2 0.0100 0.0188	5 0.0250 0.0142	5 0.0250 0.0135	7 0.0350 0.0195	0.0075	4 0.0200 0.0068	1 0.0050 0.0038	30	0.150
4.21 1.51	0.0100	0.0152	3 0.0150 0.0166	2 0.0100 0.0119	2 0.0100 0.0090	1 0.0050 0.0086	5 0.0250 0.0123	2 0.0100 0.0047	2 0.0100 0.0043	2 0.0100 0.0024	19	0.095
4.54 4.88	0.0630	0.0096	1 0.0050 0.0105	0.0075	0.0057	2 0.0100 0.0054	2 0.0100 0.0078	4 0.0200 0.0030	2 0.0100 0.0027	1 0.0050 0.0015	12	0.060
4.88 5.21	0.00205	0.0032	0.0035	0.0025	0.0019	1 0.0050 0.0018	1 0.0050 0.0026	1 0.0050 0.0010	1 0.0050 0.0009	0.0005	4	0.020
5.21 5.54	0.00105	0.0016	0.00175	0.00125	0.00095	0.00090	1 0.0050 0.0013	1 0.0050 0.0005	0.00045	0.00025	2	0.010
$_xf_a$	21	32	35	25	19	18	26	10	9	5	200	1.000
$_xf_i$	0.105	0.160	0.175	0.125	0.095	0.090	0 130	0.050	0.045	0.025	1.000	

The first row in each class field is the joint absolute frequency, the second row is the joint relative frequency and the third row is the product of relative frequencies ($_xf_i \cdot {}_yf_i$) of the marginal distributions of x and y.

the class intervals of one variable represent the rows and the class intervals of the other variable represent the columns.

The joint absolute frequency, n_{ij}, or the joint relative class frequency, f_{ij}, of a bivariate refer to the class field of the i-th row and the j-th column. The distribution of the joint frequencies over class fields is defined as the bivariate frequency surface. In the case where the frequency densities are determined for each point, or for each class mark of class fields, the bivariate frequency density surface may be obtained. If this is the case, the bivariate frequency density surface must be integrated, both over the intervals Δx and Δy, to obtain the joint frequency of a class field $\Delta x \, \Delta y$. The absolute or relative cumulative frequency of given values x and y is defined as the sum of all frequencies of class fields, f_{ij}, for which $x \leq x_i$ and $y \leq y_j$. The integral of frequency density surface from $-\infty$ to x, and $-\infty$ to y, gives the cumulative frequencies. The distribution of cumulative frequencies over class fields or over the points (x, y) is the bivariate frequency distribution.

Example 5.5. Table 5.1 gives the bivariate absolute and relative frequencies of class fields for the joint distribution of the length of two axes x and y (the largest and the intermediate) for 200 coarse gravel particles. The marginal, absolute, and relative frequencies for each class interval is given in the third line for each class field, to show that they are different from the joint relative frequencies. These differences indicate that the two variables are dependent.

Empirical distributions of multivariates and conditional variables. The uses of multivariate distributions of more than two dimensions are current in hydrologic practice. Though it is easy to visualize the same concepts of class intervals, *class space*, and class frequencies as in the bivariate case, it is extremely difficult to represent them graphically or in tabular form. As it is shown in subsequent text, the best method of analysis of multivariate distributions is either by fitted mathematical probability functions, or by their parameters in the form of sample means, sample correlation matrices, and similar statistics.

Similarly as for univariates, bivariates, and multivariates, the empirical distributions of conditional variables are obtained from observed data. In the case of simple conditional variables, $Y \mid X$ or $X \mid Y$, the distributions of the bivariate F (x, y), and of either F (x) or F (y), are used to compute $F(Y \mid X)$ or $F(X \mid Y)$, and similarly for their frequency curves. In the case of conditional variables with many dimensions, the multivariate distributions are used.

5.4. Methods of Presentation and Analysis of Empirical Distributions.

Three methods of presenting distributions. The three methods of presenting frequency curves and distributions are numerical, graphical and functional. The objectives of these three methods are to make it possible to draw hypotheses and conclusions from a sample, besides showing characteristics of a random variable.

A table giving absolute or relative class frequencies versus class intervals (with their respective class marks) is a numerical presentation of a univariate frequency curve. Similarly, a table that gives the cumulative frequencies against x is the numerical presentation of the frequency distribution. For discrete variables, the events x_i in a table are followed either by their frequencies or cumulative frequencies. Though tabular presentation does not adequately show a frequency curve or a frequency distribution, it is useful for preparing data to present distributions graphically, or to fit a function.

Graphical presentation of frequency curves and distributions consists of a plot of frequencies as ordinates against the corresponding discrete values, or class marks or any other value, as abscissas. Figures 2.2, 2.3, 2.4, 5.1, 5.2, 5.3, and 5.4 give various examples of graphical presentation of frequency graphs, curves, and distributions. The graphical presentation of distributions is very much in use, because their properties can be readily grasped. Because of the limitations of the human eye, this graphical presentation and resulting inferences must be considered, in any strict analysis, only as a procedure for advancing statistical hypotheses. Numerical and functional procedures should be used for testing the validity of these hypotheses and for drawing various conclusions.

If a function can be attached to a frequency curve, it is the frequency function. Similarly, the integral of this function from minus infinity to a value x, expressed as a function of x, is the frequency distribution function. The frequency function is the most condensed form of presenting statistical information in empirical distributions.

Three types of analysis of frequency distributions. Properties of frequency distributions, which are obtained either by observations or by experiments in hydrology, can be investigated in one of three basic ways: (a) by qualitative analysis, (b) by quantitative analysis through numbers, as parametric or non-parametric analysis, and (c) by fitting mathematical probability distribution functions to empirical distributions.

Qualitative analysis uses only descriptive words or symbols to describe properties, by telling that a distribution is bell-shaped, symmetrical, asymmetrical, uniform or rectangular, triangular, U-distributed, J-shaped, bimodal, multimodal, and so forth. It indicates whether the distribution is discrete or continuous, bounded or unbounded, and similar. This analysis is used in hydrology only as a preliminary approach, before the following two methods are applied.

The quantitative parametric or non-parametric (order-statistics) methods of analysis of frequency distributions use numbers to describe properties of distributions. Usually, each number defines a particular property, like central tendency, dispersion, asymmetry, dependence, boundaries, and similar characteristics.

The quantitative analysis by mathematical functions, as the most complete and condensed way of describing a frequency distribution, consists of the three important steps: (a) selection of the probability function to be fitted, (b) estimation of its parameters, and (c) testing its goodness of fit to the empirical distribution.

It should be stressed here that the sequential connections between observed successive values of a variable are neglected by any method of obtaining a univariate empirical distribution. A duration curve of river flows is obtained by neglecting the dependence between successive values of its time series. From the univariate distribution of x, one cannot tell anything about the connections among x-values in their sequence. To do that, the multivariate distribution of all values in a sequence must be used, considering each position in the sequence as one of numerous dimensions of the multivariate distribution, or determining a mathematical equation of dependence if all dimensions are identically distributed.

The type of distribution analysis selected depends on the problem at hand, and on the general objectives of statistical analysis. Qualitative analysis is described in the next subchapter, while quantitative analysis using numbers is treated in Chapter 6; probability distribution functions are discussed in Chapter 7.

5.5. Qualitative Analysis of Frequency Distributions.

Comparisons of distributions. Qualitative analysis is based on properties that can be determined by inspection. If only a distribution is analyzed, its general shape and other basic characteristics can be discerned regardless of the type of frequency (absolute, relative, density). If two or more distributions are to be compared qualitatively, it is desirable to reduce each frequency distribution to a common basis, say by using relative frequencies only, and some dimensionless or standard variable values. Such reduced variables are of two types. A standardized variable is $y = (x - \bar{x})/s$ for sample data, where x is the original variable, $\bar{x}$ is its mean, and s is its standard deviation. The variable y is dimensionless, and has both positive and negative values. Modular coefficients are relative values $K_i = x_i/\bar{x}$, where each value of x is divided by its mean. The new variable, K_i, is dimensionless, with the mean one. Its values oscillate about $\bar{K} = 1.00$. These reduced variables are convenient for the comparison of non-negative variables $x \geq 0$, $y \geq 0$, ..., so that they also have only positive values or zeros.

Shapes of distributions. Qualitative analysis is usually based on the shape of frequency curves or histograms. Theoretically, they can have any shape. The shapes encountered in practice or developed through theoretical considerations follow. (1) Uniform or rectangular distribution, which is not very common in hydrology; they are related to equally likely events or to an equal relative frequency. (2) Triangular distribution also is not common in hydrology. (3) Bell-shaped distribution is the most current type of distributions in the natural sciences. (4) J-shaped distribution is a highly skewed distribution, and is also common in hydrologic practice. (5) U-shaped distribution is sometimes encountered. (6) Bimodal distribution is a two peak distribution for some hydrologic variables. (7) Multimodal distribution is usually of bell-shaped type but with three or more peaks.

Other criteria of dividing distributions in their qualitative analysis. Other characteristics used in this analysis are bounded or unbounded distributions on one or both tails; only positive valued variables; discrete, continuous or mixed variables and distributions; asymptotic distributions to one or both axes, positively or negatively skewed, and similar.

Relevance of qualitative analysis. It may seem superfluous to discuss qualitative analysis, particularly the shapes of frequency distributions, but a basic knowledge about these shapes may serve some practical purposes. Often it is advantageous to know the physical conditions that predetermine one shape or another of frequency distributions.

Example 5.6. Under some conditions, the daily river flow, or monthly precipitation or runoff, may have a bimodal probability density curve. This is usually the case when their time series are composed of periodic and stochastic components.

5.6. Problems for Discussion and Solutions.

5.1. For each type of distribution, classified by shape, try to determine what the relation should be between the variable and its causative factors, in order to produce that particular shape of distribution. (Do not confuse these frequency distributions with time or line sequences.)

5.2. How may the central limit theorem infer the shape of a distribution if the individual effects of causative factors on a variable are small and additive?

5.3. If a variable x is unbounded, and if a function of this variable is proportional to x^2, what are the properties of distribution of this function?

5.4. Select a variable with a sample of at least $N = 100$ and determine the frequency density curve. Show that this curve has many wavy lines, and that the frequency histogram, or relative frequency curve, well approximates the probability density curve.

5.5. For a sample of discrete values, divide the range of the variable x into 64 class intervals, so that both x_{max} and x_{min} are included, and find class frequencies. Then use 32, 16, and 8 class intervals for the same range and find their class frequencies. Plot the four histograms and discuss the effect of the number of class intervals in this example on the shape and information of the empirical frequency histograms.

5.6. To obtain a frequency distribution, either grouped data in m class intervals, or ungrouped data of the sample of size N may be used. Discuss the dilemma of using grouped or ungrouped data for a given N, and discuss what approximately the dividing value, N_c, for these two procedures should be. How would a digital computer affect this dilemma?

5.7. How does the qualitative analysis of empirical frequency distributions help the selection of hypotheses for various tests in the quantitative analysis of distributions?

5.8. For a small sample (N = 10 to 15 years) of annual flood peak discharges, show graphically the differences, particularly at the ends, between the distribution curves obtained by the four plotting position formulas, Eqs. 5.5, 5.8, 5.9, and 5.10. Plot four curves on Gumbel graph paper and extrapolate the fitted straight lines to the return periods of 2N, 5N, and 10N. Discuss differences with regard to where they occur, the effect of the sample size, and how the four plotting positions affect the extrapolation for return periods much larger than N.

5.9. For the data in Table 5.1 of Example 5.5, find the empirical distributions of conditional variables $X \mid Y$ and $Y \mid X$. Plot the two marginal distributions and the two distributions of conditional variables and compare them.

CHAPTER 6

PARAMETERS AND ORDER-STATISTICS AS DESCRIPTORS OF DISTRIBUTIONS

Descriptors as the particular numbers of probability (frequency) distributions, in the form of moments, parameters, and order-statistics, are important in various hydrologic applications beyond probabilistic and statistical significance. These descriptors are related to populations or they are estimates from the available samples. They are discussed for the cases of univariates, bivariates, multivariates, and conditional variables.

6.1. Parameters and Order-Statistics.

Number descriptors. Descriptors of population and sample distributions in the form of parameters and order-statistics are numbers that provide insight into particular characteristics of these distributions. Each number usually refers to a specific property. Population descriptors are estimated by sample statistics. The problem opposite to the estimation of descriptors is the problem of constructing distributions from their known descriptors. Parameters and order-statistics, estimated from sample data, as descriptors of distributions, are statistics just as any other numerical property of a distribution is.

There are two types of parameters commonly used to describe distributions: (a) moments and cumulants, and (b) parameters related to geometric and other characteristics of probability distribution, which may or may not be directly related to moments.

Order-statistics are numbers derived from properties of ordered values of a sample. Parameters and order-statistics usually fall in one of four categories of distribution properties though not limited to them: (1) descriptors showing the central tendency, or values about which all other values are clustered, (2) descriptors measuring the dispersion or the spread (or in the opposite context, the concentration) of sample values about a central value, (3) descriptors measuring the asymmetry (skewness) of a frequency distribution, and (4) descriptors showing flatness (or in the opposite context, peakedness) of a distribution in comparison with the normal distribution. Descriptors may measure some other property of a distribution, like boundaries, asymptotes, etc. Moments and several parameters may be directly related, but this is not usually the case between the moments and order-statistics.

Moments. Moments of a distribution have an analogous definition to the moments used in mechanics. The frequency or the probability of a discrete random event can be considered as a mass, and the moments are taken about some convenient point or variable value. The number of moments, theoretically, can be infinite. In practice, only the first four moments are used because accuracy in estimating higher-order moments from the sample data decreases rapidly as its order, r, increases. For $r > 4$, the errors in estimating moments from the sample data are such that they are unsuitable for practical use.

Similar to moments are numbers called cumulants. The treatment of cumulants, however, is outside the scope of this text. Moment generating functions will not be discussed either. The reader is referred to classical books on statistics and probability theory for these topics.

If x_0 is a specified value of a discrete variable, the population r-th moment ζ_r about x_0 is

$$\zeta_r = \sum_{i=1}^{k} p(x_i)\,(x_i - x_0)^r, \tag{6.1}$$

in which k is the number of discrete events, x_i, and p (x) is the probability mass function, p.m.f., of x. These moments are estimated from data by the sample moments

$$s_r = \frac{1}{N} \sum_{j=1}^{N} (x_j - x_0)^r, \tag{6.2}$$

for the ungrouped sample data, x_j, $j = 1, 2, \ldots, N$, or by

$$s_r = \sum_{i=1}^{k} f(x_i)\,(x_i - x_0)^r \tag{6.3}$$

for the grouped data, or if the relative frequencies $f(x_i)$, as the sample estimates of $p(x_i)$ in Eq. 6.1, are computed for k either the number of discrete events or the number of class intervals for continuous variables.

For a continuous variable with the probability density function, f (x), and the distribution function, F(x), the population moments about x_0 are

$$\zeta_r = \int_{-\infty}^{+\infty} f(x)\,(x - x_0)^r\,dx = \int_{-\infty}^{+\infty} (x - x_0)^r\,dF(x). \tag{6.4}$$

These population moments are estimated from the sample ungrouped data by Eq. 6.2, or for the grouped data by Eq. 6.3, in which case x_i is the class mark, $f(x_i)$ is the class relative frequency, and k is the number of class intervals.

For $x_0 = 0$, the moments about the origin are those of Eqs. 6.1 through 6.4, with x_0 being replaced by zero. The population moments are designated by υ_r, or υ_1, υ_2, υ_3, and υ_4 for the first four moments.

Population central moments, designated by μ_r, $r = 1, 2, 3, 4, \ldots,$ are obtained by replacing x_0 in Eqs. 6.1 and 6.4 by the expected value Ex, while the sample central moments, designated by m_r, are obtained by replacing x_0 in Eqs. 6.2 and 6.3 by the sample mean, $\bar{x}$.

The central moments μ_r are related to moments about the origin υ_r as follows:

$$\begin{aligned} \mu_1 &= 0 \\ \mu_2 &= \upsilon_2 - \upsilon_1^2 \\ \mu_3 &= \upsilon_3 - 3\upsilon_1 \upsilon_2 + 2\upsilon_1^3 \\ \mu_4 &= \upsilon_4 - 4\upsilon_3 \upsilon_1 + 6\upsilon_2 \upsilon_1^2 - 3\upsilon_1^4, \end{aligned} \qquad 6.5$$

and the converse,

$$\begin{aligned} &\upsilon_1 \\ \upsilon_2 &= \mu_2 + \upsilon_1^2 \\ \upsilon_3 &= \mu_3 + 3\mu_2 \upsilon_1 - \upsilon_1^3 \\ \upsilon_4 &= \mu_4 + 4\mu_3 \upsilon_1 + 6\mu_2 \upsilon_1^2 + 9\upsilon_1^4. \end{aligned} \qquad 6.6$$

Population or sample moments of Eqs. 6.1 through 6.4 are used in hydrology mainly for computing various distribution parameters. In general, only four moments of a univariate are used, namely: (1) the first moment about the origin, as the expected value for the population, or the mean for the sample; (2) the second central moment, as the variance; (3) the third central moment, used for the definition of the skewness coefficient, and (4) the fourth central moment, used for the definition of the kurtosis or excess coefficients. In these last three cases, the central moments are referred either to the population or to the sample.

For moments to be meaningful, it is necessary for Eq. 6.2 to be absolutely convergent for $N \to \infty$. Although the r-th moment about any value can be computed either for the population or for the grouped or ungrouped sample data, a meaning cannot necessarily be attached to the resulting values. Probability distribution functions and empirical distributions commonly used in hydrologic practice, however, are such that all moments exist or are zeros. Moments of a function, g (x), of the random univariate x can be obtained in replacing $(x - x_0)^r$ in Eqs. 6.1 through 6.4 by $[g(x) - g(x_0)]^r$. It can be easily shown that the minimum expected value of $(x - x_0)^2$ is $\mu_2 = Dx$, or the variance of x is the smallest second moment in case $x_0 = Ex$ for the population, or in case $x_0 = \bar{x}$ for the sample.

Percentiles. The percentiles x_p are variable values that have probabilities, p, of being exceeded or not being exceeded, either for a population or a sample. For $p = 0.50$, $x_{0.50}$ is called the median. For $p = 0.25$ and $p = 0.75$, the values $x_{0.25}$ and $x_{0.75}$ are the first and the second quartiles. For $p = 0.10$ and $p = 0.90$, the values $x_{0.10}$ and $x_{0.90}$ are the first and the last deciles. The value of a discrete random variable with the greatest probability mass, or the value of a continuous random variable with the greatest probability density, is called the mode.

Order-statistics. A finite sample of N observed values of a random variable is grouped according to their magnitude, from the largest to the smallest, or the converse. This ordered set may be enumerated as 1, 2, . . . , m, . . . , N. The value x_m of a sample, which corresponds to the m-th value in this order, is an order-statistic. The largest value, the smallest value, the second largest or the second smallest value, and so forth, are all order-statistics. They are related to samples, with sample percentiles also being order-statistics.

Example 6.1. The annual flood peak discharge, the minimum daily flow in a year, and similar extremes in hydrology are typical order-statistics. They are related to the year, the day, or any other fixed sample size.

6.2. Descriptors of Central Tendency.

Definition of descriptors of central tendency. Centrally located values, either of the population or of the sample, about which other values cluster are descriptors that measure the central tendency. These descriptors are mainly the arithmetic mean (or simply the mean), the geometric mean (or its logarithm), the harmonic mean, the median, the mode, the mid-point of the range of values, and others, which are discussed later.

Arithmetic mean. Of all means, such as arithmetic, geometric, quadratic and harmonic, the arithmetic mean is the most current descriptor of central tendency. Equations 6.1 and 6.4 become for $x_0 = 0$ and $r = 1$

$$Ex = \sum_{i=1}^{k} p(x_i) \cdot x_i, \tag{6.7}$$

and

$$Ex = \int_{-\infty}^{+\infty} x\, f(x)\, dx = \int_{-\infty}^{+\infty} x\, dF(x), \tag{6.8}$$

respectively, for the populations of a discrete and a continuous variable. Similarly, Eqs. 6.2 and 6.3 become

$$\bar{x} = \frac{1}{N} \sum_{j=1}^{N} x_j, \tag{6.9}$$

and

$$\bar{x} = \sum_{i=1}^{k} x_i \cdot f(x_i), \tag{6.10}$$

respectively, for the ungrouped and grouped data of a sample. Means of Eqs. 6.9 and 6.10 are the best estimates of expected values of Eqs. 6.7 and 6.8, because the variance of the sample mean is among the lowest variances of all sample statistics. Equation 6.10 is practically a weighted mean, and it may be also used whenever values, x_i, have different weights, say different economic, safety, and similar weights. Theorems on expected value, discussed in Chapter 3, are valid for the sample arithmetic mean.

The arithmetic mean as a measure of central tendency is a concept readily grasped. The mean is easy to compute. It lends itself either to comparison or to combination with other means by virtue of its weighting property. Every item or observation enters into its computation. From a computational point of view this may appear to be a disadvantage, particularly if the total number of observations is large. The advantage of having each observation enter into the computation is, however, useful from the point of view of sampling theory. The sample mean has a high sampling stability, or it has the smallest sampling variance of all statistics measuring the central tendency, assuming that the original variables are normally distributed.

A disadvantage of using the arithmetic mean is the influence of extreme values. The small sample may contain a very large or a very small value by pure chance, which affect the accuracy of estimation. Such a value would unduly influence the computed mean of the sample, so that the sample mean may not be representative of the expected value of the population.

Example 6.2. Whenever a hydrologic variable is of a cumulative nature, such as water discharge, sediment discharge, dissolved matter in water, precipitation, and similar variables, which can be accumulated in a storage space, the arithmetic mean multiplied by the time period gives the expected total inflow into that storage space or the total outflow from it. The mean is directly related to the available water resources of a region, in time or space or both. The mean is a good measure of potential damage, cost, loss, or similar effects in various hydrologic and water resources problems. The cost of removing solids from water, or the negative effects of dissolved solids in water on the production process, and similar influences, are usually proportional to the mean content of dissolved solids of a given discharge supplied to an industry. The mean water temperature is the measure of cooling capacity of a constant flow diverted for that purpose.

Geometric mean. The population geometric mean is defined as

$$\mu_g \underset{N\to\infty}{=} \left(\prod_{i=1}^{N} x_i\right)^{1/N}, \qquad 6.11$$

while the sample geometric mean is

$$\bar{x}_g = \sqrt[N]{x_1 \cdot x_2 \cdot \ldots \cdot x_n}, \qquad 6.12$$

for the ungrouped data, and a similar expression for the grouped data. Geometric means of Eqs. 6.11 and 6.12 are often given in the form of their logarithms, such as $\mu_n = \ln \mu_g$, or $\bar{x}_n = \ln \bar{x}_g$.

In hydrologic practice, the logaritms of the positive valued observations are often used in analyses. It is useful then to apply either the geometric mean, or its logarithm as the mean of logarithms. The geometric mean is useful for many exponential or power function relations in geophysical sciences. Logarithms of variable values may follow a simple relation function. The frequency distribution of logarithms of observations often show symmetry, while the frequency distribution of observations themselves is asymmetrical. Then the geometric mean is applicable as a central tendency parameter.

Example 6.3. The lognormal distribution of the variable x in Example 2.31 has μ_y as the parameter of central tendency, which is the mean of $y = \ln x$. This parameter μ_y is the logarithm of the geometric mean of the variable x.

Harmonic mean. The population harmonic mean is defined as

$$\mu_h \underset{N\to\infty}{=} \frac{1}{\frac{1}{N}\sum_{i=1}^{N}\frac{1}{x_i}}, \qquad 6.13$$

and the sample harmonic mean is

$$\bar{x}_h = \frac{N}{\frac{1}{x_1}+\frac{1}{x_2}+\ldots+\frac{1}{x_N}}. \qquad 6.14$$

The use of the harmonic mean, as the reciprocal of the mean of reciprocals of sample values, is not widespread in hydrology, though examples of the means of reciprocal sample values can be found in some hydrologic applications.

The arithmetic mean is a greater value than the geometric mean and the geometric mean, in turn, is greater than the harmonic mean.

Example 6.4. A constant discharge, Q, along a river reach passes through cross sections with different wetted areas, A_i. These wetted areas are random variables. The mean velocity in the cross section A_i is $V_i = Q/A_i$. The mean water velocity in a reach of n cross sections is

$$V = \frac{1}{n}\left(\frac{Q}{A_1}+\frac{Q}{A_2}+\ldots+\frac{Q}{A_n}\right) = \frac{Q}{n}\sum_{i=1}^{n}\frac{1}{A_i}.$$

The harmonic mean of all wetted areas is

$$A_h = n/\sum_{i=1}^{n}(1/A_i),$$

so that the mean velocity in the reach is $\overline{V} = Q/A_h$.

Quadratic mean. The sample quadratic mean is

$$\bar{x}_g = \left(\frac{1}{N}\sum_{i=1}^{N}x_i^2\right)^{1/2}. \qquad 6.15$$

It is the mean of the function $y = x^2$, of the random variable x. This mean is not often used in hydrology, though examples may be found of its application.

Median and mode. Apart from the above means, the other central values in hydrology are the median and the mode. Because all values above the median, and all values below the median have the same probability of 0.50, the median is an attractive central value parameter. In the case of extreme value asymptotic distribution functions, the central value parameter is the mode.

A debate has been going on for a long time in hydrology as to whether it is better to use the arithmetic mean or the median as the best parameter of central tendency for any asymmetric probability distribution. For a symetrical distribution the arithmetic mean, the median, and the mode are equal. An analysis of all factors for and against the use of each of them leads to using the arithmetic mean as the best central value parameter. For very small samples, however, the use of the median in hydrology, instead of the mean, is justified because the mean is greatly affected by extreme values of small samples.

Example 6.5. When hydrologic problems are related to exceedences of random events, namely determining what is the variable value x_p that it is or is not exceeded $m = pN$ times out of N observations, the percentile x_p is a useful statistic. If the problem is to find what water level of a river is exceeded 50 percent of the time for a summer season, on the average, the median is the parameter of central tendency to use.

Central values of order-statistics. In the case of order statistics, the use of the midpoint between the observed x_{max} and x_{min}, as $(x_{max} + x_{min})/2$, has not shown any practical importance for hydrology because of the unreliability of extreme values, due either to sampling or to measurement errors, or both. Similarly, the mean of two deciles $(x_{0.10} + x_{0.90})/2$, or the mean of two quartiles $(x_{0.25}+x_{0.75})/2$, have not been used extensively.

Additional properties of arithmetic means. Besides the theorems on expected values, given in Chapter 3, some additional properties of arithmetic means are important in hydrology. The sample mean is the best estimate of the population mean or of the expected value of a variable x.

If x is multiplied by a constant C, the mean of the product is $C\bar{x}$. If a constant C is added to x, the mean of the sum is $C + \bar{x}$. For probability distributions given in the form of histograms, or frequency density curves, the arithmetic mean is the centroid of these graphs. If the mean of samples of size $N_1, N_2, \ldots, N_k$ are $\bar{x}_1, \bar{x}_2, \ldots, \bar{x}_k$, respectively, then the total sample $N = N_1 + N_2 + \ldots + N_k$ has the weighted mean $\bar{x}$ as

$$\bar{x} = \frac{\sum_{i=1}^{k} N_i x_i}{\sum_{i=1}^{k} N_i}, \quad i = 1, 2, \ldots, k. \tag{6.16}$$

Because the arithmetic mean is the most accurate sample central tendency statistic, its sampling distribution is usually the most accurately determined. As shown in Chapter 8, the distribution itself is simple. It is the fundamental statistical parameter in water resources planning and operation and always has the same dimension as the variable itself.

6.3. Descriptors of Dispersion.

Definition of descriptors of dispersion. Dispersion parameters and other dispersion statistics measure how the variable values are dispersed or spread about a central value. Usually, the reciprocals of dispersion descriptors measure the concentration of variable values about the central value. Three groups of these descriptors are: (a) statistics that are derived from the order-statistics, (b) absolute differences among sample values, and (c) parameters that are related to statistical moments. This last group comprises the three most currently used dispersion parameters: the variance, the standard deviation, and the coefficient of variation.

Dispersion descriptors of the order-statistics type. The two types of despersion statistics often found in literature are the range, R, and the relative range to the mean, $R/\bar{x}$, with

$$R = x_{max} - x_{min}, \text{ and } \frac{R}{\bar{x}} = \frac{x_{max} - x_{min}}{\bar{x}} . \qquad 6.17$$

The difference of two quartiles, D_q, and the relative difference of two quartiles to the mean, $D_q/\bar{x}$, given as

$$D_q = x_{0 \cdot 75} - x_{0 \cdot 25}, \text{ and } \frac{D_q}{\bar{x}} = \frac{x_{0 \cdot 75} - x_{0 \cdot 25}}{\bar{x}} \qquad 6.18$$

are also used as measures of dispersion. These dispersion order-statistics are not currently used in hydrology, though they are used in other statistical analyses, particularly in non-parametric statistical investigations.

Dispersion descriptors of absolute difference type. The mean absolute deviation from the sample mean is

$$D = \frac{1}{N} \sum_{i=1}^{N} | x_i - \bar{x} | , \qquad 6.19$$

and the Gini's coefficient (also a mean absolute difference) of the sample is

$$G = \frac{1}{N(N-1)} \sum_{i=1}^{N} \sum_{j=2}^{N} | x_i - x_j | . \qquad 6.20$$

These descriptors are not particularly applicable to hydrology, though they have some merit because of equal weights given to all deviations taken from the mean $\bar{x}$, or to deviations taken between the individual pairs of x-values.

Variance. Expressions for the population variance are given by Eqs. 3.38 and 3.39 for a discrete and a continuous variable. The sample variance is

$$s^2 = \frac{1}{N} \sum_{i=1}^{N} (x_i - \bar{x})^2 , \qquad 6.21$$

for ungrouped data, and

$$s^2 = \sum_{j=1}^{m} f(x_j)(x_j - \bar{x})^2, \qquad 6.22$$

for the grouped data in m class intervals, where x_j is the class mark, each with the relative frequency $f(x_j)$. When absolute frequencies are used, $f(x_j)$ in Eq. 6.22 is replaced by n_j and the sum divided by N.

As shown later in the chapters on estimation and sampling distributions, Eqs. 6.21 and 6.22 are biased estimates of the population variance Dx, also designated by σ^2 or var x. To obtain the unbiased estimate, designated by $\hat{s}^2$, the right side of Eqs. 6.21 and 6.22 must be multiplied by $N/(N-1)$, N = sample size, or by a number greater than one.

When using Eq. 6.22, the class interval width, Δx, affects the accuracy of the variance. To remove this bias, the term $(\Delta x)^2/12$ should be deducted from the right side of Eq. 6.22, which is Sheppard's correction for the second central sample moment of grouped data. To obtain the unbiased estimate of the variance, this difference must be multiplied by $N/(N - 1)$. Equation 6.21 may be rewritten as

$$s^2 = \frac{1}{N}\sum_{i=1}^{N} x_i^2 - \bar{x}^2, \qquad 6.23$$

or the sample variance is the mean of squares of variable values (quadratic mean) minus the square of its mean. Similarly, the population variance is the expected value of the variable squares minus the square of the expected value.

The use of unbiased variances, with the correction $N/(N - 1)$ in Eqs. 6.21, 6.22, and 6.23, has an advantage in hydrology only for small N, say for $N \lesssim 30$.

The variance is extensively used in many statistical analyses of hydrology. It is considered one of the basic statistics and one of the best descriptors of dispersion for the study of various problems in hydrology. The variance of the variable x has the dimension of x^2.

Example 6.6. The length of observation of a hydrologic time series, or the number of points over an area to be surveyed or observed for a hydrologic variable, in order to obtain a given accuracy of their various characteristics, are basic economic problems of any hydrologic service. These lengths or the number of points, as sample sizes, are usually proportional to the variance of random variables. At least, they are functions of the variance, besides being dependent on other properties of random variables observed or surveyed.

Example 6.7. As discussed later, the variance of a random variable often serves as the main parameter in the definition of hydrologic information that is contained in a sample. Regardless of which sample property is studied, the sample information on this property comes out to be a function of the population variance. This variance in turn is estimated by the sample variance.

Standard deviation. The standard deviation, of the same dimension as the variable x, is the dispersion parameter most currently used in hydrology. The biased sample value is designated by s, the unbiased sample value by ŝ, and the population value by σ. In its biased form, the sample standard deviation is the positive square root of Eq. 6.21, or

$$s = +\sqrt{\frac{1}{N}\sum_{i=1}^{N}(x_i - \bar{x})^2}. \qquad 6.24$$

Similarly, the standard deviation is the positive square root of Eqs. 6.22 and 6.23 for the sample data, and the positive square root of Eqs. 3.38 and 3.39 for the population. For $N \leq 30$, the unbiased form is obtained by multiplying the biased value of Eq. 6.24 by $+\sqrt{N/(N-1)}$. The standard deviation is the positive square root of the variance. This dispersion parameter, like the arithmetic mean, can be used directly in many hydrologic and water resources computations.

Example 6.8. Many problems in hydrology and water resources involve natural, artifical, or combined natural-artificial storage capacities. An input of a cumulative hydrologic variable is changed by a storage capacity. If the arithmetic mean of the input is a constant, the prescribed regulated output and the storage properties do not change with time, the capacity of storage spaces necessary to produce that required output is directly proportional to the standard deviation of the input variable.

Example 6.9. Assume that a hydroelectric power or pumping plant is installed on a three-month discharge (on the flow that is exceeded three months per year, on the average). This decision affects the average annual hydro-power that can be produced, or the average annual water volume that can be pumped. For all other properties of river flows with the demand for water and power being equal — such as their arithmetic mean, their asymmetry of flow distribution and the time dependence of discharges, the market demand, and similar — the average hydro-power that can be produced, and the average water volume that can be delivered, are functions of the population standard deviation of river flows, to be estimated by the sample (biased or unbiased)standard deviation. The greater the standard deviation, the smaller energy and water quantities available for delivery.

Coefficient of variation. The dimensionless dispersion parameter in the form of the coefficient of variation is the ratio of the standard deviation to the mean, or

$$\hat{C}_v = \frac{\hat{s}}{\bar{x}}; \; C_v = \frac{s}{\bar{x}}; \; \text{or} \; \eta_v = \frac{\sigma}{\mu} = \frac{(Dx)^{1/2}}{Ex}, \qquad 6.25$$

respectively, for the unbiased and biased parameters of the sample, and for the population parameter. Equation 6.25 is applicable only when the mean $\bar{x}$ and the expected value Ex are not zero, or near zero. Because all positive-valued hydrologic variables have values $x_i \geq 0$, the sample mean is always a positive value greater than zero.

Example 6.10. The coefficient of variation is extensively used in hydrology, particularly as a dimensionless regionalization parameter. In the case of being used as a regionalization parameter, it gives the relative variation of a variable, as that variation changes over an area, and it is usually used in conjunction with the arithmetic mean.

Comparisons between the variance, the standard deviation, and the coefficient of variation. Although the variance is the square of the standard devation, and the coefficient of variation is the ratio of the standard deviation to the mean, the properties of these three parameters differ. They are, however, adaptable to mathematical treatment for purposes of comparison, transformation, and generalization.

If a variable x is multiplied by a constant C, the variance and the standard deviation are the variance and standard deviation of x multiplied by C^2 and C, respectively, while the coefficient of variation is unaffected. If a constant C is added or subtracted from each of the variable values, neither the variance nor the standard deviation is affected, while the coefficient of variation becomes

$$C_v = \frac{s}{\bar{x} \pm C}. \qquad 6.26$$

Any change of the origin from which the variable values are measured changes the coefficient of variation.

An advantageous property of the variance is that variances from different samples can be pulled together. If $s_1^2, s_2^2, \ldots, s_k^2$ are the variances from samples of sizes $N_1, N_2, \ldots, N_k$, having means $\bar{x}_1, \bar{x}_2, \ldots, \bar{x}_k$, then the variance of the total sample $N = N_1 + N_2 + \ldots + N_k$, for the means and variances of samples being independent, is

$$s^2 \sum_{j=1}^{k} N_j = \sum_{j=1}^{k} N_j s_j^2 + \sum_{1}^{k} N_j (\bar{x}_j - \bar{x})^2, \qquad 6.27$$

in which $\bar{x}$ is defined by Eq. 6.16 as the general mean. Equation 6.27 is a weighted value of variances just as $\bar{x}$ of Eq. 6.16 is a weighted mean of means. This weighting property does not apply to the standard deviation or to the coefficient of variation, except indirectly through Eqs. 6.16 and 6.27.

Since C_v is influenced by any change in the origin, it is not a convenient parameter for comparison of distributions, when the origin is arbitrarily selected without any physical meaning. If the origin is a natural value, i.e., the measurement of sediment sizes, river discharge, angles, and similar random variables with the origin at zero, the coefficient of variation is a useful parameter for purposes of comparison. The coefficient of variation can serve for a comparison of two different populations since the influence of different means is excluded. For example, if the relative variation of sediment grain diameters between two different rivers are compared, a higher coefficient of variation means a larger relative variation.

6.4. Descriptors of Asymmetry.

Definition of descriptors of asymmetry. Skewness coefficients are descriptors of asymmetry of a distribution. Synonyms are coefficient of asymmetry, skew coefficient, and coefficient of skewness. They are usually related to the third central moment, μ_3. Even central moments ($r = 2, 4, \ldots$) are always positive; whereas, the uneven central moments (r = 3, 5, . . .) can be positive, negative, or zero. All uneven central moments are zeros for symmetrical distributions. If the uneven central moments are positive, distributions are positively skewed, and likewise if these moments are negative, distributions are negatively skewed.

The skeweness coefficient. To render the measure of asymmetry independent of the dimension of the variable, the population skewness coefficient is defined as the ratio of the third central moment to the second central moment on the power 3/2, or

$$\gamma_1 = \frac{\mu_3}{\mu_2{}^{3/2}} = \frac{\mu_3}{\sigma^3} \tag{6.28}$$

in which σ is the population standard deviation.

Pearson's skewness coefficient. Other measures of asymmetry have been proposed. One suggested by K. Pearson [1] is for the population distribution

$$\alpha_1 = \frac{\mu - m}{\sigma}, \tag{6.29}$$

in which μ is the population mean, m is its mode, and σ is its standard deviation. If the mode can be accurately determined, this is a convenient definition. In most cases, however, it cannot be determined accurately, particularly for ungrouped data.

Unbiased estimates of skewness coefficients. For small samples, the unbiased estimate of the skewness coefficient of Eq. 6.28 is

$$\hat{g}_1 = \frac{N^2}{(N-1)(N-2)} \frac{m_3}{m_2{}^{3/2}}, \tag{6.30}$$

where m_2 and m_3 are the sample second and third central moments

An approximate expression for the unbiased estimate of skewness coefficient recommended by Foster and quite frequently used in hydrology, is

$$\hat{C}_s = \frac{N}{N-1} \cdot \frac{m_3}{m_2{}^{3/2}}. \tag{6.31}$$

If N is large, $N^2/(N-1)(N-2)$ and $N/(N-1)$ are nearly unity, and the skewness coefficients, as defined by Eqs. 6.28, 6.30, and 6.31, are nearly equal, when μ_3 and μ_2 in Eq. 6.28 are estimated by sample moments m_2 and m_3.

If the variable values are multiplied by a constant or a constant is added, the skewness coefficient is unaffected.

Hydrologic significance. The skewness coefficient as a positive or negative dimensionless parameter is often used in hydrology. It is the third most desirable parameter, after the arithmetic mean and the standard deviation, for the description of probability distributions, because most hydrologic variables have asymmetrical distributions. It is used also as one of regionalization parameters.

Example 6.11. In Example 6.9, with the mean and standard deviation being constants, the total expected power production or water to be delivered depend on the skewness coefficient. They decrease with an increase of the positive skewness coefficient, but increase with a decrease of the negative skewness coefficient.

6.5. Descriptors of Flatness.

Definition of kurtosis coefficient. Although the bell-shaped frequency curves or histograms look similar, a dissimilarity exists in that their peaks are sharper or flatter. This characteristic is described by the parameter of kurtosis deriving its name from a Greek word meaning humped. It is also called the kurtosis coefficient. It is sometimes referred to as peakedness, in opposite context to flatness (though the assertion that it measures these properties is sometimes contested).

The population kurtosis coefficient is

$$\gamma_2 = \frac{\mu_4}{\mu_2^2} = \frac{\mu_4}{\sigma^4}, \qquad 6.32$$

in which μ_2 and μ_4 are the population second and the fourth central moments, and σ is the population standard deviation. This parameter γ_2 is estimated by the sample value $g_2 = m_4/m_2^2$. For very small samples of size N, the estimated value g_2 by the sample central moments m_2 and m_4 should be multiplied by $N^3/(N-1)(N-2)(N-3)$ to obtain a less biased estimate.

Geometric meaning of kurtosis. If deviations of variable values from the mean decrease, the fourth central moment tends to zero faster than the square of the second central moment, in which case the kurtosis tends rapidly to zero. This extreme corresponds to a sharp peakedness. If deviations from the mean increase, the fourth moment tends to infinity faster than the square of the second moment, in which case the kurtosis tends rapidly to infinity. This extreme corresponds to a very pronounced flatness.

Excess coefficient. It is often convenient to select a distribution function to serve as a standard. If g_0 denotes the kurtosis of that distribution, the kurtosis of any other distribution is measured relative to g_0. For the normal distribution as a standard $g_0 = 3$, so that

$$E = \gamma_2 - 3 \qquad 6.33$$

is the excess coefficient of a distribution relative to the excess of zero of the normal distribution. If E is negative, the distribution has a sharper peak than the normal distribution. If E is positive, the inverse is true. This parameter may be used in hydrology to determine the probability function of best fit to empirical distributions from the known skewness and excess coefficients.

6.6. Application Areas of Descriptors of Univariate Distributions in Hydrology and Water Resources.

Five particular areas of application. The use of descriptors in hydrology and water resources in the form of moments, parameters, order-statistics, and similar numbers are manyfold; some of them are: (1) condensation of information in the form of descriptors of univariate distributions, (2) transfer of information from one point to another, (3) applications in various water resources computations, (4) reconstruction of distributions from known descriptors, and (5) reduction of variable values to more suitable forms.

The application of descriptors to probability distributions has been already discussed. Further treatment of this application, however, is given in Chapter 7. This application represents, therefore, the condensation of information in the form of descriptors.

Transfer of information. The application of descriptors to the transfer of information from one point to another over an area is done by regionalization of hydrologic descriptors. Mainly, they are parameters rather than order-statistics.

It is a common hydrologic practice to draw maps of isolines of various parameters over an area. Because information is available only at some points, isolines are interpolations of equal parameters values between these points. The interpolation is the transfer of information from points of information to points where there is no information. In many countries, the three most common hydrologic parameters of regionalization are $\bar{x}$, C_v, and C_s, or the arithmetic mean, the coefficient of variation, and the skewness coefficient. Because C_v and C_s are dimensionless, they well describe how the distribution of a hydrologic variable changes across an area. The mean with its dimension determines the absolute values of a variable in the regionalization of three parameters, $\bar{x}$, C_v, and C_s. Other statistics — the median, the standard deviation, the first serial correlation coefficient, and similar statistics — may be also used for the regionalization. A present tendency apparent among hydrologic services in the world is to condense hydrologic information into the form of a selected number of statistical parameters regionalized over an area. Differences are in the selection of parameters that should contain as much available information as practically feasible.

Use of parameters in water resources computations. Examples for the general use of descriptors in hydrology and water resources have been given. From all parameters, hydrologists and engineers have been successful in introducing into direct computations of various water resources problems only the mean, and sometimes the standard deviation. All others are used indirectly in supplying the necessary hydrologic statistical information. The mean is used extensively, mainly as the parameter directly related to the available water resources, project costs, various losses, damage potentials,and so forth. The standard deviation is often applied indirecty; many hydrologic and water resources magnitudes are functions of it. Examples 6.8 and 6.9 are only two among many. The use of the coefficient of variation, the skewness, and excess coefficients rarely figure in practical expressions for various water resources computations.

Reconstruction of distributions from their descriptors. The use of descriptors to assess the best probability function to fit empirical distributions is given by Karl Pearson [1, p. 151]. The parmeter $\varkappa$ given by

$$\varkappa = \frac{\gamma_1^2 (\gamma_2 + 3)^2}{4(2\gamma_2 - 3\gamma_1^2 - 6)(4\gamma_2 - 3\gamma_1^2)} \qquad 6.34$$

is used as a criterion to distinguish the main types of probability functions (mainly in Pearson's family of functions), or to select the function to fit empirical distributions. In Eq. 6.34, γ_1 is given by Eq. 6.28, and γ_2 by Eq. 6.32. When estimated by sample data as C_s and $E = g_2 - 3$, Eq. 6.34 becomes

$$k = \frac{C_s^2 (E + 6)^2}{4(2E - 3C_s^2)(4E - 3C_s^2 + 12)} \qquad 6.35$$

in which C_s and E are given by Eqs. 6.30 and 6.33, in biased form for $N > 30$, and with corrections for the unbiased form in case $N \leq 30$.

With regard to the parameter k, one selects the probability function from the family of Pearson functions. If k is negative, the best function to fit is the Pearson Type I unbounded function (Beta-distribution). If the distribution is bounded by zero (positive-valued variables), the sub-case of Type I as the Pearson Type III distribution, which is often used in hydrology, should be fitted. For the Pearson Type III distribution (or Gamma distributions with two or three parameters) to fit the empirical asymmetric distributions, k should be very large or infinite but negative. For the Pearson Type VI, $k > 1$. For the Type IV, $0 < k < 1$. The Pearson method of fitting functions requires computing the first four moments of observed distributions, then computing C_s and E, and from them k, and selecting the function. Then, parameters of the function are estimated, and the test of goodness of fit performed.

Reduced variables. Descriptors are also used to reduce original variables to more convenient forms, dimensionless or otherwise. The classical reduced variable already discussed, is the standardized variable, $y = (x - \mu)/\sigma \approx (x - \bar{x})/s$, reduced to have the mean of zero and the variance (or the standard deviation) of unity. The standardization can be made for any hydrologic variable, though there is sometimes an incorrect understanding that primarily only the normally distributed variables are standardized. Many simplifications are obtained if one uses the standardized variable. However, disadvantages in their applications in hydrology are that positive-valued variables become, in reduced standardized cases, both positive and negative, so that the use of logarithms of variable values is precluded, and the coefficient of variation becomes meaningless for the mean of zero. In these cases, the use of reduced dimensionless variables in the form of modular coefficients, $K_i = x_i/\bar{x}$, is attractive, because the mean of K_i is unity, the positive-valued variables remain as such, and $s_k = C_v(x)$, and var $K_i = C_v^2(x)$.

The reduced variable of the double exponential distribution is $y_i = \alpha(x_i - \beta)$, which is a linear transformation of x. Since β is the mode and α is inversely proportional to σ, this reduced variable is similar to the standardized variable.

In using the reduced variables, once their properties are obtained the simple linear transformations, $x_i = \mu + \sigma y_i$, $x_i = \bar{x} K_i$, and $x_i = \beta + y_i/\alpha$, transform the results to their original variables.

6.7. Descriptors of Distributions of Bivariates, Multivariates, and Conditional Variables.

Covariance as the new descriptor for bivariates. For a random bivariates, (x,y), the two expected values and the two variances of its x and y components are obtained from their marginal distributions, in the same ways as described in the previous text for univariates. The new descriptor is their second central cross-product moment, of values x and y, which is called the covariance, designated by cov (x,y) or σ_{xy}. For the population of discrete bivariates, the covariance is

$$\text{cov}(x,y) = \sum_{i=1}^{n_1} \sum_{j=1}^{n_2} p_{ij}(x_i - Ex)(y_j - Ey), \qquad 6.36$$

with x_i the symbol for n_1 random discrete events of x, y_j the symbol for n_2 random discrete events of y, and p_{ij} the probability mass of the joint discrete (x_i, y_j)-event. For the continuous bivariate, with f(x,y) its joint probability density function, the population covariance is

$$\text{cov}(x,y) = \int_{-\infty}^{+\infty} \int_{-\infty}^{+\infty} (x - Ex)(y - Ey) f(x,y)\, dx\, dy. \qquad 6.37$$

It is easy to show that cov (x,y) = cov (y,x). If x and y are independent cov (x,y) = 0. However, if cov (x,y) = 0 it does not imply that x and y are independent.

In general

$$\text{cov}(x,y) = E(x - Ex)(y - Ey) = Exy - Ex\,Ey, \qquad 6.38$$

or the covariance is the expected value of cross product xy minus the product of two expected values, Ex and Ey.

The estimate of the population covariance from the ungrouped data of sample size N is by

$$\text{cov}(x,y) = s_{xy} = \frac{1}{N} \sum_{i=1}^{N} (x_i - \bar{x})(y_i - \bar{y}), \qquad 6.39$$

or from the grouped data in class fields by

$$\text{cov}(x,y) = s_{xy} = \sum_{i=1}^{m_1} \sum_{j=1}^{m_2} f_{ij}(x_i - \bar{x})(y_j - \bar{y}), \qquad 6.40$$

in which f_{ij} are relative joint class frequencies, and m_1 and m_2 are the numbers of class intervals for x and y. When the absolute frequencies, n_{ij}, are used, f_{ij} in Eq. 6.40 is replaced by n_{ij}, and the sum is divided by the total sample size, N, which is the sum of all n_{ij} values.

The r-th central moments of a bivariate, with $r = r_1 + r_2$, both for the population and the sample, respectively, are given by

$$\mu_{r_1 r_2} = E\,(x - Ex)^{r_1}\ (y - Ey)^{r_1};\ m_{r_1 r_2} = \frac{1}{N}\sum_{i=1}^{N} (x_i - \bar{x})^{r_1} (y_i - y)^{r_2}, \quad 6.41$$

with $r_1 = 0, 1, \ldots, r$, and $r_2 = r, r - 1, \ldots, 1, 0$. These moments are currently seldom used, though they may be applied in some problems of cross-correlation of two random variables, or in studying the sequential dependence of two successive outcomes of a random process.

Example 6.12. The covariance, or the correlation coefficient derived from it, are currently used in hydrology to measure the degree of linear dependence among the two components of a bivariate, or between the two variables x and y. This dependence is discussed in detail in Chapter 11 under simple linear correlation and regression methods of analysis. One group of examples are investigations of dependence between two different variables, such as precipitation and runoff of given time intervals (daily, monthly, annual, and similar), sediment discharge and water discharge, dissolved solids in water and water discharge, and so forth. The other group of examples are investigations of dependence of the same variable observed at two adjacent points, such as precipitation of two gauging stations, runoff of two rivers, evaporation at two stations, and so forth.

Correlation coefficient. The ratio of the covariance to the square root of the product of variances of two components x and y of a bivariate is defined as the population correlation coefficient

$$\rho = \frac{\text{cov}\,(x,y)}{(\text{var}\,x \cdot \text{var}\,y)^{1/2}} = \frac{\sigma_{xy}}{\sigma_x \sigma_y}. \quad 6.42$$

This coefficient, ρ, is estimated by the sample coefficient, r, when cov(x,y), var x, and var y in Eq. 6.42 are replaced by the sample estimates. The limits are $\rho = -1$ and $\rho = +1$, with ρ a dimensionless dependence parameter.

Descriptors of multivariate distributions. A multivariate in n dimensions has the components $x_1, x_2, \ldots, x_n$, and the distribution $F\,(x_1, x_2, \ldots, x_n)$. The expected values and variances of the components are obtained from their marginal distributions. Covariances [2, p. 79–83] between pairwise-taken components, x_i and x_j, $i = 1, 2, \ldots, n$ and $j = 1, 2, \ldots, n$, as cov (x_i,x_j) of this n-dimensional multivariate, form a symmetrical $n \times n$ covariance matrix, with $i = j$ producing the variances, and $i \neq j$ the covariances. This covariance matrix is represented by

$$[\sigma_{ij}] = \begin{vmatrix} \sigma_{11} & \sigma_{12} & \ldots & \sigma_{1n} \\ \sigma_{21} & \sigma_{22} & \ldots & \sigma_{2n} \\ \ldots & \ldots & \ldots & \ldots \\ \sigma_{n1} & \sigma_{n2} & \ldots & \sigma_{nn} \end{vmatrix},\quad \begin{matrix} i = 1, 2, \ldots, n \\ j = 1, 2, \ldots, n. \end{matrix} \quad 6.43$$

The expected values ($Ex_1 = \mu_1$, $Ex_2 = \mu_2$, ... , $Ex_n = \mu_n$) and the covariance matrix $[\sigma_{ij}]$ are the main descriptors. Similarly, as with the covariance matrix, the correlation matrix is defined as $[\rho_{ij}]$, $i = 1, 2, \ldots, n$ and $j = 1, 2, \ldots, n$, where ρ_{ij} are correlation coefficients between x_i and x_j components.

The r-th higher central moments (greater than two) of a random multivariate are defined in the manner similar to the way the bivariate case is defined in Eg. 6.41, extending r_1, r_2 to $r_1, \ldots, r_n$, which are the powers of various components, and with $r_1 + r_2 + \ldots + r_n = r$, and $r_i = 0, 1, \ldots, n$.

The inverse of the covariance matrix, designated by $[\sigma_{ij}]^{-1}$ or by $[\sigma^{ij}]$, has σ^{ij} as cofactors of $[\sigma_{ij}]/|\sigma_{ij}|$, where $|\sigma_{ij}|$ is the determinant of $[\sigma_{ij}]$, so that $[\sigma^{ij}]$ exists only if $|\sigma_{ij}| \neq 0$. Also, if the covariance matrix has the inverse covariance matrix, then the determinant $|\sigma^{ij}| = 1/|\sigma_{ij}|$. These properties of the covariance matrix also apply to the correlation matrix $[\rho_{ij}]$.

Descriptors of linear functions of components of multivariates. For $(x_1, \ldots, x_n)$ a random multivariate with the expected value $(\mu_1, \ldots, \mu_n)$, and the covariance matrix $[\sigma_{ij}]$, a variable obtained by a linear transformation

$$X = \sum_{i=1}^{n} c_i x_i, \quad 6.44$$

with $c_1, \ldots, c_n$ the constants, has the expected value

$$\mu(X) = \sum_{i=1}^{n} c_i \mu_i, \quad 6.45$$

and the variance

$$\sigma^2(X) = \sum_{i,j=1}^{n} c_i c_j \sigma_{ij}, \quad 6.46$$

or i and j take values 1, 2, . . . , n.

For components x_i pairwise independent, $\sigma_{ij} = 0$ for $i \neq j$, and Eq. 6.46 becomes

$$\sigma^2(X) = \sum_{i=1}^{n} c_i^2 \sigma_i^2. \quad 6.47$$

Descriptors of distributions of conditional variables. The expected value of a conditional variable $x_2 | x_1$, with the conditional distribution function (c.d.f.), $F(x_2 | x_1)$, is

$$E(x_2 | x_1) = \mu(x_2 | x_1) = \int_{-\infty}^{+\infty} x_2 \, dF(x_2 | x_1). \quad 6.48$$

Because $\mu(x_2 | x_1)$ is a function of x_1, it is called the regression function of x_2 on x_1. In this case

$$\mu(x_2 | x_1) = \alpha_1 + \beta_1 x_1. \quad 6.49$$

This is a linear regression function and α_1 and β_1 are the regression coefficients. This relation also is discussed in detail in Chapter 11.

The conditional (or residual) variance of $x_2 | x_1$ is

$$\sigma^2 (x_2 | x_1) = E\{ [x_2 - E (x_2 | x_1)]^2 | x_1\}. \tag{6.50}$$

This variance provides information about how well x_2 can be estimated with x_1 given.

Descriptors defined in Eqs. 6.48 and 6.50 are estimated by sample parameters, biased or unbiased depending on which one is appropriate for the given sample size.

6.8. Problems for Discussion and Solutions.

6.1. Select a sample of hydrologic data of $N = 40$ to 50 (say, annual precipitation, or annual runoff at any station). Compute its first four moments about the origin, and the second, third and fourth moment about the mean, $\bar{x}$. Compute all descriptors of central tendency given in Section 6.2 and compare them. To determine descriptors of central tendency based on order-statistics, plot the distribution of this variable x by using the plotting position $P(x) = m/(N + 1)$, with m the descending order from the largest to the smallest value of x.

6.2. For the sample of Problem 6.1 compute descriptors of dispersion, given in Section 6.3, and compare them by discussing the feasibility of their use in hydrology.

6.3. For the sample of Problem 6.1 compute the skewness and excess coefficients. Compute the Pearson measure of skewness, and compare it with the skewness coefficient.

6.4. The general equation of various means is

$$\bar{x} = \left(\frac{1}{N} \sum_{i=1}^{N} x_i^k \right)^{1/k},$$

with $k = -1$ giving the harmonic mean; for $k \to 0$, $\bar{x}$ converging to the geometric mean; $k = 1$ giving the arithmetic mean, and $k = 2$ giving the quadratic mean. For the sample of Problem 6.1 compare these four means. Give other examples of the use of harmonic, geometric, and quadratic means, besides those given in this chapter.

6.5. There is a controversy about whether the arithmetic mean or the median best describes the central tendency. Analyze their advantages and disadvantages for use in hydrology.

6.6. Discuss the conditions under which the coefficient of variation can be used in hydrology. Give examples of the use of this descriptor of dispersion in hydrology beyond these already given.

6.7. By using Eq. 6.35 compute the k-parameter for the sample of Problem 6.1 and show whether the Pearson Type III distribution can be fitted to the empirical distribution.

6.8. Select two adjacent precipitation stations with the concurrent annual precipitation values of at least 20 years of observations. Compute the covariance and the correlation coefficient of these two random variables.

6.9. Select a multivariate with four components from hydrology, with concurrent observations of x_1, x_2, x_3, and x_4 for at least $N = 25$, as the sample size. Find both the covariance and the correlation matrix.

6.10. For the components of Problem 6.9 find the variance of $y = 0.5 x_1 + 0.4 x_2 + 0.3x_3 + 0.2x_4$ from the covariance matrix computed in Problem 6.9. Assuming the components x_1, x_2, x_3, and x_4 pairwise independent compute also the variance of y.

CHAPTER 7

PROBABILITY DISTRIBUTION FUNCTIONS IN HYDROLOGY

Condensation of statistical information about the frequency distributions of hydrologic random variables is most feasible in the form of fitted probability distribution functions. This is often true when there are specific reasons that predetermine the conditions under which the particular probability distribution functions are applicable. The discussion in this chapter is limited to probability distribution functions of univariate, bivariate, multivariate, and conditional variables in hydrology, which are presently most often applied.

7.1. Functions for Fitting Empirical Frequency Distributions.

Selection of probability distribution functions. Nearly all distributions of hydrologic random variables are of an empirical nature, obtained from limited and observed data. Their best description is by a proper fit of probability distribution functions. Under certain conditions, particular probability distribution functions are known to apply to empirical frequency distributions. In most cases the selection of probability distribution functions for hydrologic random variables is based on experience. It is already known that some distribution functions fit empirical distributions of particular hydrologic random variables well. Often, however, two or three functions fit them equally well.

Optimization of the number of parameters of probability distribution functions. Probability distribution functions contain parameters that must be estimated from sample data. From a mathematical standpoint, the more parameters in a function the more flexible it is in fitting empirical distributions. From a statistical standpoint, to estimate parameters requires either moments, or some other statistics, or they are directly estimated. The number of moments or other statistics needed for estimating parameters increases as the number of parameters increase. The reliability of moments estimated from observed data decreases with an increase in their order. Similar loss of reliability is involved in the estimation of parameters for which higher powers of sample values are needed. The problem is this—if the number of parameters of a probability distribution function is too small, there will be insufficient flexibility of fit; but if the number of parameters is too great, some of the estimated parameters will become less reliable. Any selection of probability distribution function that is defined by a number of parameters is an optimization problem between the flexibility of fitting a function and the reliability of estimated parameters.

Sample and population probability distribution functions. Probability distribution functions of basic hydrologic variables cannot be, in general, derived theoretically. The true functions are never known. In nearly all cases it is necessary to select a function and to estimate its parameters from the sample data. If a satisfactory fit is found, the underlying assumption is that the same function applies to the population of that variable. That means, if the number of observations is continually increased the function will remain unchanged although the numerical values of the estimated parameters are altered. Assume that many small samples are available for a hydrologic variable and that the probability distribution functions of the best fit are selected separately for each sample. It is easy to be convinced that the selected probability distribution function, based only on a sample frequency distribution, may lead to a wrong conclusion about the function that best fits the population distribution. Figure 4.3 illustrated this problem. The selection of probability distribution functions to fit empirical frequency distributions, the best estimation of their parameters, the most reliable tests of these fits, and the eventual drawing of confidence limits are important aspects of statistical application to frequency distributions of hydrologic random variables. All four subjects are given attention in the subsequent text.

7.2. Properties and Classification of Probability Distribution Functions.

Fitting of probability distribution functions. Parameters of a probability distribution function that are fitted to an observed frequency distribution define properties of location (central tendency, boundary values), scale (dispersion or concentration), shape, asymmetry, peakedness, asymptotes, and other properties. A tendency exists to attach a geometrical meaning to parameters. By fitting probability distribution functions to empirical distributions, it is possible to study how the distribution of a hydrologic random variable changes by incremental changes of each parameter, or by a variation of two or more parameters, simultaneously or in sequence. It is evident that a probability distribution function defines properties of empirical frequency distributions better than any group of descriptors, exclusively.

The probability distribution function is the most concise representation of an empirical frequency distribution. A selected function does not fit exactly the observed distribution. The goodness of fit of this function depends on the sample size, the proper selection of the function, and the use of the best methods for estimating its parameters.

Comparison of empirical distributions. Two situations may arise in the comparison of two or more empirical distributions: (1) all distributions have the same probability distribution function but their parameters differ, and (2) each distribution has a different probability distribution function. In the first case parameters are directly compared. In the second case the relation between parameters of functions should be first determined before comparison is made. The use of probability distribution functions is best for comparing empirical frequency distributions.

Classification of probability distribution functions. Probability distribution functions may be classified from various standpoints. From a statistical standpoint, a function may be theoretical, semi-theoretical, empirical, or all three. If it is derived from axioms and logical derivations of the theory of probability, the function is theoretical. If conditions are such that a particular random variable must be distributed according to a specific probability distribution function, or if indications are present that the distribution of a random variable belongs to a given family of probability distribution functions, but the parameters must be estimated from the sample data, those functions may be considered semi-theoretical. If a function is simply fitted to empirical frequency curves, without any other support for its selection, the function is empirical. The same function may be theoretical, semi-theoretical, or empirical.

From the standpoint of physical properties of a random variable, a function may be also theoretical, semi-theoretical, empirical, or all three. If the cause-effect relation, established for a physical process, requires that a given function should fit the empirical distribution, the function is theoretical. If indications for a function from the physical process are relatively weak, but experience shows that such a function best fits empirical distributions, the function may be considered as semi-theoretical. If only the criterion of the best fit determines the selection of a probability distribution function for fitting observed frequency distributions, the function is empirical.

Probability distribution functions for hydrologic applications. Probability distribution functions of discrete and continuous random variables, most often used to fit frequency distributions in hydrology, are discussed in this chapter. Primary discussion is given to properties of these functions, and to various aspects of their application in hydrology. The estimation of function parameters from sample data and tests of goodness of fit of these functions are briefly discussed in this chapter when appropriate so that function properties and their applications are not separated from these estimations and tests. The estimation of parameters, the tests of goodness of fit and the confidence limits are discussed in special chapters.

It is possible to find examples of application in hydrology of all discrete univariate probability distributions functions, such as hypergeometric, binomial, Poisson, hypergeometric waiting-time distribution, binomial waiting-time distribution (also called Pascal, or the negative binominal), Neyman's type A contagious distribution, and others. Without too much manipulating, however, these discrete distributions, mainly binomial and Poisson discrete probability functions can be applied easily to hydrology; significant application, in fact, has been already found.

Nearly all existing continuous probability distribution functions may be applied to hydrology if one looks hard enough to find examples of application. Experience has shown, however, that only a limited number of functions are best applied to hydrologic empirical distributions, namely normal (Gaussian), lognormal (Galtonian), Gamma (particularly its form with two parameters, and Pearson Type III), double exponential and simple exponential functions of extreme values.

7.3. Binomial Distribution.

Definition. The binomial discrete probability distribution applies to populations that have only two discrete but complementary events. It applies to hydrology whenever there are two events with the characteristics: occurrence and non-occurrence of a property (say rainy and non-rainy day); yes-no (decisions to seed or not to seed clouds in a random seeding experimental project); either-or; zero-greater than zero; plus--minus; zero-one, and similar complementary events. If an event has probability mass p, and the complementary event $q = 1 - p$, then in a sequence of n independent trials, the expected value of the number of the first event is pn, and qn of the complementary event. Essential conditions for the application of binomial distribution is the independence of successive trials, observations or experiments, and the constant probabilities p and q at each individual trial. These are independent Bernoulli trials. Many hydrologic variables of two complementary events satisfy these conditions, at least approximately. Before a binomial distribution is applied in hydrology these basic conditions should be tested on data.

Binomial probability distribution function. The binomial probability mass function is

$$p(x) = \binom{n}{x} p^x q^{n-x} = \binom{n}{x} p^x (1 - p)^{n-x}, \qquad 7.1$$

in which n is the number of independent observations (Bernoulli trials), x is the number of occurrences of an event (with $x = 0, 1, \ldots, n$), $n - x$ is the number of occurrences of its complementary event (with $n - x = n, n - 1, \ldots, 1, 0$), p is the probability of the first event in an infinite number of trials, q is the probability of the complementary event, p (x) is the probability of x to occur in n observations, and the symbol $\binom{n}{x}$ is given by

$$\binom{n}{x} = \frac{n(n-1)\ldots(n-x+1)}{1.2\ldots(x-1)x} = \frac{n!}{x!(n-x)!}. \qquad 7.2$$

The expression n! for large n (say $n \geq 30$) is approximately $\sqrt{2\pi n}\, n^n e^{-n}$. Tables of binomial distributions are available [1].

For $p = q = 0.50$ (equal chances for the two complementary events to occur), Eq. 7.1 becomes

$$p(x) = \binom{n}{x} \frac{1}{2^n}. \qquad 7.3$$

General parameters of binomial distribution. The expected value, the variance, the skewness coefficient, and the kurtosis are considered in this text as the general parameters, and they are

$$Ex = np; \quad Dx = \text{var } x = npq; \quad C_s = \frac{q-p}{\sqrt{npq}}; \text{and } k = \frac{1 - 6pq}{npq} + 3. \qquad 7.4$$

For $p < q$, the binomial distribution is positively skewed; for $p > q$, it is negatively skewed, and for $p = q = 0.5$, it is symmetrical. It is unimodal unless pn or qn are small.

Properties of binomial distribution. The binomial distribution tends to normal probability density function, as n increases to infinity. If $np^{3/2} > 1.07$, the error in estimated probabilities in using the normal function instead of the binomial function never exceeds 0.05 for any x. It should be noted that p and n are two independent parameters. The parameter p is usually estimated by putting it equal to the frequency f of the first event, say event yes, from all the observations that are available in a sample of size N, or $f = m/N$, where m is the number of occurrences of the first of the two events in the sample. The number of trials, n, to occur or have occurred is either determined in each particular case, or is a selective parameter in the analysis. Several publications have detailed tables of binomial distribution (probability mass or cumulative probabilities) for various ranges of p and n. Sometimes, the probability of x/n may be of interest, instead of x. It is easy to transform Eqs. 7.1 through 7.4 to conform to this case.

Example 7.1. Observed data of long duration in a region of large storm activity show that in the months of this activity the probability that a large storm occurs on any individual day is 0.10, or every tenth day, on the average. Repair work is planned on a storm drainage system during this warm, storm period; the job will take 10 working days. What is the probability for no storm to occur in those 10 days? Or, what is the probability that the work must be protracted for 12 days because two storms may occur, and no work can be done for 24 hours following the beginning of each storm? It can be safely assumed that the convective summer storms are independent events and of short duration. This problem is solved by using the binomial distribution. In Eq. 7.1 the values are: $n = 10$, $p = 0.10$, $1\text{-}p = 0.90$, $x = 0$, and $n\text{-}x = 10$, and Eq. 7.2 gives one, so that $p(x = 0) = 0.32$. However, the second problem requires two days of storms in 12 days, or $n = 12$, $p = 0.10$, $1\text{-}p = 0.90$, $x = 2$, $n - x = 10$, and Eq. 7.2 gives 66, so that for $n = 12$, $p(x = 2) = 0.230$.

Example 7.2. What is the probability that a flood, which is greater than the flood of a return period of once in five years on the average, does not occur in a period of 10 years? What are probabilities that 1, 2, 3, and 4 of those floods would occur? For $n = 10$, $p = 0.20$, and x is 0, 1, 2, 3, and 4, Eq. 7.1 gives: $p(0) = 0.11$, $p(1) = 0.27$, $p(2) = 0.28$, $p(3) = 0.20$, and $p(4) = 0.11$, with the above values of $p(x)$ being rounded to two figures.

Example 7.3. The probability of a rainy day is $p = 0.30$, so that $q = 1 - p = 0.70$ is the probability of a nonrainy day, from long experience of observing precipitation at a gauging station. What is the probability that the next year would have at least 110 rainy days? The answer is

$$P(x \geq 110) = \sum_{x=110}^{365} \binom{365}{x} 0.3^x \, 0.7^{365-x}, \qquad 7.5$$

with the solution giving the probability. By using Eq. 3.24 for $a = (110 - 0.3 \times 365) / \sqrt{0.21 \times 365} = 0.0572$, and $b \approx \infty$, the approximation of Eq. 7.5 becomes $P(x \geq 110) = 0.479$. It is assumed in this solution that successive observations of rainy and nonrainy days are independent events, which in general, is not a correct hypothesis. Only the convective rain events may approximately satisfy this condition of independence.

7.4. Poisson Distribution.

Definition. If n and p of the binomial distribution are mutually dependent, and p is very small while n is very large, the binomial distribution becomes less applicable. Theoretically, if n goes to infinity while p tends to zero, but their product, np, tends to a constant λ, the distribution of x ($x = 0, 1, \ldots, \infty$) follows the Poisson discrete distribution. It is not necessary in practice that $p \to 0$ and $n \to \infty$, but that p is relatively small and n relatively large, for the Poisson function to become applicable under the condition that $\lambda = np$ stays approximately constant while n and p change.

Poisson probability distribution function. The constant λ is the only parameter of distribution that is

$$p(x) = \frac{\lambda^x e^{-\lambda}}{x!}, \qquad 7.6$$

in which x is the number of occurrences of an event of the given small probability, p, in a large number of trials, n, and p (x) is the probability of $x = 0, 1, 2, \ldots, \infty$. It is current in natural sciences that a long sample (n large) should be observed for an event of very small probability to occur. This distribution is applicable in hydrology, particularly for the occurrence of a given number of extreme values of a small probability of occurrence. It can be shown that the binomial function converges to Poisson function, or

$$\lim_{n \to \infty,\, p \to 0} \binom{n}{x} p^x (1 - p)^{n-x} = \frac{\lambda^x e^{-\lambda}}{x!}. \qquad 7.7$$

Since x are positive integers, zero included, in both distributions, the binomial is bounded by $x = 0$ and $x = n$, while the Poisson has the left lower boundary $x = 0$ but is unbounded to the right. The constant λ can be only positive, $0 < \lambda < \infty$. For $x = 0$, the value $x! = 1$. It can be easily shown that

$$\sum_{i=0}^{\infty} \frac{\lambda^{x_i} e^{-\lambda}}{x_i!} = e^{-\lambda}\left(1 + \frac{\lambda}{1!} + \frac{\lambda^2}{2!} + \ldots\right) = e^{-\lambda} e^{\lambda} = \mathbf{1}. \qquad 7.8$$

General parameters of Poisson distribution. The expected value, the variance, the skewness coefficient, and the kurtosis are

$$E\,x = \lambda;\ Dx = \text{var}\ x = \sigma^2 = \lambda;\ C_s = \lambda^{-1/2};\ k = 3 + \lambda^{-1}. \qquad 7.9$$

The expected value and the variance are equal. The skewness coefficient, $C_s > 0$, goes to zero when $\lambda \to \infty$.

Properties of Poisson distribution. Poisson distribution converges to normal, or the normal probability function fits discrete points of p (x) of Eq. 7.6, when λ goes to infinity. The values p (x) increase up to the point where $(x + 1) < \lambda$, and then they decrease toward zero. For $\lambda < 1$, the distribution is J-shaped, or a highly skewed distribution. For $\lambda = 1$, $C_s = 1$, and probabilities of $x = 0$ and $x = 1$ are $p(0) = p(1) = 1/e$, and from $x = 2$ on the probabilities decrease. For $\lambda > 1$, the distributions are unimodal. To compute probability distribution terms $P(X \leq x)$, of the sum of p (x) of the first m terms $(x = 0, 1, \ldots, m)$, it is easier to compute all p (x) of m terms from Eq. 7.6 and sum them to obtain $P(X \leq x)$, than to use expressions for cumulative probabilities, which are available in literature. Values, p (x), of the Poisson distribution for various λ, or values, P (x), obtained from the chi-square distribution, are given in Biometrica Tables for Statisticians [2]. Other tables for this distribution are also available [3].

Estimation of λ. The parameter λ is estimated from observations, or from experimental data, or it is logically constructed from the estimates of p. When λ is a function of another parameter, say the time τ of the year, or of the length position, then $d\lambda/d\tau$ can be called the rate or the density of λ. To obtain λ, the density $d\lambda/d\tau$ must be integrated over a given interval of τ.

Example 7.4. Let a sample have N = 3650 days (10 years). A rare daily event (values greater than a prescribed value) occurred five times during this period. The estimate of p is 5/3650 = 1/730. If the investigator is interested in probabilities of various numbers of this event in the period of 20 years (n = 7,300 days), then, because p is small and n is large, $\lambda = 7300/730 = 10$. This logical construction of λ has an accuracy limited by the above sample estimate, p = 1/730.

Application of Poisson distribution. Equation 7.6 implies that samples of size N, from which the probability p is estimated by the relative frequency and the sample n from which $\lambda = pn$ is determined, refer to the same population. In practice, the case may arise that a sample of size n is composed of an observation from each of n different populations, with corresponding probabilities of a given event, $p_1, p_2, \ldots, p_n$. In this case, the distribution of events x will have $\text{var}\ x = n\bar{p}(1 - \bar{p}) - n\sigma_p^2$, with σ_p^2 the variance of the above values of p_i, and var x smaller by $n\sigma_p^2$ than the variance of a binomial distribution with the same average $\bar{p}$, [4, p. 127—128]. The binomial distribution should be used for the sample of n values, each obtained from a different population, if $n\sigma_p^2$ is very small in comparison with $n\bar{p}(1 - \bar{p})$. If the converse is true, but $\text{var}\ \lambda = \text{var}\ np_i = \sigma_\lambda^2$ is small in comparison with n, the Poisson distribution should be used. Under the assumption of using the Poisson distribution, this sample has var x $= \bar{\lambda} - \sigma_\lambda^2/n$. Only if σ_λ^2/n is very small in comparison with $\bar{\lambda}$, or if σ_λ^2 is small in comparison with n, which is true for very small values p_i, is the Poisson distribution a good distribution to apply to the above sample data.

This case of the average λ, and the selection of the binomial or the Poisson distribution, may fit hydrologic applications when observations are available on a number, n, of stations but each with a small period of observation of a variable in a hydrologically homogeneous region. A particular event may be taken from data of each station, each with a probability p_i. The difficulty lies with large errors in the estimation of p_i values. If sample sizes, N_i, of various populations are different, and p_i are estimated with different accuracies, the application of the above criteria for selecting the binomial and the Poisson distribution is further complicated.

An attractive case for hydrologic applications is when each complete sample of size n is drawn from one of k different populations. An event has probabilities in each of k populations, $p_1, p_2, \ldots, p_k$. The occurrence of that event x times in the sample of n can be computed for each population by using the corresponding p_i, and the binomial distribution. It can be also computed by using the average probability of k populations, $\bar{p}$, and the sample size n. The variance of x obtained in this case would likely turn out to be smaller than the variance of x obtained among the different samples. In using the mean, $\bar{p}$, and the variance σ_p^2, of p_i, $i = 1, 2, \ldots, k$, the variance of the number x of occurrences of the event in n trials is var x $= n\bar{p}(1 - \bar{p}) + n(n - 1)\sigma_p^2$, [4, p. 128]. It would be difficult to expect the binomial distribution with parameters $\bar{p}$ and n to be applicable to this case because of the additional term $n(n - 1)\sigma_p^2$ in var x. For the Poisson distribution with the parameter $\bar{\lambda}$, as the mean of k values of $\lambda_i = np_i$, the variance of x, given by $\bar{\lambda} + \sigma_\lambda^2$, is affected by $\sigma_\lambda^2 =$ var λ_i. If, however, σ_λ^2 is small, or p_i's are very small, the Poisson distribution is still applicable.

The above example of k stations in a hydrologically homogeneous region, each station producing a sample of size n, may prove to have p_i-values which differ beyond the sampling fluctuation; thus one may reject the application of the binomial distribution with parameters n and $\bar{p}$ to this case. If there is evidence, statistical or physical or both, that the variation of p_i is produced only by the sampling fluctuation, the binomial distributions with n and $\bar{p}$ would produce more reliable probabilities of x occurrences of a particular event than the individual binomial distributions with n and p_i. If p_i-values are very small and averaged over k stations, it is evident from the above discussion that the Poisson distribution with $\bar{\lambda}$ parameter would give good results, even if the variation of p_i is not produced by sampling fluctuations only.

For var $\lambda_i = \sigma_\lambda^2$ not being too small, or for λ varying significantly among stations (for example, in the case of rare events with different λ_i for each month of the year), the distribution of λ_i is supposed to be a Gamma distribution with parameters α (shape) and β (scale). In this case [5, p. 255], the probabilities of 0, 1, 2, ... are

$$p(x) = \left(\frac{\beta}{\beta + 1}\right)^{\alpha} \left[1, \frac{\alpha}{\beta + 1}, \frac{\alpha(\alpha + 1)}{2!(\beta + 1)^2}, \frac{\alpha(\alpha + 1)(\alpha + 2)}{3!(\beta + 1)^3}, \ldots\right], \quad 7.10$$

in which the terms in brackets on the right side refer, respectively, to $x = 0, 1, 2, 3, \ldots$. The expected value and the variance of Eq. 7.10 are

$$Ex = \frac{\alpha}{\beta}, \quad \text{var } x = \alpha \frac{\beta + 1}{\beta^2}, \quad 7.11$$

where β of Eqs. 7.10 and 7.11 is equal to $1/\beta$ of Eq. 7.39. From the sample composed of mixed λ_i-values, the mean and variance enable the computation of α and β by using Ex and var x of Eq. 7.11. This case may be applicable to various regional hydrologic problems with a random variation of λ.

The application of Poisson distribution in hydrology is shown to be appropriate in cases of rare events of various types, randomly occurring either in time, along a line or across an area. For rare floods of a high return period, it may be proven easily that they occur randomly in time, and that the return period is nothing else than the average time between events, with actual times between events distributed randomly about this average time.

Example 7.5. One may ask what the probability is of flood x, with a return period of 500 years [or $p(x) = 0.002$], to occur three times in $n = 100$ years. Here $x = 3$; $\lambda = np = 100/500 = 0.2$, so that $p(x) = p(3) = 0.00101$, about one chance in 1,000, or once in every 100,000 years there would be a period of 100 years in which three 500-year floods would occur. That no 500-year flood occurs in 100 years, or $x = 0$, $p(0) = 0.82$, so that there is 80 percent chance that no flood greater than a 500-year flood occurs in a period of 100 years. Because there is still a chance of 20 percent for a flood greater than the 500-year flood to occur in 100 years, one should expect to find, on the average, one such flood among five river gauging stations of mutually independent flood events. This is sometimes overlooked by practitioners while studying the frequency distributions of observed large floods, and in their extrapolation to still larger floods.

7.5. Normal Distribution.

Normal probability distribution function. The normal (Gaussian, or Gauss-Laplacian) probability density function of a random continuous variable x is

$$f(x) = \frac{1}{\sigma\sqrt{2\pi}} e^{-(x - \mu)^2/2\sigma^2}, \quad 7.12$$

with μ the expected value (mean), and σ^2 the variance of variable x, with Eq. 7.12 often symbolized by $N(\mu,\sigma)$. These parameters are estimated from the available sample. The standardized form, $t = (x - \mu)/\sigma$, is

$$f(t) = \frac{1}{\sqrt{2\pi}} e^{-t^2/2}, \quad 7.13$$

and it represents distribution with no parameters, or with $\mu = 0$ and $\sigma^2 = 1$. It is often symbolized by N (0,1).

Figure 7.1 shows the normal probability density curves, f (x), and their probability distribution curves, F (x), with $\mu = 0$, and for three values of σ : 0.5, 1.0, and 2.0.

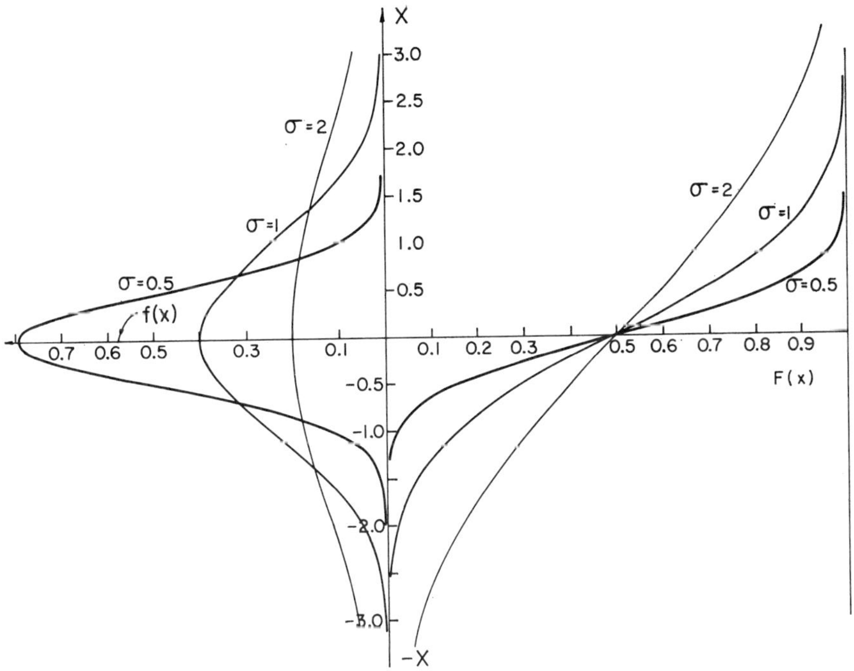

Figure 7. 1. Normal probability density curves (left side of graph), and normal probability distribution curves (right side of graph) for $\mu = 0$ and three values of σ, 0.5, 1.0 and 2.0.

Significance and properties of normal distribution. The normal distribution function, especially the standard normal, occupies a unique position among distribution functions in probability theory, mathematical statistics, and stochastic processes. It often serves as a standard or bench mark distribution for comparison with other distribution functions. Many functions converge to normal functions under particular conditions, as has been already stated for binomial and Poisson distributions. Similarly, the normal distribution serves many other purposes in hydrology which will be discussed later.

Four conditions are necessary for a variable to have a normal probability distribution: a very large number of causative factors affect the outcome of variable values, each factor taken separately has a relatively small influence on the outcome, the effect of each factor is independent of the effects of all other factors with the effects of various factors on the outcome being additive.

The normal function of Eq. 7.12 is unbounded, being defined for any real value of x. The only maximum is at $x = \mu$ with $f(\mu) = 1/\sigma\sqrt{2\pi} \approx 0.4/\sigma$. Inflection points are at $\mu \pm \sigma$ with $f(\mu \pm \sigma) = \sqrt{e}/\sigma\sqrt{2\pi} \approx 0.2420/\sigma$. The x-axis is the asymptote, and f (x) is symmetrical about $x = \mu$. The area under the curve is unity. The mean, μ, the median, M, and the mode, m, are identical. The standard normal function of Eq. 7.13 is tabulated currently as either f (t), or F (t), in many statistical books and manuals. To obtain x and f (x), the values t and f (t) are transformed : $x = \sigma t + \mu$, and $f(x) = f(t)/\sigma$.

For the normal function, the mean is the location (central value) parameter, while σ is the scale (dispersion) parameter. The skewness is zero. The kurtosis is three, or the excess iz zero. The coefficient of variation is defined only for $\mu \neq 0$, or particularly for $\mu > 0$.

The error-measuring deviation σ_p is defined as $\sigma_p = x_1 - \mu$, or $\sigma_p = \mu - x_2$, with x_1 and x_2 the values of x, symmetrically located about μ, between which the area under the probability density curve is 0.5. In this case,

$$\sigma_p = \pm 0.676\,\sigma \approx \pm 2\sigma/3. \tag{7.14}$$

In the theory of errors either σ or σ_p are used. If the absolute value $|x - \mu|$ is greater than 3σ, this deviation should be investigated as a potential mistake in observations or manipulation of data, since the probability of occurrence of values equal to or greater than $|x - \mu| = 3\sigma$ is only 0.0026 (0.26 percent).

The normal function has simple reproductive properties. If x is normal with μ_x and σ_x^2, then $y = a + bx$ (with a and b constants) is normal with $\mu_y = a + b\mu_x$ and $\sigma_y^2 = b^2\sigma_x^2$. If x and y are independent normal random variables, then $z = x \pm y$ is normal with $\mu_z = \mu_x \pm \mu_y$ and $\sigma_z^2 = \sigma_x^2 + \sigma_y^2$. If $x_1, x_2, \ldots, x_n$ are independent normal variables with μ and σ^2, then $y = \Sigma x_i$ has $\mu_y = n\mu$ and $\sigma_y^2 = n\sigma^2$. The average of n variables, $z = (\Sigma x_i)/n$, has $\mu_z = \mu$, and $\sigma_z^2 = \sigma^2/n$.

The central limit theorem indicates that the sum or the average of random variables tends to normal distribution as the number of variables in the sum increases.

Five areas of application of normal distribution in hydrology. The normal probability density function is useful in hydrology in five particular cases: (1) for fitting symmetrical empirical frequency distributions of hydrologic random variables, (2) as a probability density function in the analysis of random errors, (3) as the bench mark distribution for comparison with other distributions, (4) as many hydrologic statistical parameters may be either exactly or approximately normally distributed, the normal probability density function is used for various types of statistical inferences, and (5) to be used in Monte Carlo method (data generation) for simulating first the normal independent or dependent random numbers before they are transformed to other types of distributions.

Fitting empirical distributions. Distributions of hydrologic random variables related to large time intervals such as 2-year, 5-year, or 10-year runoff, 10-year precipitation, and similar variables are approximately normally distributed. It was found [6] that the normal function could fit relatively well empirical distributions of annual precipitation of a large number of 1614 precipitation gauging stations in western North America, and also empirical distributions of annual runoff of a large number of 446 river gauging stations in the same region. The fit is not as good for monthly rainfall and monthly runoff, and usually is not good at all for daily values.

The main theoretical difficulty in fitting normal distributions to empirical distributions of various hydrologic variables is the property that a normal variable can assume any value from $-\infty$ to $+\infty$ while most hydrologic variables are positive-valued, generally with a lower boundary zero, and sometimes also with a positive or even negative lower boundary. A standardized form of these variables cannot physically show negative values smaller than $-\mu$, or smaller than $-(\mu - x_0)$, with x_0 the lower boundary. Nevertheless, if the mean is larger than 3σ or 4σ, the probability that the value x is negative, or that it reaches the lower boundary, is very small. For many practical purposes this probability can be considered as nearly zero. Two physical factors may justify the application of normal functions in hydrology even in these cases. First, various variables may have values of zero for a large portion of time (precipitation, evaporation, sediment transport, flow of intermittent rivers, and similar). The value of zero has a finite probability mass, while the positive values have probability densities for a continuous variable for $x \geq 0$. The probability mass for $x = 0$ may be interpreted theoretically as the integral of probability densities for the negative values from $-\infty$ to zero, as shown in Fig. 7.2. The distribution may be conceived as a truncated normal function,

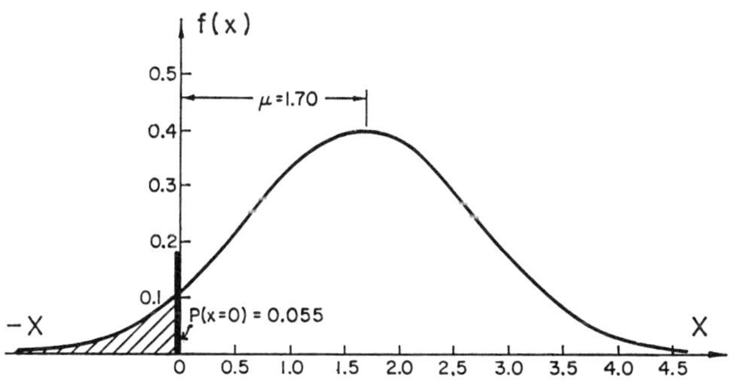

Figure 7.2. The negative part of a normal probability density function, conceived as being equal to the probability mass p $(x = 0)$, as a truncated normal distribution.

and mixed of the probability density for $x \geq 0$, and a probability mass for $x = 0$. The negative tail may be conceived as an imaginary part of the normal distribution. Second, negative values may occur for some hydrologic variables to account for a small total probability of the negative tail; this rarely happens, however. A large river with small surface slope usually has positive flows (downstream direction). However, a flash flood from a tributary may create backflow on a large river and thus the negative or upstream flows. Upstream flows are often encountered in estuaries and deltas of many rivers. The same may occur in ground water flow near rivers, lakes and reservoirs, which justify the use of negative tails of the normal probability function.

Use of the normal function for random errors. Random and systematic errors are associated with any hydrologic measurement or observation. Random errors are in most cases symmetrically distributed, and can be very often approximated by a normal probability function. The standard deviation σ, or the error σ_p, defined by Eq. 7.14, are the usual measures of these random errors.

Use of the normal function for comparison of distributions. Properties of many parameters of independent or dependent hydrologic variables may be investigated by comparing properties of corresponding parameters of normal standard independent or dependent variables. As examples, the distribution of the sample mean, the sample variance, the sample skewness coefficient, runs, ranges, the time dependence measures, and similar parameters may be investigated for nonnormal variables by comparing them with distributions of the same parameter of normal function.

Normal sampling distributions for parameters. Many statistics are either normally distributed, or their distribution may be approximated by normal distributions. The testing of hypotheses that the means or variances of two samples or two subsamples are, or are not, different from zero on a given probability level, or checking the homogeneity of data of a sample, are typical examples of the application of the normal function in hydrology. Many distributions converge to a normal function under given conditions. A typical case is the application of the central limit theorem, with the consequence that sums or averages of nonnormal variables tend to normal variables as the number of variables in the sum or in the average increases.

Use of normal function for Monte Carlo method. It is easier to generate random independent numbers normally distributed by the Monte Carlo method than the dependent random numbers of some asymmetrical distributions. Transformations may be available from these independent random normal numbers to various dependent random nonnormal numbers. The generation of random numbers of the normal probability function is an important step in the application of the data generation method to hydrology.

Graphical fitting of normal distributions. The normal function may be fitted to data either analytically or graphically. Before the estimation of parameters of normal function, and numerical tests of its goodness of fit to an empirical distribution are performed, it is often useful to check graphically whether the normal function is close to or far from a good fit. In order to facilitate this fit, a special scale for probabilities is transformed in such a way that the normal distribution function becomes a straight line. Visual inference is used to make that summary test on aligned points obtained by the selected plotting position. If the cartesian scale for the probability F (x) in Fig. 7.1, right graph, is transformed in such a way that all normal distribution curves become straight lines, this new scale is the probability scale. It is evident that the left point of F (x) = 0 goes to $-\infty$, and the right point of F (x) = 1 goes to $+\infty$.

The probability scale was first introduced to hydrology by Allen Hazen [7], with the objectives of easily extrapolating distribution curves of normal function fits and of estimating probabilities of higher and lower values than those observed. Figure 7.3 gives three probability distribution

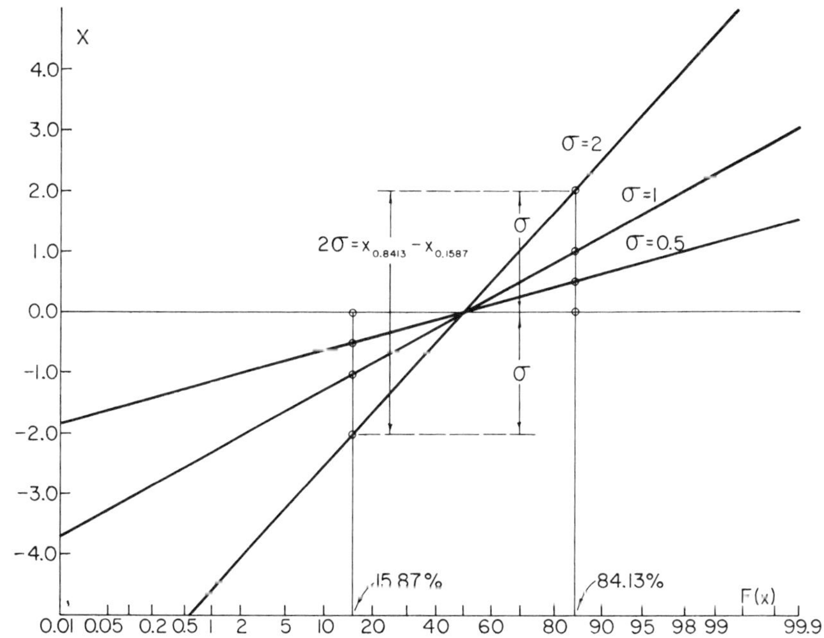

Figure 7.3. Normal probability distribution functions, with the mean of $\mu = 0$, plotted as straight lines, for $\sigma = 0.5$, 1.0, and 2.00, in cartesian-probability scales,respectively for x and F (x).

functions, with $\mu = 0$ and $\sigma = 0.5$, 1.0, and 2.0, plotted as straight lines, with the probability scale for F (x), and the cartesian scale for x. If a straight line well fits the points plotted for an empirical distribution in these scales, it is then a tentative inference drawn by inspection that the normal function fit may be also accepted by numerical tests. It then becomes justified to estimate parameters and test the goodness of fit by objective analytical methods. In summary, the cartesian-probability scales are designed in such a way that the probability distribution

$$F(x) = \frac{1}{\sigma\sqrt{2\pi}} \int_{-\infty}^{x} e^{-(x-\mu)^2/2\sigma^2} dx \qquad 7.15$$

plots as a straight line. In drawing a straight line to plotted points in cartesian-probability scales a simple rule is often observed. Points remaining on each side of the straight line should be approximately equal in number, and about uniformly spread along the line. A degree of subjectivity cannot be avoided in any visual fit.

Estimation of parameters of normal probability function. Of the four current estimation methods of parameters of distribution functions, the least square fitting method, the method of moments, the maximum likelihood method, and the graphical method, the first three give the mean and the variance of normal function approximately with the same accuracy, because the normal function is symmetrical. The mean μ is estimated by the sample mean $\bar{x}$, and the standard deviation σ is estimated by the sample standard deviation, either unbiased $\hat{s}$ for small samples, or biased s for large samples.

The graphical estimation of parameters μ and σ of the normal function is simple. The integral of f (x) from $-\infty$ to $x = \mu - \sigma$ and from $-\infty$ to $x = \mu + \sigma$ are 0.1587 and 0.8413, respectively, and from $-\infty$ to μ it is 0.50. The estimate of the mean is obtained for $F(x) = 0.5$, and the quantities $\bar{x} - s$ and $\bar{x} + s$ are at $F(\bar{x} - s) = 0.1587$ and $F(\bar{x} + s) = 0.8413$, respectively. The estimate of standard deviation, s, is equal to one half the difference of the x values corresponding to probabilities 0.8413 and 0.1587, as it is shown in Fig. 7.3. This graphical estimate by $\bar{x}$ and s are usually somewhat different from the more objective estimates by $\bar{x}$ and s (or $\hat{s}$), obtained by the method of moments.

Example 7.6. Figure 7.4 gives the annual runoff of the Willamette River, near Albany in Oregon, with N = 48 years (1893—1941), expressed in modular coefficients $x_i = Q_i/\bar{Q}$, where Q_i is the annual runoff, and $\bar{Q}$ is the mean annual runoff, plotted by the equation $P(x_i) = m/(N + 1)$. The visual straight-line fit, (1), gives the weight to points between 10 percent and 90 percent of probability, and the graphical estimates are: $\bar{x} = 1.00$ (taken as given by the modular coefficients), and $2s_x = 0.536$, or, $s_x = 0.268$. The computed value of s is 0.236 for the estimation by the method of moments. Here the values x_i with $P(x_i) < 10$ percent and $P(x_i) > 90$ percent prevail, and have rotated the straight line clockwise to the right about $\bar{x} = 1.00$ in comparison with the visual straight-line fit. There is often a significant difference between the parameters estimated graphically and those estimated analytically.

Goodness of fit of normal functions. Another criterion for preliminary testing whether the normal function would fit well an empirical distribution is by the skewness coefficient, C_s, and the excess coefficient, E, computed from data. If $-0.10 \leq C_s \leq 0.10$, or even a stronger criterion $-0.05 \leq C_s \leq 0.05$, then the empirical distribution is close to a symmetrical distribution, and the normal function may be shown to fit well the data. The negative aspect of this rough test is that a distribution can be symmetrical with C_s close to zero but still far from being well fitted by a normal function. If $-0.5 < E < 0.5$, then this second condition would usually show a good fit of the normal function.

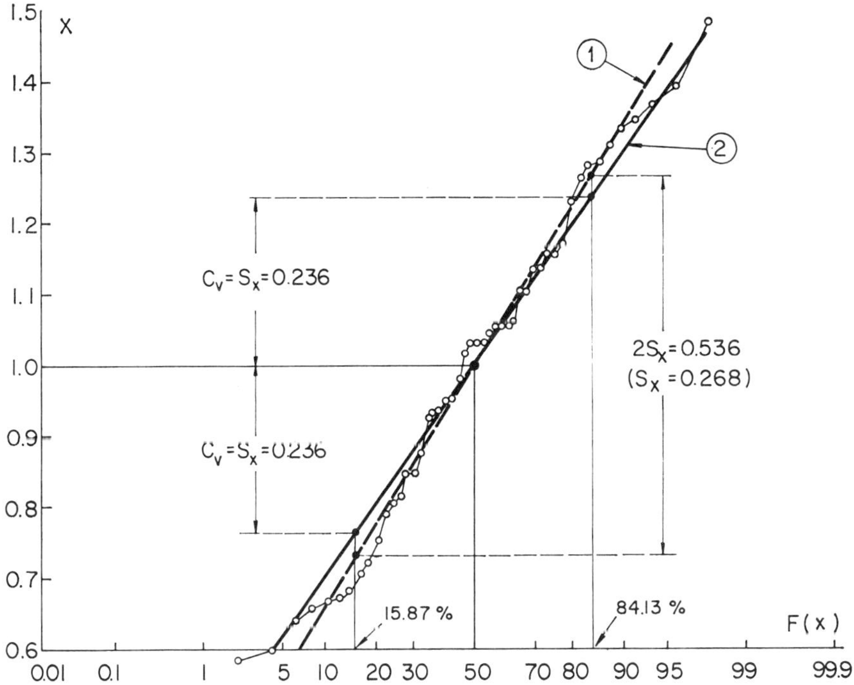

Figure 7.4. The graphical fit of normal probability function (1) to the empirical distribution of annual runoff (given in modular coefficients) of the Willamette River, near Albany in Oregon, N = 48 (1893–1941), with $\bar{x} = 1.00$ and $s_x = 0.268$, against the function (2) fitted by the method of moments, with $\bar{x} = 1.00$ and $s_x = 0.236$.

If the normal function has been found to provide an inadequate fit to a set of observations, two alternate ways are available: by fitting other distribution functions to obtain a good fit and by transforming the variable values in order to normalize them. Both procedures are empirical fits unless the transformation can be justified on some physical ground. It is often convenient to seek the normalizing transforms for any of three reasons: to utilize the simple properties of the normal function, to obtain a satisfactory fit to data, and to avoid the use of functions that cannot be justified, though they give a good fit.

Testing the goodness of fit is carried out by using the chi-square test. The Kolmogorov-Smirnov test also may be applied. They are discussed in Chapter 10.

7.6. Lognormal Distribution.

Logarithmic-normal probability distribution function. When logarithms, ln x, of a variable x are normally distributed, then the distribution of x is said to follow the logarithmic-normal probability distribution. From the discussion of distribution of functions of random univariates, it follows that $f(x)\,|dx| = \varphi(\ln x)\,|d(\ln x)|$. Because $|d(\ln x)| = |dx/x|$, then $f(x) = \varphi(\ln x)/x$, so that

$$\varphi(\ln x) = \frac{1}{\sigma_n\sqrt{2\pi}} \exp\left[-\frac{1}{2}\left(\frac{\ln x - \mu_n}{\sigma_n}\right)^2\right] \qquad 7.16$$

in which μ_n is the mean of ln x, and σ_n the standard deviation of ln x. The index n in μ_n and σ_n means that they are derived from natural logarithms. Then

$$f(x) = \frac{1}{x\,\sigma_n\sqrt{2\pi}} \exp\left[-\frac{1}{2}\left(\frac{\ln x - \mu_n}{\sigma_n}\right)^2\right], \text{ for } x \geq 0, \qquad 7.17$$

$$f(x) = 0, \text{ for } x < 0,$$

represents the logarithmic-normal probability function of the variable lnx. Since logarithms to any base can be related to natural logarithms lnx, by means of a constant, the form of Eq. 7.17 is not affected by the choice of the base. Equation 7.17 may be symbolized by $L(\mu_n, \sigma_n)$, or by $\Lambda(\mu_n, \sigma_n)$. Figure 7.5 shows several lognormal functions with various values of μ_n and variance σ_n^2 (and the corresponding values μ, or σ^2).

The mean μ_n is equal to the logarithm of the geometric mean, μ_g. McAlister and Galton [8, 9] first pointed out that the geometric mean is a better representation of the central tendency for some distributions. For this reason, the distribution is sometimes referred to as the Galton function. In the subsequent text it will be called the lognormal distribution.

A book by J. Aitchison and J. A. C. Brown [10] has the analysis of, and the necessary information about, the lognormal probability distribution function that satisfies various needs of hydrology. A large number of references are also available in the same work.

Significance and general properties of lognormal distribution. A theoretical basis is provided for the lognormal distribution by considering causative factors as having positive and multiplicative, rather than additive, effects. Hence, the logarithms of factors should satisfy the basic four conditions for normal distributions. If this is the case, then x should be a product of powers of causative factors.

The lognormal function possesses a lower limit, zero, it has a single maximum, the inflection points are at $\mu_n - (3\sigma_n^2/2) \pm (\sigma_n\sqrt{4 + \sigma_n^2}/2)$. The curve is convex upward between the inflection points and convex downward otherwise, and the mean, the median, and the mode of the function are different.

The relations between the parameters μ and σ and μ_n and σ_n permit the determination of one pair from the other. All moments exist for the lognormal distribution. The moments of x about zero are [10]

$$\gamma_r = e^{r\mu_n + r^2\sigma_n^2/2}, \qquad 7.18$$

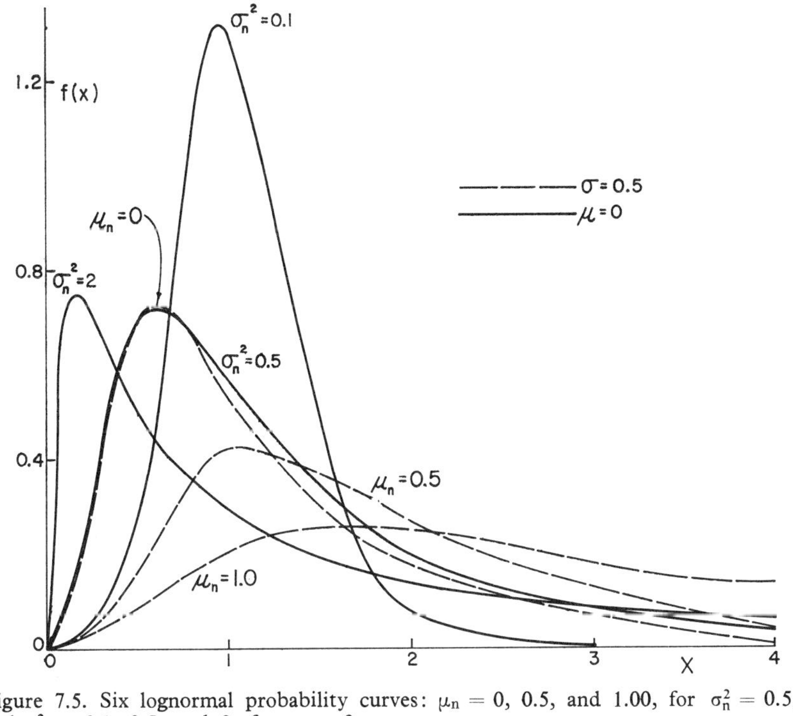

Figure 7.5. Six lognormal probability curves: μ_n = 0, 0.5, and 1.00, for σ_n^2 = 0.5, and σ_n^2 = 0.1, 0.5, and 2, for μ_n = 0.

with r the order of the moment. Thus the mean, μ, becomes for r = 1

$$\mu = e^{\mu_n + \sigma_n^2/2} \qquad 7.19$$

and is, therefore, a function of both μ_n and σ_n. The variance of x is derived from Eq. 7.18 as

$$\sigma^2 = \mu^2 (e^{\sigma_n^2} - 1). \qquad 7.20$$

Since μ is a function of μ_n and σ_n, then σ is also a function of μ_n and σ_n.

The coefficient of variation is

$$\eta = \frac{\sigma}{\mu} = (e^{\sigma_n^2} - 1)^{1/2}. \qquad 7.21$$

It is often useful to express both μ_n and σ_n in terms of μ and σ, or μ and η, and the above three equations give

$$\mu_n = \frac{1}{2} \ln \left(\frac{\mu^2}{\eta^2 + 1} \right) = \frac{1}{2} \ln \left(\frac{\mu^4}{\mu^2 + \sigma^2} \right) \qquad 7.22$$

and

$$\sigma_n^2 = \ln (\eta^2 + 1) = \ln \left(\frac{\sigma^2 + \mu^2}{\mu^2} \right). \qquad 7.23$$

The skewness and kurtosis coefficients of x are, respectively,

$$\gamma_1 = \eta^3 + 3\eta, \qquad 7.24$$

and

$$\gamma_2 = \eta^8 + 6\eta^6 + 15\eta^4 + 16\eta^2 + 3. \qquad 7.25$$

The greater σ_n^2, the greater are η, γ_1, and γ_2. Figure 7.6 gives η, γ_1, and γ_2 as functions of σ_n. The skewness and kurtosis coefficients are always positive, and are functions of η only, which in turn is defined by σ_n. The parameter σ_n is sometimes referred to in hydrology as the index of variability [11]. The greater the index of variability ($\sigma_n = I_v$), the greater the coefficient of variation, and the skewness and kurtosis coefficients.

The mode and its probability density are

$$m = e^{\mu_n - \sigma_n^2} = \frac{\mu}{(1 + \eta^2)^{3/2}}, \text{ and } f(m) = \frac{e^{-\sigma_n^2/2}}{m\sigma_n\sqrt{2\pi}}. \qquad 7.26$$

The median and its probability density are

$$M = e^{\mu_n} = \frac{\mu}{(1 + \eta^2)^{1/2}}, \text{ and } f(M) = \frac{e^{-\mu_n}}{\sigma_n\sqrt{2\pi}}. \qquad 7.27$$

Ratios of the median and mode and their probability densities are

$$\frac{M}{m} = e^{\sigma_n^2} = 1 + \eta^2, \text{ and } \frac{f(M)}{f(m)} = e^{-\sigma_n^2/2} = \frac{1}{(1 + \eta^2)^{1/2}}. \qquad 7.28$$

To determine μ_n and σ_n, two methods are available: to estimate μ and σ from the variable values and use Eqs. 7.22 and 7.23 to determine σ_n and μ_n, respectively, or by the method of moments of x, and to estimate μ_n and σ_n directly from logarithms of x, or by the maximum likelihood method of estimation. A replacement of x in Eq. 7.16 by (– x) gives the two-parameter lognormal distribution negatively skewed. This curve is the mirror image about the f (x)-axis of the curve of Eq. 7.16.

Reproductive properties of lognormal function. The lognormal probability function of Eq. 7.17 has simple reproductive properties like the normal function, but the multiplicative properties replace the additive properties. If x is lognormal, the $y = a x^b$, with a and b constants, is lognormal with $\mu_n(y) = \ln a + b\ \mu_n(x)$ and $\sigma_n^2(y) = b^2\ \sigma_n^2(x)$. If x and y are independent lognormal random variables, then $z = xy$ is lognormal with $\mu_n(z) = \mu_n(x) + \mu_n(y)$, and $\sigma_n^2(z) = \sigma_n^2(x) + \sigma_n^2(y)$, or additive properties are valid for logarithms of these variables. The converse is also true. For $z = x/y$, and x and y are mutually independent, z is lognormal with $\mu_n(z) = \mu_n(x) - \mu_n(y)$, and $\sigma_n^2(z) = \sigma_n^2(x) + \sigma_n^2(y)$. If $x_1, x_2, \ldots, x_n$ are independent lognormal variables with the same parameters μ_n, and σ_n, then the geometric mean of n variables, $\bar{x}_g = (\Pi\, x_i)^{1/n}$, is also lognormal with the mean μ_n and the variance σ_n^2/n.

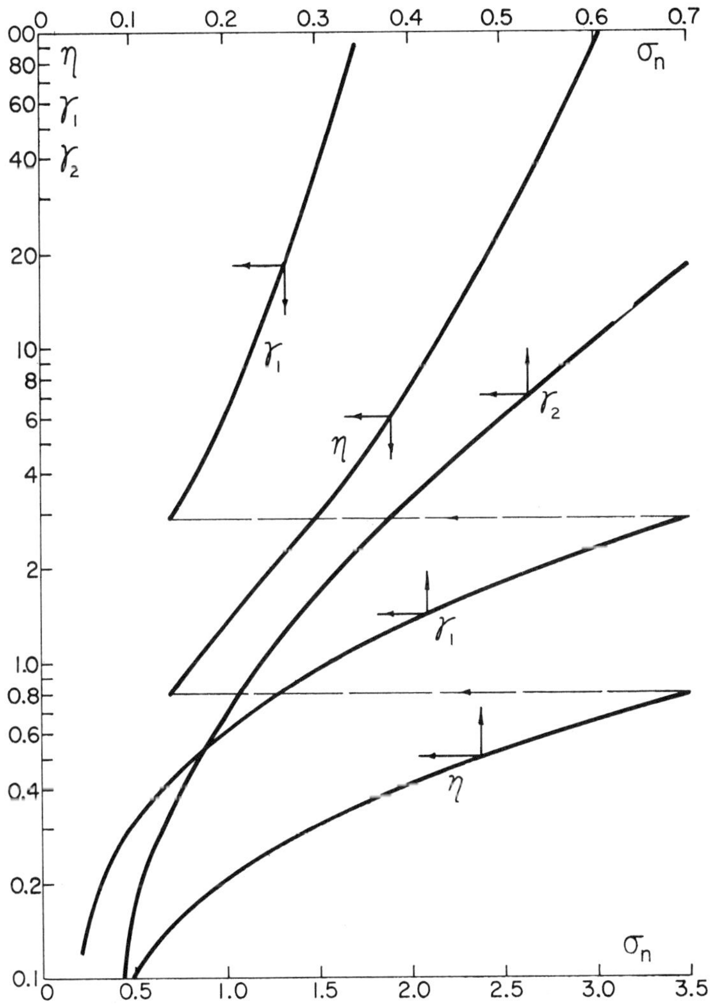

Figure 7.6. Relations of the coefficient of variation η, the skewness coefficient γ_1, and the kurtosis γ_2 to the standard deviation of logarithms σ_n.

The product of n variables, Πx_i, i = 1, 2, . . . , n is lognormal with the mean $n\mu_n$, and the variance $n\sigma_n^2$. The central limit theorem implies that the product of n skewed random variables, each with μ_n and σ_n^2 of their logarithms, is asymptotically lognormally distributed with the mean $n\mu_n$ and the variance $n\sigma_n^2$. A product, or a geometric mean of random, non-lognormal variables, independent or dependent, with various values of μ_n and σ_n^2 (various skewness coefficients also), converge to the lognormal distribution as the number of variables increases.

Lognormal probability density function with three and four parameters. If the lower boundary of distribution is not zero, it is necessary to modify the lognormal distribution function by introducing the third parameter, the lower boundary. The three-parameter lognormal function is

$$f(x) = \frac{1}{(x - \beta)\,\sigma_n\sqrt{2\pi}} \exp\left\{-\frac{[\ln(x-\beta) - \mu_n]^2}{2\sigma_n^2}\right\}, \qquad 7.29$$

in which β represents the lower boundary. Equation 7.29 is Eq. 7.17 with its variable x replaced by $(x - \beta)$.

In practice, two situations can arise concerning the lower boundary. The first is when the lower boundary is known, and the transformed variable $(x - \beta)$ can be determined. In this case, the relations and the procedures for computing the parameters given for the two-parameter lognormal function are used. The second situation, which most often arises, is when the lower boundary is not known a priori. In this case, the relations for parametes given for the two-parameter lognormal functions cannot be applied directly.

The addition or subtraction of a constant from a variable does not change its variance, σ^2, but it does change the mean, μ. Hence, the change of μ implies a change of the coefficient of variation, η. Those parameters that depend on η and σ_n are also changed. If β is negative, the values of x can be negative; however, $(x + \beta) > 0$ for all x.

The easiest procedure for expressing the parameters β, μ_n, and σ_n of Eq. 7.29 is by using the moments of x, namely, m_1, μ_2, and μ_3, where m_1 and μ_2 define μ_x, σ and η_x. As previously stated, μ_x and η_x are now different from the corresponding parameters of the two-parameter lognormal function.

Parameters μ_n and σ_n are defined for $(x - \beta)$ by Eqs. 7.22 and 7.23, respectively. To find β, assumed here a positive value, the third central moment of x, μ_3, must be used. The coefficient of variation $\eta_x = \sigma/\mu_x$, pertaining to x, and the coefficient of variation $\eta = \sigma/\mu$, pertaining to $(x - \beta)$, with the mean of $(x - \beta)$ being $\mu = \mu_x - \beta$, give β by eliminating σ, as

$$\beta = \mu_x\left(1 - \frac{\eta_x}{\eta}\right). \qquad 7.30$$

This represents the method of moments in estimating the lower boundary, which is not as efficient as the method of maximum likelihood discussed in Chapter 8.

Since the second and third central moments of Eq. 7.29 are independent of β, Eq. 7.24 provides a way to compute η based on γ_1, computed for x. By introducing η into Eq. 7.30, the parameter β is determined. To express η in terms of γ_1 requires the solution of a cubic equation. The analytical solution of Eq. 7.24 is

$$\eta = \left[\frac{\sqrt{\gamma_1^2 + 4} + \gamma_1}{2}\right]^{1/3} - \left[\frac{-\sqrt{\gamma_1^2 + 4} + \gamma_1}{2}\right]^{1/3}. \qquad 7.31$$

Figure 7.6 makes it possible to determine η graphically once γ_1 is computed.

Forms of the three-parameter functions other than that given by Eq. 7.29 have been introduced in hydrologic literature. One such form, proposed by R. Gibrat [12], is

$$F(x) = \frac{1}{\sqrt{2\pi}} \int_{-\infty}^{z} e^{-z^2}\, dz, \qquad 7.32$$

in which

$$z = a \log (x - q_0) + b, \qquad 7.33$$

with a, b, and q_0 representing the three parameters, and q_0 denoting the lower boundary. Gibrat called the lognormal distribution the law of proportionate effect, and did not consider it as being exclusively empirical.

When $(x - \beta)$ in Eq. 7.29 is replaced by $(\beta - x)$, the negatively skewed lognormal distribution with three parameters is produced. If x in Eq. 7.17 is replaced by

$$x = \frac{X - \beta}{\alpha - X}, \qquad 7.34$$

the new variable X has both the upper and the lower boundary, α and β, respectively, or it is confined to the range $\alpha < X < \beta$, and μ_n, σ_n are parameters of x. This four-parameter lognormal bounded distribution function (μ_n, σ_n, α, β) is skewed and may be used only for those phenomena bounded on both ends by their physical properties.

Reproductive properties of the two-parameters lognormal function are not applicable to the three- and four-parameter lognormal functions. Because of difficulties and inaccuracies in estimating parameters, the four-parameter distribution has not found an application in hydrology.

Estimation of parameters of lognormal functions. A main issue in estimating parameters of lognormal distributions is the reliability of alternatives to the method of maximum likelihood. The estimation of μ_n and σ_n of Eq. 7.17 by the sample means $\bar{x}_n$ and s_n of ln x gives the maximum likelihood estimates. The maximum likelihood method in the treatment of ungrouped data requires determining logarithms, ln x, of all values of a sample, before the moments are computed, which may be laborious except when a computer is used. The procedure for the grouped data is even more laborious. Methods in using sample moments of x-values, and the graphical procedures, are simpler but less accurate and involve, therefore, a loss of information when compared to the method of maximum likelihood. The estimation procedures are even more complex when the three- or four-parameter lognormal functions are involved, when distributions are truncated, or when they contain zero values. The estimation of parameters of lognormal distributions is discussed at length in the literature [10, and 13 through 19].

Visual test to determine whether the lognormal function fits an empirical frequency distribution as the hypothesis, and whether the graphical estimates of parameters μ_n and σ_n in Eq. 7.17 are best, is carried out by plotting the empirical cumulative distributions in graphs with logarithmic-probability scales. The log-probability graph paper has the property that the distribution functions, F (x), of Eq. 7.17 plot as straight lines. If points $P(x) = m/(N + 1)$ of a sorted sample according to ascending or descending values follow approximately a straight line in this graph, the hypothesis is attractive that the variable x follows approximately the lognormal distribution. The value $P(x) = 50$ percent gives $\bar{x}_n$ as the estimate of μ_n, and the value $(\ln x_1 - \ln x_2)/2$, with $P(x_1) = 84.13$ percent and $P(x_2) = 15.87$ percent, gives s_n as the estimate of σ_n. To test how good is the fit, either the Kolmogorov-Smirnov statistic, $\Delta_{max} = P(x) - F(x)$, is used, where P (x) is the plotted point and F (x) is the corresponding point at the straight line, or the chi-square test is applied.

If the three-parameter lognormal function is fitted, and if the lower boundary β in Eq. 7.29 is known a priori, the above graphical procedure is used for the transformed variable $(x - \beta)$. On the other hand, if β is not known a priori, a trial-and-error graphical method may be applied. For several values of β assumed, various sets of $(x - \beta)$ samples are determined. These sets are then plotted on log-probability paper. The set that most nearly plots as a straight line defines the boundary β. With the boundary β thus determined, the other parameters are determined graphically as stated above. The accuracy of these estimates are, however, lower than in the analytical estimation by Eqs. 7.30 and 7.31. The best estimates of the three parameters of Eq. 7.29 are determined analytically by the maximum likelihood method [6].

Applications of lognormal function in hydrology. Many hydrologic variables such as precipitation, discharge, grain size of sediments, and others, are measured from the base zero and possess no upper bound. Since the logarithm of zero is minus infinity, the logarithmic transformation of these variables vary from $-\infty$ to $+\infty$. The application of the normal function to such transformed variables is currently practiced in hydrology.

It has been found through extensive experience that the lognormal probability function fits well a large number of distributions in hydrology, because they are asymmetrical and because of other properties. It can be assumed in many cases, that the physical causative factors of a stochastic hydrologic process are interdependent, with their effects multiplicative. This implies the theoretical background for lognormal function, though some factors may be more important than others. By advancing several hypotheses of physical processes, it can be shown in some cases that the distributions should be lognormal. Kolmogorov [20], and Rosumovsky [21], have shown that the grain size of ground rock under given conditions is lognormally distributed. These studies, and others, imply also that the sediment grain size in a natural mixture of sand is expected to be lognormally distributed under a given set of assumptions.

Example 7.7. It has been shown [6] that among the probability functions that best fit the empirical distributions of annual runoff and annual precipitation of 2060 stations in the western part of North America, the lognormal probability density function of Eq. 7.17, with the two parameters μ_n and σ_n is one of the best. The three-parameter distribution of Eq. 7.29 was not, on the average, any better than the two-parameter distribution function.

Example 7.8. Figure 7.7 shows the points of empirical distribution of annual river flow of the Rhine River at Basle, Switzerland, for 150 years (1808 – 1957), expressed in modular coefficients, $K_i = x_i/\bar{x}$, and plotted on log-probability graph paper, with a straight line fit. Figure 7.8 shows the frequency density polygon and the fitted lognormal probability density function. The graphical estimate of σ_n is $s_n = 0.165$, while the maximum likelihood method (or the method of moments applied to logarithms) produces $s_n = 0.162$. Visually the fit seems good, so that the basic hypothesis of a lognormal function fitting the empirical curve well is attractive.

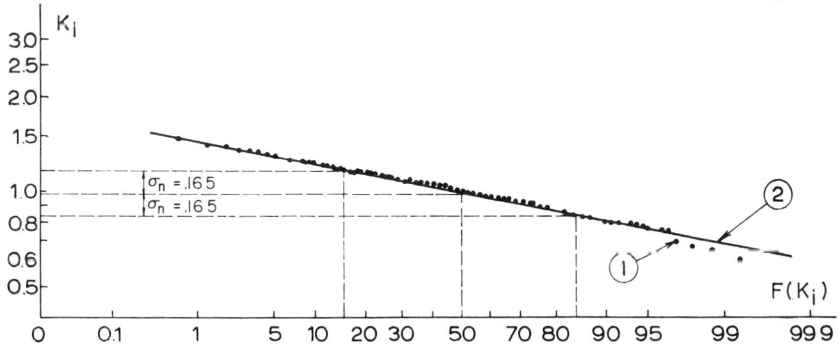

Figure 7.7. Visual fitting of the straight line lognormal function in log-probability scales to the empirical distribution of annual river flows of the Rhine River at Basle, Switzerland, for the period 1808 – 1957 (N = 150 years): (1) points plotted by $P(K_i) = m/(N+1)$, and (2) fitted straight line.

Example 7.9. Figure 7.9 gives the graphical fitting of the straight line in log-probability scales to the empirical distribution of annual one-day low-flow discharges of the Kings River at Piedra, California, for 60 years (1897–1956); three curves are

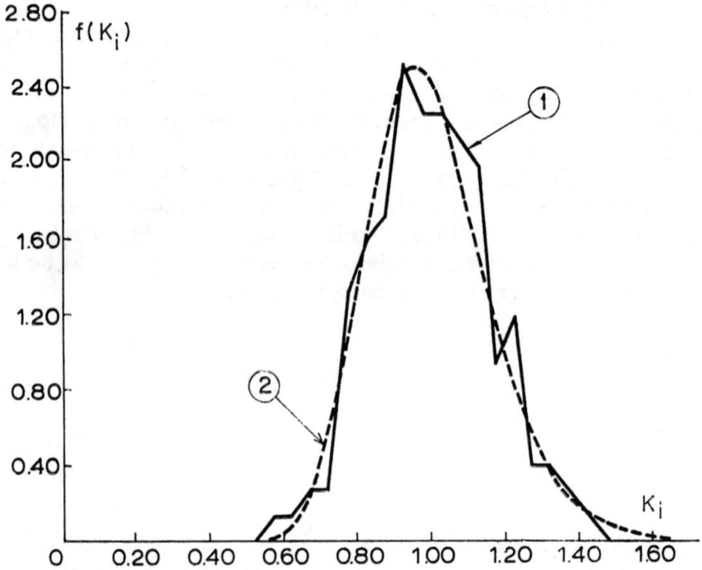

Figure 7.8. Comparison between (1) the empirical frequency density curve and (2) the fitted lognormal probability density function of Fig. 7.7 for annual river flows of the Rhine River.

given: (1) observed frequency distribution, (2) visual straight-line fit of the two-parameter lognormal function, and (3) analytically estimated three-parameter lognormal function of the best fit. Analytical estimates of three parameters of Eq. 7.29 are: $m_n = 5.24$ as the estimate of μ_n, $s_n = 0.33$ as the estimate of σ_n, and $\hat{\beta} = -19.0$ cfs as the estimate of the lower boundary, β.

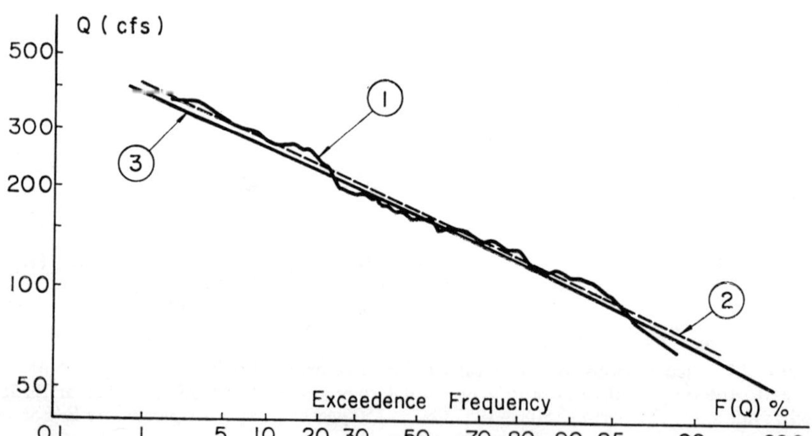

Figure 7.9. The fitting of a lognormal function to the empirical distribution of the annual one-day low-flow discharge of the Kings River at Piedra, California: (a) observed distribution, (2) visual straight-line fit of two-parameter function, and (3) fit of three--parameter function with numerically estimated parameters.

7.7. Gamma Distributions.

Incomplete gamma function. The integral

$$\Gamma(\alpha) = \int_0^\infty x^{\alpha-1} e^{-x} dx \qquad 7.35$$

converges if $\alpha > 0$, and is called the gamma function of α. Its simple properties are: $\Gamma(\alpha) = (\alpha - 1)!$, if α is a positive integer; for $\alpha > 0$ but not an integer, $\Gamma(\alpha)$ is computed by series expansion and numerical integration: $\Gamma(0) = \infty$, $\Gamma(1/2) = \sqrt{\pi}$, $\Gamma(1) = \Gamma(2) = 1$, and various functions can be reduced to gamma function by simple transformations [an example is $-\int_0^1 (\ln x)^{1/5} dx = \int_0^\infty y^{1/5} e^{-y} dy = \Gamma(6/5)$, with $y = \ln x$, $\alpha = 6/5$].

By putting $\alpha = p + 1$, and integrating Eq. 7.35 from 0 to $u\sqrt{p+1}$, then the function

$$I(u,p) = \frac{1}{\Gamma(p+1)} \int_0^{u\sqrt{p+1}} x^p e^{-x} dx \qquad 7.36$$

is called the incomplete gamma function, a term introduced by K. Pearson [22]. He published tables of this function with u ranging from 0 to 12, and p ranging from 0 to 50 [23]. Usually, Eq. 7.35, with limits 0 and x, is called the incomplete gamma function, $\Gamma(x, \alpha)$.

Gamma probability density function with one parameter. The continuous variable x, which has the probability density function

$$f(x) = \frac{1}{\Gamma(\alpha)} x^{\alpha-1} e^{-x}, \text{ for } x > 0 \qquad 7.37$$

$$f(x) = 0, \text{ for } x < 0,$$

or of the range $0 \leq x \leq \infty$, is said to follow one-parameter gamma distribution, with α the shape parameter. Figure 7.10 shows this probability function for various values of α.

Properties of one-parameter gamma function. The properties of this distribution are: the density curves for $0 \leq \alpha \leq 1$ are J-shaped, with both axes being asymptotes; the curve for $\alpha = 1$ is the simple exponential function $f(x) = e^{-x}/\Gamma(a)$; the bell-shaped curves for $1 < \alpha \leq 2$ have an f(x)-axis as the tangent at $x = 0$, while the x-axis is the asymptote; the curves for $\alpha > 2$ start at $x = 0$ by touching the origin while the x-axis is the asymptote, and the mode for $\alpha > 1$ is at $m = \alpha - 1$. The gamma probability density function can be integrated exactly if α is an integer.

The expected value (mean), variance, skewness coefficient, and kurtosis are, respectively,

$$E\,x = \alpha; \quad \text{var } x = \alpha; \quad \gamma_1 = \frac{2}{\sqrt{\alpha}}; \quad \gamma_2 = \frac{6}{\alpha} + 3. \qquad 7.38$$

If x is normally distributed, N (μ, σ), then

$$y = \frac{1}{2}\left(\frac{x - \mu}{\sigma}\right)^2$$

follows the one-parameter gamma distribution with $\alpha = 1/2$. This equation gives a physical justification for using the gamma probability function. If the variable values as effects depend on squares of causative factors, and all other conditions of normal functions for cause-effect relations are also satisfied, the variable is gamma distributed. The sum of two independent one-parameter gamma variables, with parameters α_1 and α_2, follows one-parameter gamma distribution with $\alpha = \alpha_1 + \alpha_2$. The converse property is also valid. In general, a sum of n independent gamma variables

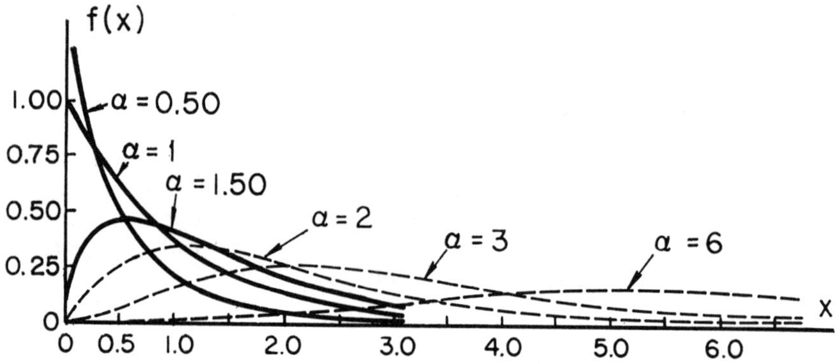

Figure 7.10. One-parameter gamma probability density curves for six values of α (0.50, 1, 1.50, 2, 3, and 6).

with parameters $\alpha_1, \alpha_2, \ldots, \alpha_n$ is the gamma variable with the parameter $\alpha = \Sigma\, \alpha_1$. Since $(x - \mu)/\sigma$ is a standard normal variable, t, the above property states that the sum $\frac{1}{2} \sum_{i=1}^{n} t_i^2$ is a one-parameter gamma distribution with $\alpha = n/2$, when t_i's are independent. This makes it possible to compute the gamma variable values of α for any multiple of 1/2 by using the sum of squares of the values t normally distributed. This property may serve to generate gamma independent random numbers, of α multiples of 1/2, by using the standard normal independent random numbers. The gamma distribution converges for large α to normal distribution (say $\alpha > 30$).

Hydrologic application of one-parameter gamma function. The application of one-parameter gamma probability function in hydrology is limited because of the relatively small flexibility of all functions with one parameter to fit properly hydrologic empirical frequency distributions. The chi-square parameter in testing many hypotheses is, however, distributed as a one-parameter gamma function with the parameter n/2, and n is the number of degrees of freedom used in testing a hypothesis. This general application of a one-parameter gamma distribution in the form of a chi-square distribution is widely used in hydrology.

Gamma probability function with two parameters. If x in Eq. 7.37 is replaced by a new variable x/β, then

$$f(x) = \frac{1}{\beta^{\alpha}\,\Gamma(\alpha)}\, x^{\alpha-1}\, e^{-x/\beta} \qquad 7.39$$

is the two-parameter gamma probability density function, with $0 \leq x < \infty$, and $f(x) = 0$ for $x < 0$. The shape parameter is $\alpha > 0$, and the scale parameter is $\beta > 0$. Equation 7.39 is identical to Eq. 7.37 for $\beta = 1$. By the transformation $x/\beta = y$ and $\beta f(x) = \varphi(y)$, Eq. 7.39 is reduced to Eq. 7.37 in case $\beta \neq 1$.

Properties of two-parameter gamma function. The mode is $m = \beta^{\alpha}(\alpha - 1)$ for $\alpha \geq 1$. The expected value and the variance of Eq. 7.29 are

$$Ex = \mu = \alpha\beta; \quad \text{var } x = \alpha\beta^2, \qquad 7.40$$

while the skewness coefficient $\gamma_1 = 2/\sqrt{\alpha}$ and the kurtosis $\gamma_2 = 3 + 6/\alpha$ are independent of β. Equation 7.39 sometimes is written with $m = 1/\beta$, or β is replaced by β^{-1}, which are only various ways of representing the same two-parameter gamma distribution. Figure 7.11 gives three curves of Eq. 7.39 for $\alpha = 2$, and $\beta = 0.5$, 1.0, and 2.0.

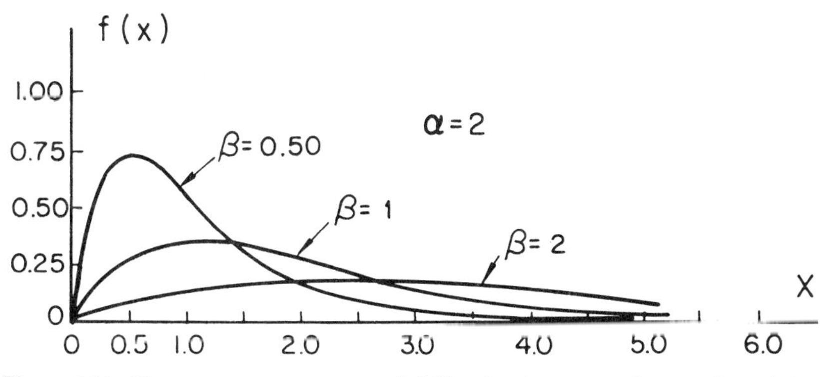

Figure 7.11. Two-parameter gamma probability density curves for $\alpha = 2$, and three values of β (0.50, 1, and 2).

Gamma probability density function with three parameters. Similarly, as for Eq. 7.39, if x in Eq. 7.37 is replaced by a new variable $(x - \gamma)/\beta$, then

$$f(x) = \frac{1}{\beta^{\alpha}\,\Gamma(\alpha)}\,(x - \gamma)^{\alpha-1}\, e^{-(x-\gamma)/\beta} \qquad 7.41$$

is the three-parameter gamma probability density function. It has the only new property that γ is the location parameter, the lower boundary for x, so that $\gamma \leq x < \infty$, with $\gamma \neq 0$. The three parameters are: α (shape), β (scale), and γ (location). Equation 7.41 has the mean $\mu = \gamma + \alpha\beta$, while σ^2, γ_1, and γ_2 are the same as for Eq. 7.39. The mode is at $m = \gamma + \beta^\alpha (\alpha - 1)$.

The transformation $y = x - \gamma$ reduces Eq. 7.41 to Eq. 7.39, and $z = (x - \gamma)/\beta$ reduces Eq. 7.41 to Eq. 7.37. The opposite transformations are useful in Monte Carlo experiments (generating data) in transforming independent random numbers of one-parameter gamma to the two-parameter and three-parameter gamma distributions, with the same α, and with the given β, or β and γ, respectively, for the two latter distributions.

Sometimes the gamma distribution is given in the form of a Pearson Type III probability density function:

$$f(x) = p_0 \cdot e^{-x/d} \left(1 + \frac{x}{b}\right)^{b/d}, \qquad 7.42$$

in which p_0 is the probability density of the mode; d is the difference between the mean and the mode, and b is the difference between the mode and the lower boundary of $f(x) = 0$. In Eq. 7.42 the mode is $m = 0$, or the peak of the function is at $x = 0$, with $f(x) = p_0$; the lower boundary is at $x = -b$; and $2d = C_v C_s (= \eta\gamma_1)$. The two functions of Eqs. 7.39 and 7.42 are the same gamma distributions; the following are the relations of parameters

$$p_0 = \frac{\beta^{\alpha(\alpha-2)}(\alpha - 1)^{\alpha-1}}{\Gamma(\alpha)} e^{-(\alpha-1)\beta^{\alpha-1}} ; d = \alpha\beta - \beta^\alpha(\alpha - 1); b = \beta^\alpha(\alpha - 1). \qquad 7.43$$

Once b and d are selected, p_0 is also determined.

Estimation of parameters in gamma distributions. The parameters α, or α and β, or α, β, and γ, respectively, for Eqs. 7.37, 7.39, and 7.41, are estimated from the sample data. For a highly skewed distribution, a maximum likelihood estimation is recommended, and even necessary, as will be shown later. For the two-parameter gamma distribution, $\mu = \alpha\beta$, $\sigma^2 = \alpha\beta^2$, so that $\beta = \sigma^2/\mu$ and $\alpha = \mu^2/\sigma^2 = \eta^2$. In estimating μ and σ by $\bar{x}$ and $\hat{s}$ of a sample, the estimates a and b of α and β are obtained by the method of moments. This may be satisfactory if the skewness coefficient $\gamma_1 = 2/\sqrt{\alpha}$ is small, or α is relatively large, and if the computed values of C_s as an estimate of γ_1, and k as an estimate of γ_2, are approximately equal to $2/\sqrt{\alpha}$ and $3 + 6/\alpha$, respectively. These last conditions may be used for the above estimate of α as a first and rough test of goodness of fit for the two-parameter gamma function before a more efficient method of estimation (the maximum likelihood) and the test of goodness of fit (chi-square test) are undertaken. Similarly, by using $\mu = \gamma + \alpha\beta$, $\sigma = \alpha\beta^2$, and $\gamma_1 = 2/\sqrt{\alpha}$, the estimates a, b, and g, of α, β, and γ, may be obtained from the sample estimates $\bar{x}$, $\hat{s}$, and C_s of the parameters μ, σ, and γ_1 as

$a = 4/C_s^2$, $b = \hat{s}\, C_s/2$, and $g = \bar{x} - 2\hat{s}/C_s$. However, for C_s sufficiently large, say $C_s > 0.5$, the maximum likelihood estimates of α, β and γ should be used, and the goodness of fit tested by the chi-square method.

Applications of two- and three-parameter gamma distributions in hydrology. The use of two-parameter and three-parameter gamma distributions in hydrology is as common as the use of the lognormal distribution. Unfortunately, it is not easy to transform the coordinate scales in such a way that all cumulative gamma distributions plot as straight lines. This disadvantage makes the use of the gamma distribution less attractive to some practitioners than the lognormal.

It was shown by Marković [6] that there is no practical difference between a fitting of lognormal and of gamma distributions to empirical frequency distributions of annual precipitation and annual runoff of 2060 gauging stations. It was also shown that the gamma and the lognormal distributions, with three-parameters, have no significant advantage in fitting the above empirical distributions when compared with the two-parameter gamma and the two-parameter lognormal distributions. This is an important conclusion, not expected by most practicing hydrologists.

Example 7.10. Figure 7.12 shows the fitting of both the two-parameter gamma function ($\hat{\alpha} = 1.727$ and $\hat{\beta} = 0.579$) and the three-parameter gamma function ($\hat{\gamma} = -1.500$, $\hat{\alpha} = 12.360$, and $\hat{\beta} = 0.202$) to the mean annual river flows of the Weldon River at Mill Grove, Missouri, USA, with flows given in modular coefficients $K_i = Q_i/\bar{Q}$, with $\bar{Q} = 256.7$ cfs, and with the parameters estimated by the maximum likelihood method [6].

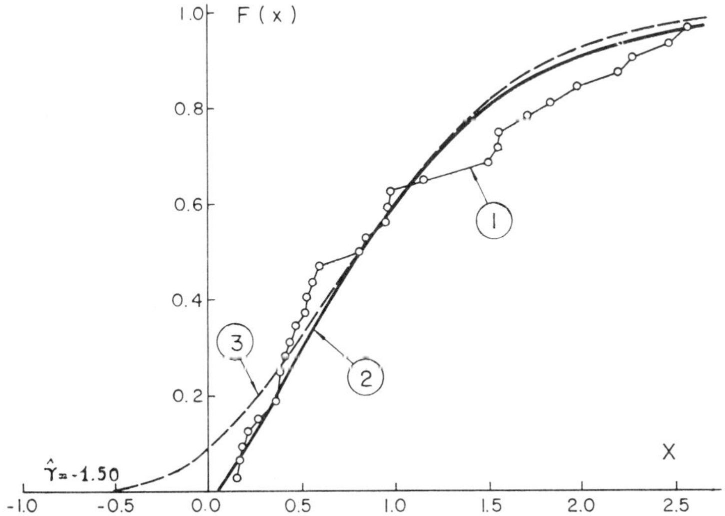

Figure 7.12. Fitting the two-parameter and three-parameter gamma distribution functions to the empirical distribution of annual flows of the Weldon River at Mill Grove, Missouri, with flows given in modular coefficient $x_i = Q_i/\bar{Q}$: (1) observed distribution, (2) fitting the two-parameter gamma function, and (3) fitting the three-parameter gamma function.

Example 7.11. Figure 7.13 gives a fit of the three-parameter gamma function of Eq. 7.41 to the empirical distribution of the annual minimum 9-month low water volume of the Little Arkansas River at Valley Center, Kansas, for a period of 33 years (1923–1955).

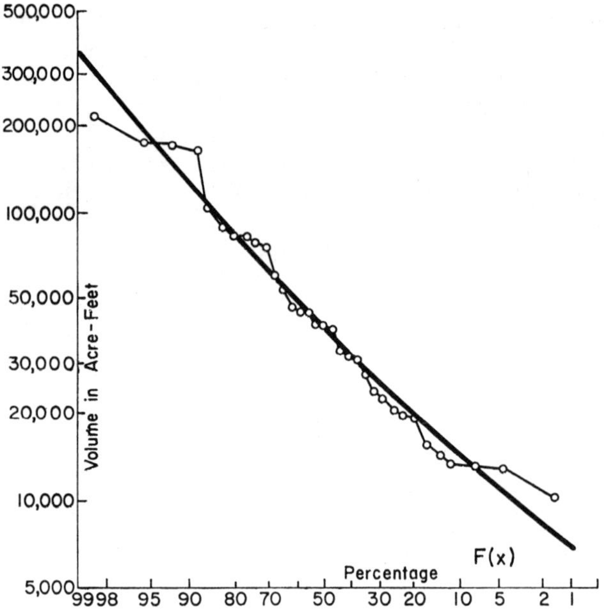

Figure 7.13. The fitting of the three-parameter gamma function (Pearson Type III) to the empirical distribution of the annual minimum 9-month low-flow water volume of the Little Arkansas River at Valley Center, Kansas.

Beta probability density functions. The family of beta probability distribution functions is related to gamma distributions. The beta function has two parameters and is defined as

$$B(\alpha,\beta) = \int_0^1 x^{\alpha-1} (1 - x)^{\beta-1} dx = \frac{\Gamma(\alpha)\,\Gamma(\beta)}{\Gamma(\alpha+\beta)}, \qquad 7.44$$

in which $\alpha > 0$, and $\beta > 0$, which shows how the beta function is related to gamma functions. From Eq. 7.44 it follows that α and β are interchangeable. The special values are: $B\left(\frac{1}{2}, \frac{1}{2}\right) = \pi$, $B(1,1) = 1$, $B(2,2) = \frac{1}{6}$, and $B(1,2) = \frac{1}{2}$. The probability density function

$$f(x) = \frac{x^{\alpha-1}(1-x)^{\beta-1}}{B(\alpha,\beta)} \qquad 7.45$$

is a beta distribution with boundaries 0 and 1, and two parameters α and β, often called the B_1-distribution. For $\alpha > 0$ and $\beta > 1$, the mode is

$m = (\alpha - 1)(\alpha + \beta - 2)$. For $\alpha > 2$ the curve touches the x-axis at the origin. For $1 < \alpha < 2$ the curve is tangent to the f (x)-axis at the origin. For $0 < \alpha < 1$, the curve has the f (x)-axis as asymptote. Similar properties are valid at $x = 1$ as related to β. For $0 < \alpha < 1$ and $0 < \beta < 1$, the distribution is a U-distribution. Equation 7.45 has $E\,x = \alpha/(\alpha + \beta)$, and $\text{var}\,x = \alpha\beta/(\alpha + \beta)^2 (\alpha + \beta + 1)$. In another form, the function

$$f(x) = \frac{x^{\alpha-1}(1+x)^{(\alpha+\beta)}}{B(\alpha,\beta)} \qquad 7.46$$

is also a beta distribution but with boundaries 0 and ∞, and with two parameters α and β, often called the B_2-distribution. For $\alpha > 1$, the mode is $m = (\alpha - 1)/(\beta + 1)$. The conditions at $x = 0$ are the same as for B_1. For B_2, the mean is $E\,x = \alpha/(\beta - 1)$, and the variance is $\text{var}\ x = \alpha(\alpha + \beta - 1)/(\beta - 1)^2(\beta - 2)$.

Furthermore, the B_1-distribution may well be used when a distribution in hydrology has two boundaries, either 0 and 1, or by simply transforming any other two boundaries. If boundaries are different from 0 and 1, and must be estimated from sample data, then the B_1-distribution becomes a four-parameter function.

Example 7.12. Whenever there are two fixed boundaries for a random variable, like the highest and the lowest level of a reservoir, the probability density curve of the levels in between may be different from a uniform (rectangular) distribution, and the B_1-distribution may be appropriately used. If this is the case, the two boundaries are physically defined and only two parameters α and β must be estimated. Similarly, a ground water table may fluctuate between the impervious base and the ground surface, with two well defined boundaries. Because the probabilities of the levels near the empty aquifer and near the overspilling aquifer to free surface are small, the use of B_1-distribution for the ground water table may be justified in many cases.

7.8. Extreme Value Distributions.

Sampling extremes. A discrete series of size N, or a continuous series of the length T, are given as samples of a random variable x. These samples are divided into n subsamples each of size m, so that $N = nm$, or $T = nm$. From each of these subsamples, a selection of variable values is made according to a given property of the subsamples. These selected subsample statistics may be either a certain magnitude of the variable, a property, a cumulative value, or a variable value having a certain order of magnitude. By such a selection a new random variable, y, with its new population is formed.

Selections according to the largest, smallest, or central values in each subsample are of particular interest and importance in hydrology. The largest and the smallest value are commonly referred to as extreme values, and are often associated with floods and droughts, surplus or deficit, and similar economic and safety connotations. Many water resources problems require random variables to refer to subsamples of a given size, such as the minimum 10-day flow in a season, the maximum one-month flood volume in a year, and similar statistics. One can select a large number of these random variables derived from the basic series of a random variable x.

Sometimes, the selection rule is such that values greater than, or smaller than, a particular x value are selected from a series. If only peak flood discharges are selected from each separate flood hydrograph, which are greater than a given value x_0, and all occur in a subsample, then 0, 1, 2, . . . , values result from each subsample. In this sampling, the total number of selected values may be greater than, equal to, or smaller than the number n of the subsamples. New samples selected in such a way are often called the partial series in hydrology. For partial series the concept of subsamples is not necessary, except for the number of events to occur in the subsamples.

In this chapter, only selections referring to extreme values of subsamples are considered. Thus, from the sample of size $N = nm$, n subsamples each of size m provide a new sample of size n, consisting of the largest and/or smallest value in each subsample. For further analysis the n extreme values selected are arranged in order of magnitude, largest to smallest or smallest to largest.

Asymptotic distributions of extreme values. A useful concept in analyzing extreme values is that of the number of exceedences, defined as the number of variable values that exceed a certain magnitude. This definition is analogous to that of the absolute cumulative frequency. The general feeling that prevails among practitioners is that the variance of extreme values is so large as to limit their applications. R. A. Fisher, R. von Mises, and E. J. Gumbel, have shown, however, that under certain conditions the variances of extremes are less than those of some central values. Extremes may be more stable, from one subsample to another, than some central values. To gain insight into the stability of extremes, it is necessary to consider distributions of the number of exceedences. The reader is referred to the book Statistics of Extremes by E. J. Gumbel, for a detailed analysis of these distributions [24].

According to R. von Mises, a continuous distribution may be either bounded or unbounded in either direction. If the distribution is unbounded, the moments may or may not exist. Thus the distributions may be classified as (a) distributions which are unbounded and where all moments exist, (b) distributions which are unbounded but where only a finite number of moments exist, and (c) bounded distributions where all moments exist. With respect to each class of distribution, the distributions of the extremes tend to the same asymptotic form as the sample size N tends to infinity. These asymptotic distributions are called: (1) exponential-initial distributions, unbounded and all moments exist; (2) Cauchy-intial distributions, unbounded but only a finite number of moments exist, and (3) limited-initial distributions, bounded.

If the largest and smallest extremes are considered, their asymptotic probability distribution functions for each of these three types are given in Table 7.1.

Table 7.1
Asymptotic Distributions of Extreme Values

Type	Largest Value	Conditions	Smallest Value	Conditions
Exponential	$F(x) = \exp[-e^{-\alpha(x-\beta)}]$	$\alpha > 0$	$F(x) = \exp[-e^{-\alpha(x-\beta)}]$	$\alpha > 0$
Cauchy	$F(x) = \exp\left[-\left(\frac{\beta}{x}\right)^{\alpha}\right]$	$x \geq 0$; $\alpha > 0, \beta > 0$	$F(x) = \exp\left[-\left(\frac{\beta}{x}\right)^{\alpha}\right]$	$x \leq 0$; $\alpha > 0, \beta < 0$
Limited	$F(x) = \exp\left[-\left(\frac{\gamma-x}{\gamma-\beta}\right)^{\alpha}\right]$	$x \leq \gamma$; $\alpha > 0, \beta < \gamma$	$F(x) = \exp\left[-\left(\frac{x-\gamma}{\beta-\gamma}\right)^{\alpha}\right]$	$x \geq \gamma$; $\alpha > 0, \beta > \gamma$

Asymptotic distributions of extreme values in hydrology. Gumbel has shown in several works that the exponential probability function of largest value fits asymptotically the distributions of annual flood peak discharge, while the limited probability function of smallest value fits asymptotically the distributions of annual lowest discharge. These two functions are discussed in subsequent text. As the selection refers to discrete ordered values, the extremes are the highest and the smallest value in this ordered set. If a hydrologic variable is unbounded, then the first row functions of Table 7.1 fit asymptotically distributions of both extremes. The Cauchy function did not find an application in hydrology. Some practitioners try to use these asymptotic extreme value functions to cumulative variables such as 10-day smallest total flow in a year, one-month largest volume in a season, and others. This cannot be justified by the theory of exceedences. If one sums up ten values, in a sequence of small values, it means the sum of the smallest, second smallest, . . . , and the tenth value in this sequence of ten values, or something similar. According to the theory of exceedences, this is a sum of 10 variables each with a different distribution. The central limit theory implies that the distribution should converge asymptotically to normal rather than to distribution functions of Table 7.1. In a continuous series, the largest and the smallest instantaneous values of a subsample should be selected.

Functions in Table 7.1 are given in their cumulative forms as F (x) because they are mathematically of simpler forms than their f (x) functions. The selection of the subsample size is a matter of practical need or convention. The year is often selected in time series as the subsample for two reasons. First, the astronomical cycle of a year introduces the annual periodic movement into many hydrologic phenomena (precipitation, evaporation, runoff, ground water level, sediment transport, water quality variables, etc.), so that the subsample of the size of a year encompasses that period. Second, the same astronomical cycle has been reason for the establishment of the year as the basic time unit, particularly as the economic and social unit (one measures benefit, cost, risk, and similar water resources characteristics by referring to the year as a unit). Because of hydrologic seasonal variations, individual seasons or months may also be selected as subsamples. Sometimes the day, because of 24-hour periodicity, is the subsample. The selection of the subsample size for sequences other than time series is a matter of convenience. The parameters α, β, and γ of equations in Table 7.1 are functions of the subsample size.

It should be clear that the equations in Table 7.1 are asymptotic probability distributions for both the large samples and the large subsamples. If they are small, the deviations between the true and asymptotic distributions may be significant. This is often a point of controversy in applying distribution functions of extreme values in hydrology, because the sample and subsample sizes may be small. Even if the sample and subsample look large, such as N = 50 – 80 years of daily river flows as a sample, and 365 days as a subsample, the high time dependence of the series would reduce the effective sizes in comparison with an independent series. It is likely that there are as many independent daily flows in a year as there are independent storms, so that the effective sample and subsample sizes are greatly reduced.

Double exponential distribution function of the largest value. The double exponential distribution function of the largest value in its cumulative form, called in Table 7.1 the exponential of largest value, is defined as

$$F(x) = e^{-e^{-\alpha(x-\beta)}}, \qquad 7.47$$

in which F (x) is the probability distribution of the largest value x of subsamples, and α and β are parameters, with β the location (central value) and α the scale parameter. This function may be also written as $F(x) = \exp\{-\exp[-\alpha(x-\beta)]\}$. It was introduced for extreme values by R. A. Fisher and L. H. C. Tippett. The function is sometimes known in hydrology as the Gumbel function because of Gumbel's various studies of distributions of extreme flood peak discharge.

Equation 7.47 can be reduced to a function involving no parameters by means of the transformation

$$y = \alpha(x - \beta), \qquad 7.48$$

whereby

$$F(x) = \exp[-\exp(-y)] = e^{-e^{-y}}. \qquad 7.49$$

Properties of double exponential function. Equation 7.49 plays the same role among the family of double exponential curves as the standard normal does among the family of normal curves.

The first derivative, f (x), of Eq. 7.47 is

$$f(x) = \alpha \exp\{-\alpha(x-\beta) - \exp[-\alpha(x-\beta)]\}, \qquad 7.50$$

so that for the reduced variable y,

$$f(x) = \exp[-y - \exp(-y)] = e^{-y-e^{-y}}. \qquad 7.51$$

As x or y tend to $+\infty$ or $-\infty$, F (x) tends to 1 or 0, respectively. Hence, Eqs. 7.50 and 7.51 are unbounded probability density functions with the x-axis as asymptote. The second derivatives of Eq. 7.47 and Eq. 7.49 are zero at $x = \beta$ and $y = 0$, respectively, so that the mode is at $m = \beta$ for Eq. 7.47, and $m = 0$ for Eq. 7.49. For Eq. 7.47, the probability of the mode F (m) and the modal density f (m) are:

$$F(m) = F(\beta) = e^{-1} = 0.368;\ f(m) = f(\beta) = \alpha e^{-1} = 0.368\alpha. \qquad 7.52$$

The mean, the standard deviation, and the median are, respectively,

$$\mu = \beta + \frac{0.5772}{\alpha};\ \sigma = \frac{\pi}{\alpha\sqrt{6}} = \frac{1.281}{\alpha};\ M = \beta + \frac{0.3665}{\alpha}. \qquad 7.53$$

For the mean, μ, and the standard deviation, σ, given, the parameters α and β become

$$\beta = \mu - 0.450\sigma;\ \alpha = \frac{1.281}{\sigma}. \qquad 7.54$$

From the quartiles $F(x_1) = 0.25$ and $F(x_2) = 0.75$, the parameter α becomes

$$\frac{1}{\alpha} = 0.6359\,(x_2 - x_1). \qquad 7.55$$

This relation may be used as a first analytical estimation for α. The skewness and kurtosis coefficients are constants, or $\gamma_1 = 1.29857 \approx 1.3$, and $\gamma_2 = 5.4$. Since $\gamma_2 > 3.0$, the peakedness of Eq. 7.47 is different from the normal function.

For large values of cumulative frequency, $F(x) \geq 0.99$(or $y \geq 4.60$), Eq. 7.47 is well approximated by the function $\Phi(x) = e^{-y}$, which is the simple exponential function, and it can be considered as an asymptotic tail function of Eq. 7.47. By developing Eq. 7.49 in a power series and using $\Phi = e^{-y}$, then

$$F(x) = e^{-\Phi} = 1 - \Phi + \frac{\Phi^2}{2} + 0\,(\Phi^3). \qquad 7.56$$

By neglecting all terms of higher order than the 2nd, Eq. 7.49 is approximated for values of F (x) close to unity by

$$F(x) = 1 - e^{-y}. \qquad 7.57$$

Expressing the probabilities of very large values of x, not by values close to 1 but by values close to zero, then $F(x) = e^{-y} = \exp[-\alpha(x - \beta)]$. For F (x) replaced by P, this latter expression gives

$$x = \beta\,[1 - \log_{10}P/\alpha\beta \log_{10}e], \qquad 7.58$$

which gives x for a given probability of occurrence, P. The approximation of Eq. 7.58 is a theoretical justification of a simple expression for extreme flood peaks in the form of $x = \beta\,(1 + k \log_{10}P)$, given by Fuller [25].

Estimation of parameters of double exponential function. Taking twice the logarithm of Eq. 7.47,

$$\alpha(x - \beta) = - \ln[- \ln F(x)], \qquad 7.59$$

so that the double logarithmic scale $-\ln[-\ln F(x)]$ for $F(x)$ and the cartesian scale for x produce a straight line probability distribution plot. The graph paper with these scales is not commercially available, but many hydrologic offices have them, often under the name of extremal-probability graph paper, or Gumbel graph paper. This paper has the property that for $F(x) = 0.368$ it gives the mode $m = \beta$, or the estimate of β, and by using the two quartiles, x_2 and x_1, α can be estimated by Eq. 7.55.

The analytical estimation of α and β by the method of moments uses $\bar{x}$ and s as estimators of μ and σ in Eq. 7.54, from which then the estimations $\hat{\alpha}$ and $\hat{\beta}$, of α and β, are obtained. Because $\gamma_1 = 1.3$ approximately, the double exponential distribution is sufficiently skewed warranting the use of the maximum likelihood method in estimating α and β.

Applications in hydrology. An objection is often made to the application of the double exponential function to phenomena with a finite lower boundary because the function is unbounded. When this function is applied to positive-valued variables, such as flood peak discharge which cannot be less than zero, the function is not strictly applicable. Although floods have a lower limit of zero, to achieve a flood of zero in a given year requires that the year be dry from the beginning to the end. This is not unusual in some regions having arid climates. Nevertheless, most rivers are not of this type. If a region is very dry for several years and rivers have no runoff, the probability mass for discharge zero is finite, which may justify the use of truncated double exponential function. The part of the probability distribution for very large values close to the probability of unity are, however, of the greatest practical interest in the analysis of floods. In this case, the lower part of the probability distribution function loses practical importance, regardless of whether it is bounded or unbounded.

The cumulative frequency for variable values less than zero may be taken as a criterion for determining the applicability of the double exponential function. Suppose that the variable x is positive-valued (including zero) and this function is fitted. For $x = 0$, Eq. 7.47 gives $F(x = 0) = \exp(-e^{\alpha\beta})$, where $\alpha\beta = (\alpha/\eta) - 0.5772$. Two extreme conditions may arise: (1) $\eta = C_v = 0$, and (2) $\eta = \infty$. If $\eta = 0$, Eq. 7.54 gives $\alpha\beta = \infty$, and $F(x = 0) = 0$; and if $\eta = \infty$, $\alpha\beta = - 0.5772$ and $F(x = 0) = 0.168$. Thus it is seen that for the nonexistent values of x (negative tail of x), the maximum cumulative probability of $x \leq 0$ is about 0.17.

Other conditions for applying the double exponential function concern the constant values of skewness and kurtosis. If the skewness and kurtosis coefficients of data are estimated by moments, they can be compared with the theoretical constant values for the double exponential function. If the observed values are near to theoretical values, at least as close as may be expected because of sampling errors, the hypothesis is legitimate that the double exponential distribution may well fit the empirical frequency distributions of extreme values. This enables checking the applicability of the double exponential distribution to floods. By gathering records of floods of many rivers in a region, a set of skewness and kurtosis coefficients may be computed, and their respective means obtained. If these means are close to the respective theoretical values of 1.3 and 4.5, the double exponential function may be applicable to floods of that region.

Example 7.13. Figure 7.14 gives the distribution of flood peak discharge of the Licking River at Tobosco, Ohio, for a period of 26 years (1922–1947), in the graph with cartesian-extremal probability scales, with a visual straight-line fit of double exponential distribution function. The estimates of α and β of Eq. 7.47 from this graph are: m = 10,200 cfs as the estimate of β, and $x_1 = 8{,}100$ cfs and $x_2 = 18{,}400$ cfs, as the estimates of the two quartiles x_1 and x_2, from which Eq. 7.55 gives the estimate $\alpha = 0.000152$. The sample mean and standard deviation are x = 13,740 cfs and s = 7,000 cfs, as estimates of μ and σ. Then Eq. 7.54 produces m = 10,550 cfs, and $\alpha = 0.000183$. This shows that $1/\alpha$, rather than α, is a practical parameter to be used. These two estimates by the method of moments are somewhat different from the graphical estimates. As $C_s = 1.19$ and k = 6, the straight-line fit for N = 26 seems acceptable.

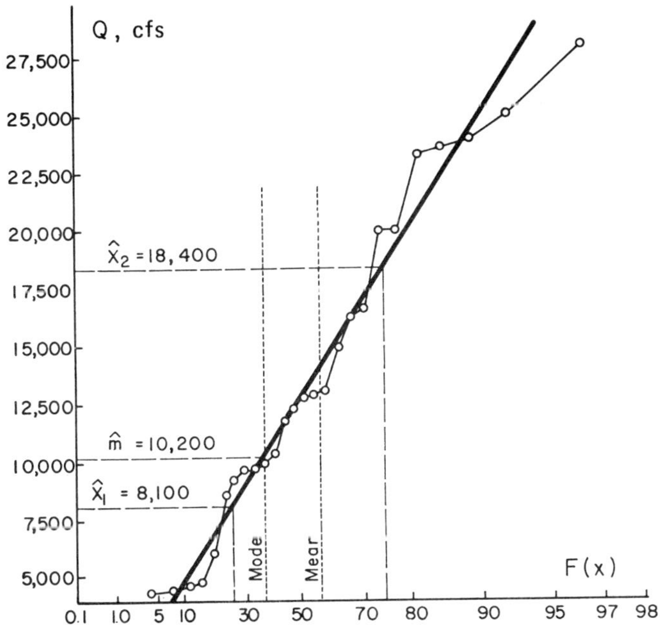

Figure 7.14. Fitting the double exponential probability function to the distribution of annual flood peak discharge of the Licking River at Tobosco, Ohio.

Example 7.14. Figure 4.5 represents the distribution of flood peak discharge of the Danube River at Orshova, for N = 110 years (1840–1950), both for the non-homogeneous and homogeneous series. It shows that the double exponential function fits well these empirical distributions, at least in the range of probabilities 5 to 95 percent.

Bounded exponential distribution of the smallest value. The bounded exponential distribution function of the smallest value is

$$F(x) = \exp\left[-\left(\frac{x-\gamma}{\beta-\gamma}\right)^{\alpha}\right], \tag{7.60}$$

in which F (x) is the probability distribution of the smallest value x of a given subsample, and α, β, and γ are parameters. It was also introduced by R. A. Fisher and L. H. C. Tippett, and was used and proposed by E. J. Gumbel for the analysis of the smallest discharge during a given time interval (generally a year, or a season as subsamples).

The application of this function requires four conditions: $x \geq \gamma$; $\alpha > 0$, $\beta > \gamma$, and $\gamma \geq 0$. Equation 7.60 may be transformed to a parameterless function by means of the transformation

$$y = \left(\frac{x-\gamma}{\beta-\gamma}\right)^{\alpha}, \tag{7.61}$$

which gives

$$F(x) = e^{-y}. \tag{7.62}$$

This is the simple exponential probability function.

The first derivative of Eq. 7.60 gives

$$f(x) = \frac{\alpha}{\beta-\gamma}\left(\frac{x-\gamma}{\beta-\gamma}\right)^{\alpha-1} \exp\left[-\left(\frac{x-\gamma}{\beta-\gamma}\right)^{\alpha}\right] = -\frac{\alpha}{\beta-\gamma} y^{(\alpha-1)/\alpha} e^{-y}. \tag{7.63}$$

Properties of bounded exponential distribution. From Eq. 7.63 it is seen that as $(x-\gamma)$ tends to zero, f (x) also tends to zero, and as $(x - \gamma)$ tends to infinity, f (x) tends to zero. From Eq. 7.60 it follows that for $x = \gamma$, or the smallest value of x, $F(x) = 1$, while for $x \to \infty$, $F(x) \to 0$, so that F (x) gives probabilities of $X \geq x$. This exponential function of the smallest value is bounded to the left and unbounded to the right, with x in the range γ to $+\infty$. The parameter γ is the lower boundary, while β is the location or central value parameter, and α is the scale parameter. For $x = \beta$, $F(x) = e^{-1} = 0.368$, so that the values $x \geq \beta$ have the probability 0.368. The median, M, and mode, m, are:

$$M = \gamma + \beta(\ln 2)^{1/\alpha} \tag{7.64}$$

and

$$m = \gamma + \beta\left(1 - \frac{1}{\alpha}\right)^{1/\alpha}. \tag{7.65}$$

A mode exists only if $\alpha > 1$, and $m < M$ if $(1/\alpha) > 0.30685$, $M = m$ if $(1/\alpha) = 0.30685$, and $M < m$ if $(1/\alpha) < 0.30685$. When $(1/\alpha) = 0.30685$, the function is nearly symmetrical but does not approach the normal function.

Estimation of parameters of bounded exponential distribution. Equation 7.60 gives

$$\ln[-\ln F(x)] = \alpha \ln(x - \gamma) - \alpha \ln(\beta - \gamma), \qquad 7.66$$

so that only $(x - \gamma)$ versus $F(x)$ plotted in graphs with logaritmic-extremal probability (or log-double log) scales becomes a straight line. In general $\gamma > 0$, so that it is unknown and must be estimated from sample data. Only for $\gamma = 0$ (the lower boundary is zero) does x and F (x) follow a straight line if plotted in log-extremal probability scales (or log-double log scales). In this case, the graphical estimates of α and β are obtained by using Eqs. 7.64 and 7.65. For $F(x) = 0.50$ and $F(x) = 0.368$, the fitted straight line gives the estimates of M and m, respectively. For $\gamma = 0$, Eqs. 7.64 and 7.65 give the estimates $\hat{\alpha}$ and $\hat{\beta}$ of α and β. The only objection to this procedure is the closeness of 0.50 and 0.368 on the F (x) scale, so that small errors in reading M and m affect the accuracy of results.

For $\gamma > 0$, the method of moments gives estimates of α, β, and γ. The following equations enable this estimation

$$\beta = \mu + \sigma A(\alpha); \text{ and } \gamma = \beta - \sigma B(\alpha), \qquad 7.67$$

where

$$A(\alpha) = \left[1 - \Gamma\left(1 + \frac{1}{\alpha}\right)\right] B(\alpha), \qquad 7.68$$

$$B(\alpha) = \left[\Gamma\left(1 + \frac{2}{\alpha}\right) - \Gamma^2\left(1 + \frac{1}{\alpha}\right)\right]^{-1/2}, \qquad 7.69$$

and

$$\gamma_1 = C_s = \left[\Gamma\left(1 + \frac{3}{\alpha}\right) - 3\Gamma\left(1 + \frac{2}{\alpha}\right)\Gamma\left(1 + \frac{1}{\alpha}\right) + \right.$$

$$\left. + 2\Gamma^3\left(1 + \frac{1}{\alpha}\right)\right] \cdot B^3(\alpha). \qquad 7.70$$

For the computed skewness coefficient, C_s, of the sample as the estimate of γ_1, Eq. 7.70 gives the estimate $\hat{\alpha}$ of α. With it $B(\hat{\alpha})$ is determined, and then $A(\hat{\alpha})$. For the sample estimates $\bar{x}$ and $\hat{s}$, of μ and σ, Eq. 7.67 gives first the estimate $\hat{\beta}$ of β, by substituting $\bar{x}$, $\hat{s}$, and $A(\hat{\alpha})$ for μ, σ and $A(\alpha)$, and second $\hat{\gamma}$ as the estimate of γ by using $\hat{\beta}$, $\hat{s}$, and $B(\hat{\alpha})$. The easiest way to do computations is by using tables [24, pp. 282—284] of $A(\alpha)$, $B(\alpha)$, and $C_s(\alpha)$ against $1/\alpha$, or by using the corresponding curves. Figure 7.15 gives these three curves: $A(\alpha)$, $B(\alpha)$, and $C_s(\alpha)$. The value $A(\alpha) = (\beta - \mu)/\sigma$ may be called the standardized difference of the central value β from the mean μ, and the value $B(\alpha) = (\beta - \gamma)/\sigma$ the standardized difference of the central value β from the lower boundary. This requires $\gamma \geq 0$ for $\mu/\sigma > [B(\alpha) - A(\alpha)]$. For $\alpha = 1$, the simple exponential function has $\beta = \mu$ and $\gamma = \mu - \sigma$.

Example 7.15. Figure 7.16 gives the example of the empirical distribution curve of annual lowest daily discharge of the Kings River at Piedra, California, for N = 60

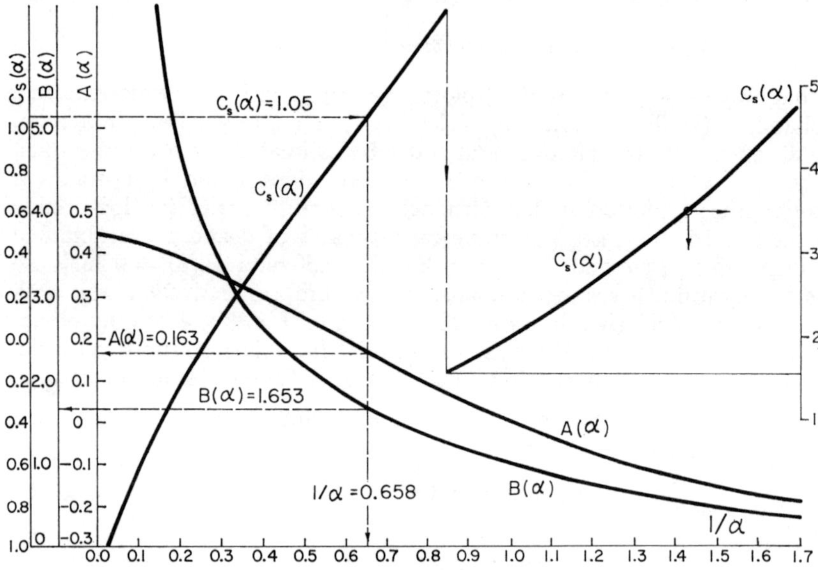

Figure 7.15. The functions A (α), B (α), and Cs (α) for the estimation of parameters of bounded exponential distribution function of smallest values.

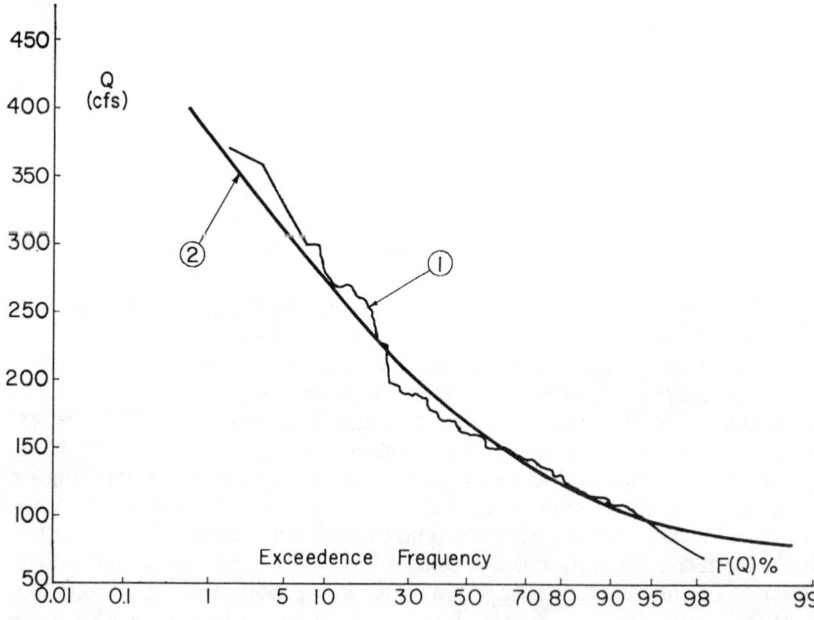

Figure 7.16. Fitting the bounded exponential distribution function of smallest values to the empirical distribution of annual one-day lowest discharge of the Kings River at Piedra, California.

years (1897–1956), with the lower boundary different from zero. The curve is plotted on the graph with log-extremal probability scales, with the plotting position $F(x) = m/(N+1)$. The sample estimates are $\bar{x} = 180$ cfs, $\hat{s} = 68$ cfs, and $C_s = 1.05$. Figure 7.15 for $C_s = 1.05$ gives $1/\hat{\alpha} = 0.658$, $\hat{\alpha} = 1.52$, $A(\hat{\alpha}) = 0.163$, and $B(\hat{\alpha}) = 1.653$. By using Eq. 7.67, $\hat{\beta} = 191.0$ cfs and $\hat{\gamma} = 79.5$ cfs. The distribution function of the low-flow discharge is then

$$F(Q) = \exp\left[-\left(\frac{Q-79.5}{111.5}\right)^{1.52}\right].$$

The observed lowest points in Fig. 7.16 are somewhat smaller than the estimated lower boundary, $\hat{\gamma} = 79.5$ cfs, which is a disadvantage of fitting this probability function to empirical distributions. This example and Fig. 7.16 clearly show that the visual straight-line fit would be highly in error, and that forcing the lower boundary to be zero, as the likely lowest flow discharge in a very long period of observations, should be tested first in log-double log scales for a straight-line fit.

7.9. Other Univariate Distribution Functions, and Concluding Remarks.

Discrete distribution functions. Apart from the binomial and Poisson discrete distribution functions, some other discrete distributions may be used in hydrology under special conditions. The hypergeometric distribution applies to samples obtained from a finite population without replacement. A finite population is N, and there are m items of type A. All items of the population are equally likely to occur on the next observation or draw. For a sample of size n, the value x is the number of events of type A, and the distribution of x is called hypergeometric. It is

$$p(x) = \frac{{}_mC_x \; {}_{N-m}C_{n-x}}{{}_NC_n} = \frac{\binom{m}{x}\binom{N-m}{n-x}}{\binom{N}{n}}, \qquad 7.71$$

with ${}_mC_x$, ${}_{N-n}C_{N-x}$, and ${}_NC_n$, defined by Eq. 3.4, and $x = 0, 1, 2, \ldots, [n,m]$, or whichever is smaller of n or m. The three parameters are N, m, n, and $Ex = nm/N$, and $\operatorname{var} x = nm(N-m)\cdot(N-n)/N^2(N-1)$. For various applications see the literature [26, p. 59–66]. For $N \to \infty$, Eq. 7.71 converges to binomial distribution. For $n < 0.1\,N$, the binomial approximation is satisfactory.

The geometric distribution gives the number of trials, x, that are necessary for an event, A, of the constant probability, p, to occur for the first time in a sequence of Bernoulli trials, or

$$p(x) = p(1-p)^{x-1}; \; F(x) = \sum_{i=1}^{x} p(i) = 1-(1-p)^x, \qquad 7.72$$

in which $x = 0, 1, 2, \ldots$. This means that $x - 1$ other events must occur before the event A occurs on the x-th trial [26, p. 66–67]. For this distribution, $Ex = (1-p)/p$, and $\operatorname{var} x = (1-p)/p^2$.

The Pascal distribution function gives the probabilities of the number x of Bernoulli trials of an event A, with the constant probability p of its occurrence on any trial or observation, up to and including the r-th occurrence of the event A, or

$$p(x) = {}_{x-1}C_{r-1} \cdot p^r (1-p)^{x-r} = \frac{(x-1)!}{(r-1)!\,(x-r)!}\, p^r (1-p)^{x-r}, \qquad 7.73$$

in which $x = r, r+1, \ldots$. For $r = 1$, this distribution reduces to a geometric probability mass function.

Continuous distribution functions. Several other continuous distribution functions may be used in hydrology under appropriate conditions. The Rayleigh distribution with parameter $\alpha > 0$ is given by

$$f(x) = \frac{1}{\alpha^2}\, x\, e^{-x^2/2\alpha^2}, \text{ for } x > 0, \qquad 7.74$$

and zero for $x < 0$.

The Maxwell distribution with parameter $\alpha > 0$ is given by

$$f(x) = \frac{4}{\sqrt{2\pi}\,\alpha^3}\, x^2 e^{-x^2/\alpha^2}, \text{ for } x > 0, \qquad 7.75$$

and zero for $x < 0$. It can be shown that these two distributions are special cases of the χ^2-distribution which is given and discussed in Chapter 9.

The uniform or rectangular distribution is given by

$$F(x) = \frac{x-a}{b-a}, \text{ or } f(x) = \frac{1}{b-a}, \text{ for } a \leq x \leq b, \qquad 7.76$$

and zeros for $x < a$ and $x > b$.

If the derivative of a probability density function is expressed as

$$\frac{df(x)}{dx} = \frac{f(x)(x+a)}{F(x)}, \qquad 7.77$$

and the function $F(x)$ expanded in Maclaurin's power series, then

$$\frac{df(x)}{dx} = \frac{f(x)(x+a)}{b_0 + b_1 x + b_2 x^2 + \ldots} \qquad 7.78$$

represents the general expression for the study of Karl Pearson's distribution functions. Assuming some of parameters, a, b_0, b_1, b_2, , . . , being different from zero, various Pearson's functions are obtained. The study of these functions is outside the scope of this text; the reader is referred to the literature [4, p. 148—154 and p. 175].

For fitting distributions by a flexible type of series, the Gram-Charlier series of Type A may be used [4, pp. 156—157], or the Edgeworth's form of the Type A series [4, pp. 157—160]. To better fit empirical functions, sometimes the distributions of variables, which are functions of random variables having simple distribution functions, are used. An example has alredy been shown, Example 2.32, with the Pearson Type III function applied to logarithms of variable values. Several such functions are known under the names of authors who applied them to particular cases.

Concluding remarks on univariate probability distribution functions. Table 7.2 gives several probability functions, discussed previously, with the summary of their various characteristics. The table also includes three probability functions used for statistical inferences: chi--square ($\chi^2 = \xi$), Student-t, and Fisher-z distributions, to be discussed in detail in Chapter 9. The table is intended for rapid access to properties of various functions currently used in hydrology.

Using computers in hydrologic statistical analysis is another reason for giving particular attention to fitting probability functions to empirical distributions. With the digital computer replacing manual computations and desk calculators, the input of statistical information into a computer is simplest in the form of probability distribution functions. The sampling variation along an empirical distribution does not justify its detailed reproduction. The fit of a probability function is equivalent to smoothing an empirical distribution. Experience shows that a continuous increase of the sample size slowly removes these small variations on an empirical distribution, making it progressively smoother.

Information on physical properties of a random variable and its probability distribution function. Two approaches are feasible in obtaining or selecting the most reliable probability functions for fitting distributions of hydrologic variables: (a) the use of observed statistical data only by selecting probability functions of the best fit, often supported by experience from previous fits, and (b) the use of additional information on the physical properties related to a hydrologic random variable by determining the probability function that the variable should follow.

A random variable may be normally distributed in its population while the sample distribution could be highly asymmetrical. If it is known from the physical properties of the variable that a normal distribution should be the population distribution, it is then best to fit it to the empirical distribution, regardless of its asymmetry. The majority of probability functions previously discussed have particular conditions under which they apply. It is often difficult, however, to determine the physical properties of a hydrologic process to ascertain whether particular types of probability functions are applicable to the empirical distribution of its variables.

If a variable is the effect of a very large number of additive causative factors, each being of a small effect, and these small effects are mutually independent, as already stated, the probability distribution should be normal. Whenever hydrologic conditions satisfy these requirements, the normal distribution should be used regardless of empirical distributions.

TABLE 7.2

DISTRIBUTION FUNCTION	PROBABILITY DISTRIBUTION, OR PROBABILITY DENSITY, OR MASS FUNCTION	RANGE OF VARIABLE VALUES	PARAMETERS OF THE FUNCTION	MEAN μ	VARIANCE σ^2	3rd CENTRAL MOMENT μ_3
BINOMIAL	$f(x) = \binom{n}{x} p^x q^{n-x}$ $q = 1 - p,$	Integers $0 \leq x \leq n$	n, p	n p	n p q	npq (1−2p)
POISSON	$f(x) = \frac{\lambda^x e^{-\lambda}}{x!}$	Integers $0 \leq x \leq \infty$	λ	λ	λ	λ
STANDARD NORMAL	$f(t) = \frac{1}{\sqrt{2\pi}} e^{-t^2/2}$	$-\infty$ to $+\infty$		0	1	0
NORMAL	$f(x) = \frac{1}{\sigma\sqrt{2\pi}} e^{-\frac{(x-\mu)^2}{2\sigma^2}}$	$-\infty$ to $+\infty$	μ, σ	μ	σ^2	0
LOGNORMAL 2-PARAMETER	$f(x) = \frac{1}{x\sigma_n\sqrt{2\pi}} e^{-\frac{(\ln x - \mu_n)^2}{2\sigma_n^2}}$	0 to $+\infty$	μ_n, σ_n	$e^{\mu_n + \sigma_n^2/2}$	$\mu^2 (e^{\sigma_n^2} - 1)$	$\mu^3(C_v^6 + 3C_v^4)$
GAMMA 1-PARAMETER	$f(x) = \frac{x^{\alpha-1} e^{-x}}{\Gamma(\alpha)}$	0 to $+\infty$	α α = Shape Parameter $\alpha > 0$	α	α	2α
GAMMA 2-PARAMETER	$f(x) = \frac{x^{\alpha-1} e^{-x/\beta}}{\beta^\alpha \Gamma(\alpha)}$	0 to $+\infty$	α, β α = Shape Parameter, $\alpha > 0$ β = Scale Parameter, $\beta > 0$	$\alpha\beta$	$\alpha\beta^2$	$2\alpha\beta^2$
GAMMA 3-PARAMETER	$f(x) = \frac{1}{\beta\Gamma(\alpha)}\left(\frac{x-\gamma}{\beta}\right)^{\alpha-1} e^{-\left(\frac{x-\gamma}{\beta}\right)}$	Bounded on Left by γ	α, β, γ α = Shape Parameter β = Scale Parameter γ = Lower Boundary	$\gamma + \alpha\beta$	$\alpha\beta^2$	$2\alpha\beta^2$
BETA I 2-PARAMETER	$f(x) = \frac{x^{\alpha-1}(1-x)^{\beta-1}}{B_1(\alpha, \beta)}$ $0 \leq x \leq 1$	0 to 1	α, β $\alpha > 0$; $\beta > 0$	$\frac{\alpha}{\alpha+\beta}$	$\frac{\alpha\beta}{(\alpha+\beta)^2(\alpha+\beta+1)}$	
BETA II 2-PARAMETER	$f(y) = \frac{x^{\alpha-1}(1+x)^{-(\alpha+\beta)}}{B_2(\alpha, \beta)}$ $0 \leq x \leq \infty$	0 to $+\infty$	α, β $\alpha > 0$; $\beta > 0$	$\frac{\alpha}{\beta - 1}$ for $\beta > 1$	$\frac{\alpha(\alpha+\beta-1)}{(\beta-1)^2(\beta-2)}$ for $\beta > 2$	
DOUBLE EXPONENTIAL 2-PARAMETER	$F(x) = e^{-e^{-\alpha(x-\beta)}}$ for $y = \alpha(x-\beta)$; $F(x) = e^{-e^{-y}}$	$-\infty$ to $+\infty$	α, β α = Shape Parameter β = Scale Parameter	$\beta + \frac{0.5772}{\alpha}$	$\frac{\pi^2}{6\alpha^2}$	
BOUNDED EXPONENTIAL 3-PARAMETER	$F(x) = e^{-\left(\frac{x-\gamma}{\beta-\gamma}\right)^\alpha}$	Bounded on Left by γ, Unbounded on Right	α, β γ $\quad x \geq \gamma$ β = Location Param. $\beta > \gamma$ α = Scale Param. $\alpha > 0$ γ = Location Param. $\gamma \geq 0$	$\gamma + (\beta-\gamma) \cdot \Gamma\left(1 + \frac{1}{\alpha}\right)$	$(\beta-\gamma)^2\left[\Gamma\left(1 + \frac{2}{\alpha}\right) - \Gamma^2\left(1+\frac{1}{\alpha}\right)\right]$	
CHI-SQUARE $\chi^2 = \xi$	$f(\xi) = \frac{1}{2^{n/2}\Gamma(n/2)} \xi^{\frac{n}{2}-1} e^{-\frac{\xi}{2}}$	0 to $+\infty$	n n = Degrees of Freedom	n	2n	8n
STUDENT-t	$f(t) = \frac{1}{\sqrt{n\pi}} \frac{\Gamma\left(\frac{n+1}{2}\right)}{\Gamma(n/2)} \left(1 + \frac{t^2}{n}\right)^{-\frac{n+1}{2}}$	$-\infty$ to $+\infty$	n n = Degrees of Freedom	0	$\frac{n}{n-2}$	0
FISHER-z	$f(z) = \frac{2m^{m/2} n^{n/2}}{B\left(\frac{m}{2}, \frac{n}{2}\right)} \frac{e^{mz}}{(me^{2z} + n)^{(m+n)/2}}$	$-\infty$ to $+\infty$	m, n	$\frac{n}{n-2}$ for $n > 2$	$\frac{2n^2(m+n-2)}{m(n-2)^2(n-4)}$ for $n > 4$	

TABLE 7.2 (continued)

DISTRIBUTION FUNCTION	4 th CENTRAL MOMENT μ_4	SKEWNESS $\frac{\mu_3}{\sigma^3}$ γ_1 or C_s	EXCESS $\frac{\mu_4}{\sigma^4}-3$ $\gamma_2 - 3$, or E	MODE m, and p (m)	MEDIAN M, and p (M)	REMARKS
BINOMIAL	$3n^2p^2q^2 +$ $+ npq(1-6pq)$	$\frac{1-2p}{\sqrt{npq}}$	$\frac{1-6pq}{npq}$			p and n must be known and $\binom{n}{x} = \frac{n!}{x!\,(n-x))!}$ Events must be independent.
POISSON	$\lambda(3\lambda+1)$	$\frac{1}{\sqrt{\lambda}}$	$\frac{1}{\lambda}$	Mode is Largest Integer Less Than Or Equal to λ		Skewness is always positive. $\lambda = pn$, and $\lambda > 0$, must be known. Events must be independent.
STANDARD NORMAL	3	0	0	$m = 0$ $p(m) = \frac{1}{\sqrt{2\pi}}$	$M = 0$ $p(M) = \frac{1}{\sqrt{2\pi}}$	Mean is zero, and variance in unity. Standardization by $t = (x-\mu)/\sigma$.
NORMAL	$3\sigma^4$	0	0	$m = \mu$ $p(m) = \frac{1}{\sigma\sqrt{2\pi}}$	$M = \mu$ $p(M) = \frac{1}{\sigma\sqrt{2\pi}}$	μ = any real number, σ = any positive number, a change in μ shifts curve horizontally; a change in σ raises or lowers the peak.
LOGNORMAL 2-PARAMETER	$\mu^4(C_v^{12}+6C_v^{10}+$ $+15C_v^8+$ $+16C_v^6+3C_v^4)$	$C_v^3 + 3C_v$	$C_v^8+6C_v^6+$ $+15C_v^4+$ $+16C_v^2$	$m = \frac{\mu}{(1+C_v^2)^{3/2}}$ $p(m) = \frac{(1+C_v^2)^{1/2}}{\mu\sqrt{2\pi\ln(1+C_v^2)}}$	$M = \frac{\mu}{\sqrt{1+C_v^2}} = e^{\mu_n}$ $p(M) = \frac{\sqrt{1+C_v^2}}{\mu\sqrt{2\pi\ln(1+C_v^2)}}$	$\infty \le \mu_n \le \infty$; $0 < \sigma_n \le \infty$. μ_n = mean of ln x; σ_n = standard deviation of ln x. Skewness is always positive.
GAMMA 1-PARAMETER	$3\alpha(\alpha+2)$	$\frac{2}{\sqrt{\alpha}}$	$\frac{6}{\alpha}$	$m = \alpha - 1$ for $\alpha > 1$		Definition of Γ-function: $\Gamma(\alpha) = \int_0^\infty x^{\alpha-1}e^{-x}dx$
GAMMA 2-PARAMETER	$3\alpha\beta^4(\alpha+2)$	$\frac{2}{\sqrt{\alpha}}$	$\frac{6}{\alpha}$	$m = \beta^2(\alpha-1)$ for $\alpha > 1$		
GAMMA 3-PARAMETER	$3\alpha\beta^4(\alpha+2)$	$\frac{2}{\sqrt{\alpha}}$	$\frac{6}{\alpha}$			$f(x) = \frac{1}{\beta\Gamma(\alpha)}\left(\frac{x}{\beta}\right)^{\alpha-1}e^{-x/\beta}$ for $\gamma = 0$; $\gamma = \mu - \sigma\sqrt{\alpha}$; $\beta = \sigma/\sqrt{\alpha}$; $\alpha = 4/C_s^2$
BETA I 2-PARAMETER				Mode is: $=(\alpha-1)(\alpha+\beta-2)$ for $\alpha > 0$, $\beta > 0$		Definition of B-function $B(\alpha,\beta) = \int_0^1 x^{\alpha-1}(1-x)^{\beta-1}dx = \frac{\Gamma(\alpha)\Gamma(\beta)}{\Gamma(\alpha+\beta)}$ β in this case is parameter.
BETA II 2-PARAMETER				Mode is: $m = \frac{\alpha-1}{\beta-1}$ for $\alpha > 1$		β in this case is parameter.
DOUBLE EXPONENTIAL 2-PARAMETER		Constant 1.29857 ≈ ≈ 1.3	Constant 1.500	$m = \mu - 0.5772\alpha^{-1}$ $f(m) = \alpha e^{-1} = 0.368\alpha$ $F(m) = e^{-1} = 0.368$	$M = \frac{\beta + 0.3665}{\alpha}$	$\sigma = \frac{\pi}{\alpha\sqrt{6}}$; $\beta = \mu - 0.450\sigma$; $\alpha = 1.281\sigma^{-1}$; $f(x) =$ $= \alpha\exp[-\alpha(x-\beta)e^{-\alpha(x-\beta)}]$.
BOUNDED EXPONENTIAL 3-PARAMETER				$m = \gamma + \beta\left(1-\frac{1}{\alpha}\right)^{1/\alpha}$ for $1/\alpha < 1$	$M = \gamma + \beta(\ln 2)^{1/\alpha}$ for $\frac{1}{\alpha} < 1$	$m < M$ for $1/\alpha > 0.30685$ $m = M$ for $1/\alpha = 0.30685$ $m > M$ for $1/\alpha < 0.30685$ for $x = \beta$; $F(\beta) = e^{-1} = 0.3678$
CHI-SQUARE $\chi^2 = \xi$	$12n(n+4)$	$\sqrt{\frac{8}{n}}$	$\frac{12}{n}$			$\chi^2 = \sum_1^n \frac{(x_i-\mu)^2}{\sigma^2} = \xi$
STUDENT-t		0		$m = \mu = 0$	$M = \mu = 0$	As $n \to \infty$ distribution converges to normal distribution.
FISHER-z				$m = \frac{m-2}{m}\;\frac{n}{n+2}$ for $m > 0$		This function is independent of σ. $z = \frac{1}{2}\ln F$; F = Fisher's statistics = ratio of variance estimates for two samples, with m and n degrees of freedom, respectively.

If logarithms of a variable satisfy these conditions, the lognormal probability distribution should be used.

If causative factors are random variables, and are approximately normally distributed, though this condition may not be necessary, and if their individual effects on a random variable are proportional to squares of causative factors, and additive, the gamma distribution should be used. Because there are several hydraulic and other laws under which a random variable is a quadratic function of another random variable, or it is a sum of these quadratic functions, it is justified to fit the gamma distribution to empirical distributions of many hydrologic variables.

It is often rightfully claimed by hydrologists that the distribution of extreme values (highest or lowest) of a sample of size N of the variable x is a function of the underlying distribution, F (x), and the sample size, N. These exact distributions are discussed in Chapter 9 in the text dealing with sampling distributions. It should not be forgotten, however, that the population distributions are relevant for the study of extreme value distributions, and that the sample size N should go to infinity to find these asymptotic distributions. Therefore, if extreme high values are independent events, like the largest annual flood discharges, the double exponential distribution is applicable as its asymptotic distribution. Similarly, for extreme low values, the bounded exponential function is applicable as the asymptotic distribution.

In conclusion, the hydrologist can add significant information when selecting the appropriate probability function to fit empirical distributions by studying types of cause-effect relations of hydrologic processes. If a random variable must be distributed according to its predetermined population probability function, a good deal of information is already added from the physical information. This approach represents a combination of statistical and physical information, which promises some significant improvements in obtaining hydrologic information beyond the simple treatments of statistical data.

Example 7.16. Random errors of repeated measurements of a magnitude under proper operational procedures may be shown to satisfy conditions of the normal distribution. If one is sure that the procedure of measurements in hydrology does not change, and does not introduce systematic errors, it is safe to assume random errors to be approximately normally distributed.

Example 7.17. The distribution of the grain size of large samples of river bedload is often close to a straight line on graphs with log-probability scales. If it can be shown that the effects of causative factors are multiplicative for a particular process of sediment transport (or the logarithms of causative factors are additive), the grain size should follow the lognormal distribution.

Example 7.18. Experiments [27] by the Monte Carlo method, generating very large samples of hydrologic data, show that the double exponential function is approximately applicable in hydrology under a variety of conditions, rather than just to normal independent variables. These conditions include the asymmetrical independent variables, normal variables with the linear autoregressive dependence, and normal independent variables with added periodic components.

7.10 Distribution Functions for Bivariates, Multivariates, and Conditional Variables.

Normal bivariate distribution function. For a bivariate (x,y) with the simple correlation coefficient, ρ (x,y), of components x and y, the bivariate normal probability density function is

$$f(x,y) = \frac{1}{2\pi\sigma_x\sigma_y\sqrt{1-\rho^2}} \exp\left\{-\frac{1}{2(1-\rho^2)}\left[\frac{(x-\mu_x)^2}{\sigma_x^2} + \frac{(y-\mu_y)^2}{\sigma_y^2} - 2\rho\frac{(x-\mu_x)(y-\mu_y)}{\sigma_x\sigma_y}\right]\right\} = \frac{e^{-Q/2(1-\rho^2)}}{2\pi\sigma_x\sigma_y\sqrt{1-\rho^2}}, \quad 7.79$$

in which μ_x and σ_x are the mean and the standard deviation of x, and μ_y and σ_y of y, and Q is the term in brackets [ ]. The area under the surface is unity. The marginal distributions of x and y are, respectively,

$$f(x) = \frac{1}{\sigma_x\sqrt{2\pi}} e^{-(x-\mu_x)^2/2\sigma_x^2}; \qquad f(y) = \frac{1}{\sigma_y\sqrt{2\pi}} e^{-(y-\mu_y)^2/2\sigma_y^2}; \quad 7.80$$

they are obtained by integrating Eq. 7.79 in respect to y from $-\infty$ to $+\infty$ for the first marginal distribution of Eq. 7.80, and in respect to x, in the same limits, for the second marginal distribution of Eq. 7.80. The density function of the conditional variable X | Y is

$$f(x \mid y) = \frac{f(x,y)}{f(y)} = \frac{1}{\sigma_x\sqrt{2\pi(1-\rho^2)}} e^{-Q/2(1-\rho^2)+(y-\mu_y)^2/2\sigma_y^2}, \quad 7.81$$

and similarly for the conditional variable Y | X by interchanging x and y in Eq. 7.81. The expected value of X | Y is $E(x \mid y) = \rho\sigma_x/\sigma_y$, and its variance $D(x \mid y) = \sigma_x^2(1-\rho^2)$; similarly for Y | X, the expected value and the variance are $E(y \mid x) = \rho\sigma_y/\sigma_x$ and $D(y \mid x) = \sigma_y^2(1-\rho^2)$, respectively.

If $f(x,y) = f(x) \cdot f(y)$, the two variables x and y are said to be independent, and ρ in Eq. 7.79 is zero. It is not sufficient for the independence of x and y that f (x,y) can be factored, but the two factors must also equal the respective marginal density functions. It will be shown in Chapter 11 that the two variables x and y may be dependent even if they are linearly uncorrelated.

Lognormal bivariate distribution function. It is easy to apply Eq. 7.79 to the lognormal case in replacing x by ln x, y by ln y, μ_x and σ_x by $\mu_n(x)$, and $\sigma_n(x)$, and μ_y and σ_y by $\mu_n(y)$ and $\sigma_n(y)$, or by the means and standard deviations of logarithms, and in further dividing Eq. 7.79 by xy, so that

$$f(x,y) = \frac{1}{2\pi xy\,\sigma_n(x)\,\sigma_n(y)\sqrt{1-\rho_n^2}} e^{-Q_n/2(1-\rho_n^2)}, \quad 7.82$$

in which f (x,y) is bivariate lognormal density function, Q_n is Q of Eq. 7.79 but related to logarithms, and ρ_n is the correlation coefficient of Eq. 6.42 between ln x and ln y. Similarly as in Eqs. 7.80 and 7.81, the density functions of marginal and conditional distributions are obtained by corresponding replacements.

Example 7.19. A bivariate distribution is given as

$$f(x, y) = \beta^2 x e^{-\beta x(y+1)},$$

with $0 \leq x \leq \infty$, and $0 \leq y \leq \infty$. Integrating f (x, y) in respect to y from 0 to ∞, the marginal x-density function is $f(x) = \beta e^{-\beta x}$, $0 \leq x \leq \infty$, and, in respect to x from o to ∞, the marginal y-density function is $f(y) = \beta/(1 + y)^2$, $0 \leq y \leq \infty$. As $f(x, y) \neq f(x) \cdot f(y)$, the two variables x and y are dependent. The density functions of conditional variables are: $f(x|y) = \beta(1 + y^2) x e^{-\beta x(y+1)}$, and $f(y|x) = \beta x e^{-\beta xy}$, with both in the range 0 to ∞. This shows another criterion for independence. Whenever the distribution of conditional variables depend on both variables x and y, or x on y for X | Y, or y on x for Y | X, the two variables x and y are depedent. This dependence on the other variable should not be either explicit by equations of conditional distributions, nor implicit by the range of one variable being dependent on the other variable.

Normal and lognormal multivariate distributions. The multivariate distributions of three or more components have theoretical and practical importance in mathematical statistics, particularly when applied to hydrology. They are not handy to write or to treat, except for simple cases of multivariate normal or lognormal distributions with mutually independent components. In the case of more complex multivariate distributions, their investigation is often reduced to studying marginal distributions of components of multidimensional variables, and covariances or autocorrelation coefficients between the pairs of these components.

A multivariate distribution of the vector X is normal if all its components $x_1, x_2, \ldots, x_n$ are normal. A multivariate distribution of the vector Y is lognormal if all its components $y_1, y_2, \ldots, y_n$ have the property that $x = \ln y$ give the vector X whose components $x_1, x_2, \ldots, x_n$ are normally distributed. The multivariate lognormal distribution requires that the multivariate distribution of logarithms of its components is a multivariate normal distribution, regardless of whether the components are independent or dependent.

A multivariate distribution of n variables $(x_1, x_2, \ldots, x_n)$ has the correlation matrix $[\rho_{ij}]$ of the correlation coefficients, ρ_{ij}, as

$$[\rho_{ij}] = \begin{vmatrix} \rho_{11}, & \rho_{12}, & \rho_{13}, & \ldots, & \rho_{1n} \\ \rho_{21}, & \rho_{22}, & \rho_{23}, & \ldots, & \rho_{2n} \\ \ldots & \ldots & \ldots & \ldots & \ldots \\ \rho_{n1}, & \rho_{n2}, & \rho_{n3}, & \ldots, & \rho_{nn} \end{vmatrix} . \qquad 7.83$$

It is a symmetrical matrix about ρ_{ii}-diagonal with values $\rho_{ii} = 1$, and $i = 1, 2, \ldots, n$. If all variables $x_1, x_2, \ldots, x_n$ are standardized, then the covariance matrix and correlation matrix are identical. If it is known that the distribution is normal, or lognormal, then the n expected values, μ_i, the n variances, σ_i^2, and the above correlation matrix give sufficient information about a multivariate normal or lognormal distribution, provided the components are linearly dependent. In the case of lognormal multivariate distribution, the expected values, variances, and ρ_{ij}-coefficients of the matrix are those of logarithms of components $y_1, y_2, \ldots, y_n$, of the Y vector.

Example 7.20. Assume that n variables, as components of a multivariable, are lognormally distributed, and that the relationship $y_1 = a_1 y_2^{a_2} y_3^{a_3} \ldots y_n^{a_n}$ is found among them. Their multivariate distribution is lognormal with the correlation matrix of components $x_i = \ln y_i$ giving the main information about the linear dependence among the logarithms of components. The above relation proves that the pairwise relations among $x_i = \ln y_i$ are linear, so that the matrix of correlation coefficients $[\rho_{ij}]$ of $\ln y_i$ is a sufficient measure of dependence.

Example 7.21. One may ask in hydrology what is the probability that a flood Q_a occurs at river A, when a flood Q_b occurs at river B. If this simultaneous occurrence is very important for flood control works, the probability distribution of this conditional variable is in the form $F(Q_a \mid Q_b)$. Because it is easier to determine analytically the probability density function of a conditional variable from the joint distribution function and the marginal distribution functions of the two variables, fitting the lognormal function to empirical distributions of annual flood peak discharges may be more attractive than fitting the double exponential function. This property of bivariate lognormal probability function, to allow the analytical derivation of distributions of conditional variables, is an advantage to fitting the lognormal function to the distribution of flood peaks.

Example 7.22. The question can be also asked what is the probability that a severe drought year occurs in an area A while (or after) the severe drought has occurred in the adjacent area B? This problem is best treated by using conditional variables and their probability distributions. Fortunately, in most cases, the annual precipitation and runoff variables may be well fitted by normal or lognormal probability functions, which permits the use of their bivariate or multivariate distribution functions, and from them, derive the marginal distributions and the distributions of conditional variables.

7.11. Problems for Discussion and Solutions.

7.1. Give four examples of the application of binomial discrete distribution in hydrology. Discuss particularly whether the conditions of application are fulfilled in each example. Why is it difficult to completely satisfy the conditions of binomial distribution in geophysical sciences like hydrology, meteorology, oceanography, geology, etc.?

7.2. From a record of 85 years of observations of river flows at a gauging station, the probability that a flood event Q_i would be exceeded is $P(Q \geq Q_i) = 0.02326$. Assuming the annual flood as independent events, compute the probability for four floods greater than Q_i to occur in the next 10 years, as well as the probability that none would occur.

7.3. Give four examples of the application of Poisson distribution in hydrology. Discuss difficulties that may arise in applying this probability function to an experimental science, or a science with historical observational data.

7.4. It was found by experiments that $\lambda = \lim np = 1$, for $n \to \infty$ and $p \to 0$, where n is the sample size and p is the probability of an event. What is the probability that a rare event: (a) such as a 10-year rare event would occur three times in a future period of 10 years? (b) such as a 20-year rare event would not occur at all in the period of 20 years?

7.5. Use the Poisson distribution for Problem 7.2 and compare results with those obtained from the binomial distribution.

7.6. Use the binomial distribution for Problem 7.4 and compare results with those obtained from the Poisson distribution.

7.7. From any publication of hydrologic data select a river with daily flows, at least $N = 30$ years. From this data determine four new samples: (a) annual average discharge, (b) January discharge, (c) maximum daily flow of each year, and (d) minimum daily flow of each year. Sort each sample in ascending order of magnitude. By using the plotting position $m/(N + 1)$, plot the four distributions: (a) annual flows on cartesian-probability graph paper, (b) January flows on log-probability graph paper, (c) maximum values in cartesian-extremal probability graph paper, and (d) minimum values in log-extremal probability graph paper. Fit straight lines and estimate parameters graphically.

7.8. For n stations of river flows in a region, compute the skewness and kurtosis coefficients of series of annual flood peak discharges, as well as the means of each set of these n coefficients. Find differences between these means and the constant theoretical values of the double exponential function, $\gamma_1 = 1.29857$ and $\gamma_2 = 4.50$. Discuss if, on the average, the double exponential function is applicable to the above stations. Since the condition of independent peak discharges among these stations is not satisfied, the sample parameters C_s are mutually dependent, which is also the case with the sample parameters k. Is the number n then really the effective number of independent samples of flood discharges?

7.9. Put into Eq. 7.79 the following substitutions: $\mu_x = \mu_y = 0$, $\sigma_x = \sigma_y = \alpha\sqrt{2}$, and $\rho = 1/2$, and find the marginal density functions and the density functions of the two conditional variables $X \mid Y$ and $Y \mid X$.

7.10. Determine if any of the univariate functions studied in this chapter, excluding the normal and lognormal functions, can be easily expressed as a bivariate distribution function by using the correlation coefficient ρ between the components x and y, or between their transformations to new variables $u = f(x)$ and $v = g(y)$.

CHAPTER 8

ESTIMATION METHODS

One of the main objectives of inferential statistics is the reliable estimation of population characteristics of random variables. The procedures used are called estimation methods or estimation techniques. Various properties of estimates are first defined and then the estimation methods currently used in hydrology are reviewed and discussed in a condensed form. The better the estimates inferred from a sample about the population characteristics, the larger is the information on random variables extracted from the sample data.

8.1. Definitions and Properties of Estimates.

Definition of estimates. The population properties of random variables are usually characterized by parameters, percentiles, or other similar numbers. Because populations are rarely known in sciences like hydrology, the population properties of random variables must be estimated from the available sample data. Any estimate obtained from a sample is a statistic. The statistic also may be defined as a function of sample values, $x_1, x_2, \ldots, x_N$ and this function refers to some property of the population of a random variable.

Methods derived from mathematical statistics for estimating various parameters, percentiles, or other population properties of a variable x from sample values in such a way that they depart to a minimum from the population values are called estimation methods. If a probability distribution has three parameters, α, β, and γ, the estimates of these values denoted here as a, b, and c should be as close as practically possible to the population values. The measure of goodness of these estimation methods are the variances of differences $(\alpha - a)$, $(\beta - b)$, and $(\gamma - c)$. The smaller these variances the better an estimation method. The estimated value a, b, and c, are called the estimates or estimators of α, β, and γ. There is an infinite number of possible estimates of any property. Therefore, only some of them are acceptable as best estimates. The concept of best estimate must be clearly defined. Though the best estimates, designated by $\hat{\alpha}$, $\hat{\beta}$, and $\hat{\gamma}$, are those that fall closest to α, β, and γ, there may be a large number of these estimates. Because the various samples of the same size drawn from a given population have different estimates a, b, c, ..., of $\alpha, \beta, \gamma, \ldots$, the sampling distributions of estimates a, b, c, ... about their population values $\alpha, \beta, \gamma, \ldots$ are important properties of estimation methods. Variances of these sampling distributions are measures of the best estimation. The smaller the variance of an estimate, the better that estimate in comparison with the others.

An estimate is often classified as biased or unbiased, regular or irregular, efficient or inefficient, sufficient or insufficient, and consistent or inconsistent [1, p. 473—497]. To define these five concepts of estimates, assume that the estimation is needed for α in the one-parameter probability density function, $f(x; \alpha)$, of the random variable, x; or for α and β in its two-parameter probability density function, $f(x; \alpha, \beta)$. The variance of partial derivatives in α of logarithms of these functions, designated by D (a), is given in one of these three forms:

$$D(a) = \int_{-\infty}^{+\infty} \left[\frac{\partial \ln f(x; \alpha)}{\partial \alpha}\right]^2 f(x; \alpha)dx = E\left(\frac{\partial \ln f}{\partial \alpha}\right)^2 = \int_{-\infty}^{+\infty} \frac{1}{f}\left(\frac{\partial f}{\partial \alpha}\right)^2 dx. \qquad 8.1$$

In the case of $f(x; \alpha, \beta)$, the corresponding expressions of Eq. 8.1 give D (a) and D (b). The corresponding covariance is C (a, b).

Unbiased estimates. If the expected value of an estimate, a, of α is

$$E\,a = \alpha + \upsilon(\alpha), \qquad 8.2$$

in which $\upsilon(\alpha) \neq 0$, the estimate, a, is considered biased. In the case $\upsilon(\alpha) = 0$, or $E\,a = \alpha$, the estimate, a, is unbiased. The expected value of an unbiased estimate, $\hat{\alpha}$, must be the population value, α.

Regular estimates. The difference between the sample and population values is $a - \alpha$. If the variance of this difference satisfies the condition that it is greater than a certain value (given by Eq. 8.3), the estimate is called regular. This inequality is [1, p. 480]

$$\text{var}(a - \alpha) = E(a - \alpha)^2 \geq \frac{\left[1 + \frac{d\upsilon(\alpha)}{d\alpha}\right]^2}{N\,D(a)}, \qquad 8.3$$

in which N is the sample size, D (a) is defined by Eq. 8.1, and $\upsilon(\alpha)$ is defined by Eq. 8.2. For an unbiased estimate, $\hat{\alpha}$, Eq. 8.3 becomes $\text{var}(\hat{\alpha} - \alpha) \geq 1/N\,D(\hat{\alpha})$.

Efficient estimates. If an estimate is unbiased and regular, and if

$$\text{var}\ \hat{\alpha} = E(\hat{\alpha} - \alpha)^2 = \frac{1}{N\,D(\hat{\alpha})}, \qquad 8.4$$

the estimate is efficient. It has the smallest possible variance of deviations $(\hat{\alpha} - \alpha)$ for given N and $f(x; \alpha)$, and is usually understood as the best estimate.

For any other estimate a, different from $\hat{\alpha}$, the variance is $E(a - \alpha)^2 > E(\hat{\alpha} - \alpha)^2$, and the ratio

$$e_a = \frac{E(\hat{\alpha} - \alpha)^2}{E(a - \alpha)^2} \leq 1, \qquad 8.5$$

in which the numerator is given by Eq. 8.4, is called the efficiency of an estimation method which produces the variance given by the denominator of Eq. 8.5. Only an efficient estimate has $e_a = 1$. Two efficient estimates, α_1 and α_2, of the parameter α have the same mean, α, the same variance given by Eq. 8.4, and their correlation coefficient is $\rho = 1$. When an efficient estimate exists, it can always be determined by the Fisher method of maximum likelihood.

Sufficient estimates. If α_1 and α_2 are two independent estimates of α, α_1 is considered a sufficient estimate if the joint probability distribution of α_1 and α_2 has the property

$$F(\alpha_1, \alpha_2) = F(\alpha_1) \cdot F(\alpha_2 \mid \alpha_1) = F(\alpha_1)\, K(x_1, x_2, \ldots, x_n), \qquad 8.6$$

in which $F(\alpha_1)$ is the distribution of α_1, $F(\alpha_2 \mid \alpha_1)$ is the conditional probability distribution of α_2 given α_1, and $K(x_1, x_2, \ldots, x_n)$ is not a function of α. When Eq. 8.6 holds, α_2 does not produce any new information about α, which is not already contained in α_1, regardless of the estimation method used to produce the estimate α_2. In this case, α_1 is optimum, or it is also a sufficient estimate.

Consistent estimates. If sample size N is large, $\hat{\alpha}$ is called a consistent estimate of α if it converges with the probability unity to α as N tends to infinity. Because many unbiased estimates have the variances of the type var $\hat{\alpha} \simeq c/\sqrt{N}$, in most cases the condition of consistency is satisfied. If this is the case, and if the efficiency e_a tends to unity as N goes to infinity, the estimate $\hat{\alpha}$ is called the asymptotically efficient estimate.

Joint efficient estimates of two parameters. If a population density function of x has both parameters unknown, $f(x; \alpha, \beta)$, then the efficient estimates, $\hat{\alpha}$ and $\hat{\beta}$, have the following variances

$$\text{var}\, \hat{\alpha} = \frac{1}{1 - \rho^2(\hat{\alpha}, \hat{\beta})} \cdot \frac{1}{N D(\hat{\alpha})}, \quad \text{and} \quad \text{var}\, \hat{\beta} = \frac{1}{1 - \rho^2(\hat{\alpha}, \hat{\beta})} \cdot \frac{1}{N D(\hat{\beta})}, \qquad 8.7$$

where

$$\rho(\hat{\alpha}, \hat{\beta}) = \frac{C(\hat{\alpha}, \hat{\beta})}{[D(\hat{\alpha})\, D(\hat{\beta})]^{1/2}},$$

$$\text{with } C(\hat{\alpha}, \hat{\beta}) = E\left[\frac{\partial \ln f(x; \alpha, \beta)}{\partial \alpha}\right] \left[\frac{\partial \ln f(x; \alpha, \beta)}{\partial \beta}\right]. \qquad 8.8$$

For $\hat{\alpha}$ of Eq. 8.7 to be equal to $\hat{\alpha}$ of Eq. 8.4, the expressions of Eq. 8.8 should be zeros, or $\hat{\alpha}$ and $\hat{\beta}$ should be uncorrelated. In any other case, the estimate $\hat{\alpha}$ is less efficient by Eq. 8.7 than by Eq. 8.4.

Practical significance of various properties of estimates. The above five properties of estimates have different significance when applied. In the majority of hydrologic cases of estimation, the most important properties are to have unbiased (and therefore regular), and efficient (and usually sufficient and consistent) estimates. These two properties of unbiased and efficient estimates are usually required in hydrology when the requirement is to extract the maximum information from the sample data.

The theory of estimation was mainly developed and shaped by R. A. Fisher. Though there are several other ideas on this subject, Fischer's approach seems to have shown some advantages in its use in hydrology, especially his estimation method of maximum likelihood.

Some of these advantages are manifest in different ways. Hydrologic data are often scarce. In arid and semi-arid regions, when the hydrologic variables have the largest variation and often have distributions which are highly skewed, the irony is that sample data are shortest in length, and the amount of observational points is relatively small. It is a mistaken attitude of many practitioners in hydrology that simple estimation methods, like the graphical method or method of moments, are sufficiently accurate because of scarce data. The approach should be quite the opposite. For those variables in hydrology with great variation, a highly skewed distribution, and a short series of observation, the unbiased and efficient estimates should be always searched for. The main problem is selecting the estimation method in each particular case that gives the unbiased and efficient estimates with the least computational cost.

Methods of estimation. Various methods of estimation used currently in hydrology, given in their ascending order of efficiency, from the least to the most efficient, are the graphical method, the least-squares method and/or the method of moments, and the maximum likelihood method.

8.2. Graphical Estimation Method.

Description. If one draws the normal distribution in the form of a straight line in cartesian-probability scales to the plotted empirical distribution, as already described in Chapter 7, the 50 percent probability gives the estimate $\bar{x}$, and the difference of values of x for 84.13 percent and 15.87 percent gives 2s, or the double standard deviation. These values, $\bar{x}$ and s, are said to be graphically estimated, because the straight line fit of the normal distribution in these scales is usually made visually in drawing a straight line. The accuracy, however, is very limited. It is sufficient to estimate $\bar{x}$ and s by the least-squares method, or by the method of moments, to ascertain the unreliability of visual fits for estimating parameters. The preceding chapter gives several examples of graphical estimation of hydrologic distribution parameters. The basic condition for the use of this method is for the coordinate scales to be transformed in such a way that a given probability distribution function, $F(x; \alpha, \beta, \ldots)$, plots as a straight line.

Use of m points and graphical fit of curves. The graphical fit may, however, be used for the fitting of functions

$$y = F(x; \alpha, \beta, \ldots), \tag{8.9}$$

without a transformation of coordinate scales. This requires a visual tracing of a curve, selecting m points for m parameters α, β, ..., in the various parts of the curve, through which Eq. 8.9 should pass. The coordinates y_i and $F_i(x; \alpha, \beta, \ldots)$ of m points give m equations from which the parameters may be estimated. The accuracy of this method is highly subjective, and depends on experience and good judgment. It does not have a particular advantage in comparison with the graphical visual straight-line fits in transformed coordinates.

8.3. Least-Squares Estimation Method.

Description. This method consists of the estimation of parameters by fitting a theoretical function to an empirical distribution, or any other empirical curve. The sum of all squares of deviations of observed points from the fitted function is minimized to produce least-squares.

In the case a function $y = f(x; \alpha, \beta, \gamma, \ldots)$ should be fitted to data by determining the best estimates a, b, c, ..., of α, β, γ, ..., the analytical method of least-squares minimizes the sum

$$S = \sum_{1}^{N} e_i^2 = \sum_{1}^{N} (y_i - y)^2 = \sum_{1}^{N} [y_i - f(x_i; \alpha, \beta, \gamma, \ldots)]^2, \tag{8.10}$$

in which x_i and y_i are coordinates of observed points, α, β, and γ are replaced by their estimates a, b, and c, and N is the sample size. Instead of minimizing $\Sigma (y_i - y)^2$, often $\Sigma (x_i - x)^2$, or Σe_n^2, where e_n are perpendicular deviations from the function, $f(x; \alpha, \beta, \gamma, \ldots)$, also may be mini mized. So that the line $f(x; a, b, c, \ldots)$ may have the minimum of Eq. 8.10, all partial derivatives of this sum with respect to parameters a, b, c, ... should be zero, so that

$$\frac{\partial \sum_{i=1}^{N} (y_i - y)^2}{\partial a} = 0; \quad \frac{\partial \sum_{i=1}^{N} (y_i - y)^2}{\partial b} = 0; \ldots \quad . \tag{8.11}$$

These partial derivatives give m equations for the determination of m parameters. The simplest functions $f(x; \alpha, \beta, \gamma, \ldots)$ are those which are linear with respect to parameters. Among such functions the polynomial function for x is a general case, since other functions can be approximated by polynomials if they are developed in power-series form. There must be more points than parameters, or $N > m$. Generally, N should be much greater than m, if the derived line is to be used for purposes of prediction.

Fitting of a quadratic parabola. The equation for fitting a quadratic parabola is

$$y = a + bx + cx^2, \tag{8.12}$$

and it gives three equations by using Eq. 8.11:

$$aN + b\,\Sigma x_i + c\Sigma x_i^2 = \Sigma y_i,$$
$$a\Sigma x_i + b\Sigma x_i^2 + c\Sigma x_i^3 = \Sigma x_i y_i, \text{ and} \qquad 8.13$$
$$a\Sigma x_i^2 + b\Sigma x_i^3 + c\Sigma x_i^4 = \Sigma x_i^2 y_i,$$

with the summations taken from $i = 1$ to $i = N$. The solution of these three linear equations gives the estimates a, b, and c.

Efficient estimation. Three conditions should be satisfied for the least-squares method to be an efficient estimation method: (a) that the deviations e_i of Eq. 8.10 are normally, or at least symmetrically, distributed; (2) that the population variance of e_i is independent of the position of a given group of values e_i, or the deviations are mutually independent along the line, and (3) that the population variance of e_i along the curve is a constant. These conditions are rarely satisfied in hydrology, especially the latter two, because often the deviations e_i depend on y; usually they increase with an increase of y. In this case, by minimizing S of Eq. 8.10 through Eq. 8.11, greater weights are given to larger y-values than to smaller ones. Therefore, a bias is introduced. To alleviate this, various transformations of coordinates are used in an attempt to make the above conditions satisfied as closely as practically feasible. The logarithmic transformation is currently used in hydrology for this purpose, though it has the bias of giving larger weights to smaller y-values than to larger ones. Other transformations are given in Table 11.7.

Least-square method as the χ^2-minimum method. The least--squares method is also called χ^2-minimum method [1, p. 506]. If used for probability distributions, it is applicable to continuous distributions with grouped data, expressed as absolute frequencies of class intervals, or to absolute frequencies of discrete distributions. The χ^2-method as the method of testing various hypotheses is discussed in Chapter 10.

If an equation for χ^2 of the following type is used

$$\chi^2 = \sum_{i=1}^{m} \frac{(f_i - np_i)^2}{np_i}, \qquad 8.14$$

in which m is the number of class intervals for continuous variables, or the number of events of discrete variables, f_i are the observed absolute frequencies of class intervals, or absolute frequencies of discrete events, and p_i are the probabilities of class intervals or of discrete events, of the probability function $p(x; \alpha, \beta, \gamma, \ldots)$. Then the partial derivatives $\partial\chi^2/\partial\alpha = 0$, $\partial\chi^2/\partial\beta = 0$, $\partial\chi^2/\partial\gamma = 0$, ... give as many equations as there are parameters $\alpha, \beta, \gamma, \ldots$ to be estimated.

For the first derivative,

$$-\frac{1}{2}\frac{\partial\chi^2}{\partial\alpha} = \sum_{i=1}^{m}\left[\frac{f_i - np_i}{p_i} + \frac{(f_i - np_i)^2}{2np_i^2}\right]\frac{\partial p_i}{\partial\alpha} = 0; \qquad 8.15$$

the second term inside the brackets on the right side is small in comparison with the first term, when n is large, so that

$$\sum_{i=1}^{m} \frac{f_i - np_i}{p_i} \frac{\partial p_i}{\partial \alpha} = 0. \tag{8.16}$$

The use of Eq. 8.16 is called the modifed χ^2-minimum method of estimation. However, it can be shown that Eq. 8.16 for α, β, γ, ... gives the estimates a, b, c, ... which are identical with the estimates obtained by the maximum likelihood method.

Example 8.1. Table 8.1 gives the results of fitting a parabolic rating curve to measured discharges, Q, and stages, H. The sums in Eq. 8.13 are given at the bottom of Table 8.1. The computation must be carried out to the same decimal point for all sums, because the sums may be of the same order of magnitude as the differences involved in the use of large numbers. The rating curve by the least squares method is found to be $Q_1 = 29.9 + 0.567\ H + 0.00720\ H^2$. The parabola through three selected points (points 6, 11, and 13 in Table 8.1) is $Q_2 = 38.6 + 0.396\ H + 0.007597\ H^2$. The last two columns of Table 8.1 give e_1^2 and e_2^2, which are the respective measures of goodness of fit in the case the least-squares method, and the method of three points are used in fitting the parabola to the rating curve.

TABLE 8.1. Fitting a Parabola to a Rating Curve

	H (stage) cm*	Q (discharge) m³/sec**	Estimates of Parameters		e_1^2	e_2^2
			By least-squares	By three-point		
1	−23	15.55	20.7	35.5	1.25	9.55
2	−22	15.46	20.9	33.6	1.39	9.55
3	−16	20.07	23.6	34.2	0.26	5.00
4	−16	21.99	23.6	34.2	0.11	4.33
5	14	36.11	39.3	45.6	0.26	1.98
6	33	59.82	56.6	59.9	0.19	0.00
7	46	86.58	70.9	72.9	3.50	2.57
8	69	110.96	103.3	102.1	0.57	0.77
9	88	136.52	135.6	132.3	0.01	0.13
10	120	201.40	200.7	190.5	0.07	1.00
11	136	232.87	240.2	233.0	0.22	0.00
12	220	492.50	503.2	493.4	0.23	0.00
13	400	1,412.48	1,409.0	1,412.5	0.01	0.00
					8.07	35.12

* 1 cm = 0.3937 in.
** 1 m³/sec = 35.314 cfs

N = 13; $\Sigma x_i = 1{,}049.00$; $\Sigma x_i^2 = 258{,}727.00$; $\Sigma x_i^3 = 80{,}006{,}477.00$; $\Sigma x_i^4 = 28{,}580{,}940{,}099.00$; $\Sigma y_i = 2{,}846.31$; $\Sigma (x_i y_i) = 754{,}258.87$; $\Sigma (x_i^2 y_i) = 258{,}971.73$.

8.4. Method of Moments.

Description. This method was introduced by K. Pearson as one of the oldest proposed estimation method. The basic principle is that relations of parameters to moments are established for each probability distribution function in the form

$$\alpha = f_1(\mu_i, \mu_{i+1}, \ldots); \ \beta = f_2(\mu_j, \mu_{j+1}, \ldots); \ \gamma = f_3(\mu_k, \mu_{k+1}, \ldots); \ \ldots, \quad 8.17$$

in which $\mu_i, \mu_{i+1}, \ldots, \mu_j, \mu_{j+1}, \ldots, \mu_k, \mu_{k+1}, \ldots$, are the population central moments; the moments about the origin may also be used. As these moments are estimated by the sample moments, either as biased or unbiased estimates, then Eqs. 8.17 give a, b, c, ..., or $\hat{a}$, $\hat{b}$, $\hat{c}$, ..., as biased or unbiased estimates of α, β, γ, ... As it was shown in Chapter 7 on probability distribution functions, there should be as many equations 8.17 as there are parameters, and as many sample moment estimates as there are parameters. The moments μ_i, μ_j, μ_k, ... may be the same moments in Eqs. 8.17. For small samples, say $N \leq 30$, the unbiased moments should be used, otherwise the sample moments are used directly to replace the population moments.

As it was shown in examples in Chapter 7, the computations of estimates by the method of moments are comparatively simple. This may be the reason for the extensive use of this method of estimation. The method also facilitates computation of distributions of various statistics. Because many parameters estimated by moments are asymptotically normal, the means of these parameters may defer from the population values by a quantity of the order of $1/N$, and their variances are proportional to $1/N$. This bias of the order of $1/N$ can easily be corrected by the unbiased estimates of parameters.

Efficiency of the method of moments. It was shown by R. A. Fisher that the estimates by moments are asymptotically efficient, and that their efficiency is usually smaller than unity; for skewed distributions, it may be considerably smaller than unity. Very often, the estimates by moments are considered only as first approximations. If a probability distribution is symmetrical, and particularly if it is normal, it can be shown that the efficiency of the method of moments is either unity or close to it, so that in this case the other methods, the maximum likelihood or the modified χ^2-minimum method, do not have any preference over the method of moments. Unfortunately, the majority of hydrologic variables are more or less skewed, so that the use of estimation method by moments always represents a small or large loss of efficiency in estimation.

Example 8.2. In an earlier discussion the question of hydrologic information in arid and semi-arid regions was mentioned. Besides the scarcity of data and large variances of most hydrologic variables in arid regions, the high skewness is rather the rule than an exception. So, if the estimation method by moments is used instead of the maximum likelihood method, a large amount of information in this scarce data may not be extracted.

8.5. Maximum Likelihood Method.

Description. Let

$$f(x;\ \alpha,\ \beta,\ \ldots) \tag{8.18}$$

be a probability density function of x, with α, β, ..., the parameters to be estimated. The product,

$$L = \prod_{i=1}^{N} f(x_i;\ \alpha,\ \beta,\ \ldots), \tag{8.19}$$

is called the likelihood function of a sample of N values from a population of the continuous variable, x. In the case of discrete variable with probability mass, $p_i(x; \alpha, \beta, \ldots)$, the likelihood function is

$$L = \prod_{i=1}^{N} p_i(x_i;\ \alpha,\ \beta,\ \ldots), \tag{8.20}$$

in which N is the sample size and α, β, ..., are replaced by the estimates a, b, ... The estimation of α, β, γ, ... consists of determining a, b, c, ... from sample data in such a way as to obtain the value L as great as possible. Since ln L attains the maximum values at given a, b, c, ... as L does, the likelihood equation is

$$\ln L = \ln \prod_{i=1}^{N} f(x_i;\ a,\ b,\ \ldots) = \sum_{i=1}^{N} \ln\ f(x_i,\ ;\ a,\ b,\ \ldots), \tag{8.21}$$

while its partial derivatives in a, b, ..., equated to zero and called the maximum likelihood equations, are

$$\frac{\partial \ln L}{\partial a} = 0;\quad \frac{\partial \ln L}{\partial b} = 0;\ \ldots. \tag{8.22}$$

These equations, of the same number as the number of parameters, allow the computation of estimates a, b, c, ... Therefore, the maximum likelihood equations are functions of parameters to be estimated for a given sample. In this case the estimates a, b, c, ... are the efficient estimates of α, β, , ..., if the efficient estimates exist.

Properties of the method. The solution of Eqs. 8.22 must exclude all estimates which are constants, and retain only those roots of these equations that depend on sample values, x_1, x_2, ..., x_N. The estimates by the maximum likelihood method have these properties: (a) they are asymptotically unbiased , (b) if efficient estimates exist for the parameters, α, β, γ, ... the maximum likelihood method will produce them, and (c) if sufficient estimates exist, the solutions of the likelihood equations will be functions of these sufficient estimates.

Often the likelihood equation, Eq. 8.21, gives a solution for an estimate that converges to the true population value of the parameter as N goes to infinity, with the probability of unity, so that the estimate is consistent.

Example 8.3. For the normal distribution function, Eq. 8.22 gives the estimate, $\bar{x}$, of the mean, μ, as $\frac{1}{N}\sum_{i=1}^{N} x_i$, and the estimate s of the standard deviation, σ, as $\left[\frac{1}{N}\sum_{i=1}^{N}(x_i-\bar{x})^2\right]^{1/2}$. These estimates are asymptotically efficient.

Example 8.4. For the two-parameter lognormal function, the mean and the standard deviation of logarithms, $\bar{x}_n$ and s_n, are the maximum likelihood estimates of μ_n and σ_n of Eqs. 7.16 and 7.17. For the three-parameter lognormal function, the mean and standard deviation of $\ln(x_i-\gamma)$ are the maximum likelihood estimates of μ_n and σ_n in Eq. 7.29, while the estimate $\hat{\gamma}$, of the lower boundary γ, is given by

$$\sum_{i=1}^{N}\frac{1}{x_i-\hat{\gamma}}\left\{\frac{1}{N}\sum_{i=1}^{N}\ln^2(x_i-\hat{\gamma})-\left[\frac{1}{N}\sum_{i=1}^{N}\ln(x_i-\hat{\gamma})\right]^2-\frac{1}{N}\sum_{i=1}^{N}\ln(x_i-\hat{\gamma})\right\}+ + \sum_{i=1}^{N}\frac{\ln(x_i-\hat{\gamma})}{x_i-\hat{\gamma}}=0, \quad 8.23$$

in which $\hat{\gamma}$ is the maximum likelihood estimate of the population lower boundary γ. It may be solved only by an iterative computational procedure.

Example 8.5. For the one-parameter gamma distribution, Eq. 7.35, the maximum likelihood estimation method gives the lower limit of the variance of a regular and unbiased estimate of α, for a sample of size N, and is determined by Eq. 8.22 as

$$\frac{1}{N}\frac{\partial \ln L}{\partial\alpha}=\frac{1}{N}\Sigma\ln x_i-\frac{\partial[\ln\Gamma(\alpha)]}{\partial\alpha}=0, \quad 8.24$$

in which the expression $\partial[\ln\Gamma(\alpha)]/\partial\alpha=\psi(\alpha)$ is called the digamma function of α. The estimate $\hat{\alpha}$, as the solution of Eq. 8.24, gives an asymptotically efficient estimate of α, so that $\hat{\alpha}$ is normally distributed with the mean, α, and the standard deviation $\{N\partial^2[\ln\Gamma(\alpha)]/\partial\alpha^2\}^{-1/2}$. If α is estimated from the mean, $\hat{\alpha}=\bar{x}$, or from the variance as $\hat{\alpha}=\hat{s}^2$, as shown in Table 7.2, it can be seen that the efficiency is smaller than unity. For the case of the two-parameter gamma distribution (with the parameters α and β) of Eq. 7.39, the two equations that follow from Eq. 8.22 are

$$\ln\hat{\alpha}-\frac{\partial[\ln\Gamma(\hat{\alpha})]}{\partial\hat{\alpha}}=\ln\frac{1}{N}\sum_{i=1}^{N}x_i-\frac{1}{N}\sum_{i=1}^{N}\ln x_i, \quad 8.25$$

with $\hat{\alpha}$ as the estimate of α, and

$$\hat{\beta}=\frac{1}{\hat{\alpha}}\cdot\frac{1}{N}\sum_{i=1}^{N}x_i, \quad 8.26$$

with $\hat{\beta}$ as the estimate of β. Davis [2] has provided tables for the $\psi(\alpha)$-digamma function. Masuyama and Kuroiwa [3] prepared tables of $\ln\hat{\alpha}-\psi(\hat{\alpha})$ from tables of logarithms and tables of the digamma function.

Nörlund [4] gives the digamma function as an asymptotic expression

$$\psi(\hat{\alpha})=\ln\hat{\alpha}-\frac{1}{2\hat{\alpha}}-\sum_{i=1}^{n}\frac{(-1)^{i-1}B_i}{2i\hat{\alpha}^{2i}}+R_n, \quad 8.27$$

in which B_i are the Bernoulli numbers: $B_1=\frac{1}{6}$, $B_2=\frac{1}{30}$, ..., and R_n = the remainder after n terms. For $n=1$, the equation obtained gives

$$\hat{\alpha}=\frac{1+\sqrt{1+\frac{4}{3}\left(\ln\bar{x}-\frac{1}{N}\sum_{i=1}^{N}\ln x_i\right)}}{4\left(\ln\bar{x}-\frac{1}{N}\sum_{i=1}^{N}\ln x_i\right)}-\Delta\hat{\alpha}, \quad 8.28$$

as an approximate estimate of α. However, the correction $\Delta\hat{\alpha}$ is necessary [5], which is given in Table 8.2.

For the three-parameter gamma distribution (α, β, γ) of Eq. 7.41, Eq. 8.28 is valid for the estimate of α, if x_i in that equation is replaced by $(x_i - \hat{\gamma})$, where $\hat{\gamma}$ is the estimate of γ. Similarly, Eq. 8.26 gives the estimate of β if $\bar{x}$ is replaced by $(\bar{x} - \hat{\gamma})$. The estimate $\hat{\gamma}$ of γ is given by the following equation

$$\frac{1 + \sqrt{1 + \frac{4}{3} A}}{1 + \sqrt{1 + \frac{4}{3} A - 4A}} - \frac{(\bar{x} - \hat{\gamma})}{N} \sum_{i=1}^{N} \frac{1}{(x_i - \hat{\gamma})} = 0, \qquad 8.29$$

in which

$$A = \left[\ln (\bar{x} - \hat{\gamma}) - \frac{1}{N} \sum_{i=1}^{N} \ln (x_i - \hat{\gamma}) \right]. \qquad 8.30$$

TABLE 8.2. Correction Factor $\Delta\alpha$ for Computation of Maximum Likelihood Estimates of the Shape Parameter of Gamma Functions with two and three Parameters.

α	$\Delta\alpha$	α	$\Delta\alpha$
0.200	0.034	1.400	0.006
0.300	0.029	1.500	0.005
0.400	0.025	1.600	0.005
0.500	0.021	1.700	0.004
0.600	0.017	1.800	0.004
0.700	0.014	1.900	0.003
0.800	0.012	2.200	0.003
0.900	0.011	2.300	0.002
1.000	0.009	3.100	0.002
1.100	0.008	3.200	0.001
1.200	0.007	5.500	0.001
1.300	0.006	5.600	0.000

Example 8.6. The annual flows of the Weldon River at Mill Grove, Missouri, are given as the example of estimates of parameters of five distribution functions (Chapter 7, ref. 6). Table 8.3 gives the data, the original Q_i, and in modular coefficients $K_i = Q_i/\bar{Q} = Q_i/256.7$. Multiplying estimates of K_i-distribution by $\bar{Q}$, the estimates of Q_i-distribution are obtained. The various sums of K_i are: $(K_i - 1)^2 = 17.018$; $\Sigma \ln K_i = -9.810$; $\bar{x}_n = \Sigma \ln K_i/N$; $(\ln K_i - \bar{x}_n)^2 = 21.837$; $\Sigma (K_i - \hat{\gamma}_1) = 29.450$, for $\hat{\gamma}_1 = 0.050$; $\Sigma [1/(K_i - \hat{\gamma}_1)] = 73.867$; $\Sigma \ln (K_i - \hat{\gamma}_1) = -13.130$; $\Sigma \ln^2 (K_i - \hat{\gamma}_1) = 32.492$; $\Sigma [\ln (K_i - \hat{\gamma}_1)/(K_i - \hat{\gamma}_1)] = -95.407$; $\Sigma (K_i - \hat{\gamma}_2) = 77.500$ for $\hat{\gamma}_2 = -1.500$; $\Sigma [1/(K_i - \hat{\gamma}_2)] = 13.418$; and $\Sigma \ln (K_i - \hat{\gamma}_2) = 27.124$.

With the above sums the estimates are: (a) for the normal distribution, $\bar{K} = 1$ and $s_h = 0.741$; (b) for the two-parameter lognormal distribution, $\bar{x}_n = -0.317$ and $s_n = 0.840$; (c) for the three-parameter lognormal distribution, $\hat{\gamma}_1 = 0.050$; $\bar{x}_n = -0.424$ and $s_n = 0.933$; (d) for the two-parameter gamma distribution, Eq. 8.28 gives $\hat{\alpha} = 1.727$ with $\Delta\hat{\alpha} = 0.004$, and $\hat{\beta} = 0.579$; and (e) for the three-parameter gamma distribution, Eqs. 8.29 and 8.30 give $\hat{\gamma} = -1.500$, $\hat{\alpha} = 12.360$, and $\hat{\beta} = 0.202$.

When the method of moment is used for the two-parameter gamma function, $\hat{\alpha}\hat{\beta} = \bar{K}$, $\hat{\alpha}\hat{\beta}^2 = s_K^2$, so that $\hat{\alpha} = \bar{K}^2/s_K^2 = 1/0.741^2 = 1.821$, and $\hat{\beta} = 0.741^2 = 0.549$ The differences in estimates by the method of maximum likelihood and the method of moments are of some measurable importance.

If $\alpha < 10$, the maximum likelihood method gives much more efficient estimates than the method of moments. For $\alpha = 1$, by using the method of moments instead of maximum likelibood, the sample size 2N is needed to obtain the same information in the estimation of β, and 5N/2 in the estimation of α, in comparison with the estimates a and b by Eqs. 8.25 and 8.26 [5]. This is an important aspect of estimation for the asymmetrical distributions. For symmetrical distributions, the methods of moments, the least-squares, and the maximum likelihood give the same results. Because many hydrologic variables have extremely skewed distributions (say, daily flows, daily precipitation), it is clear the method of moments or the method of least-squares for the estimation of distribution parameters may represent a significant loss of valuable hydrologic information.

TABLE 8.3. Annual Flows, Q_i, of the Weldon River at Mill Grove, Missouri.

Year	Annual River Flow Q_i (cfs)	$K_i = \frac{Q_i}{\bar{Q}}$	Year	Annual River Flow Q_i (cfs)	$K_i = \frac{Q_i}{\bar{Q}}$
1930	108.0	0.421	1945	441.0	1.718
1	53.6	0.209	6	386.0	1.504
2	585.0	2.279	7	567.0	2.209
3	98.1	0.382	8	122.0	0.475
4	40.6	0.158	9	151.0	0.588
1935	472.0	1.839	1950	244.0	0.951
6	96.5	0.376	1	400.0	1.558
7	217.0	0.845	2	245.0	0.955
8	42.7	0.166	3	114.0	0.444
9	208.0	0.810	4	659.0	2.567
1940	143.0	0.557	1955	132.0	0.514
1	93.7	0.365	6	44.0	0.171
2	398.0	1.551	7	72.5	0.282
3	298.0	1.161	8	135.0	0.526
4	248.0	0.966	9	635.0	2.474
			1960	508.0	1.979

8.6. **Problems for Discussion and Solutions.**

8.1. For the annual flows in Problem 7.7, (a), estimate the parameters $\bar{x}$ and s by the method of moments, and compare them with the graphical estimates obtained in Problem 7.7.

8.2. For the empirical distribution in Problem 7.7, (a), fit the two-parameter and three-parameter lognormal functions on the graph with the log-probability scales. For the graphical fit of the three-parameter distribution use several values of the lower boundary, and estimate it by selecting the lognormal fit closest to a straight line. Estimate parameters by the method of moments of logarithms, and compare them with the estimate obtained from graphical fits.

8.3. For the January flows of Problem 7.7, (b), estimate parameters of the two-parameter gamma distribution function, both by the maximum likelihood method and the method of moments, and compare them.

8.4. For the maximum and minimum annual daily flows of Problems 7.7, (c), and 7.7, (d), fit the lognormal functions with two parameters by the method of moments of logarithms, then write the bivariate distribution function of these two extremes. Are these flows mutually independent or dependent?

8.5. Select at least 10 measurements of stage and discharge for the stage-discharge relation. Assuming that a second order polynomial of the form $Q = a + bH + cH^2$ is a suitable mathematical expression for the rating curve, determine the parameters a, b, and c for the best fit of data using the least squares estimation procedure. State the underlying assumptions of the least squares approach and discuss their validity in this case.

8.6. For Problem 8.5, fit also the equation $Q = m(H - H_0)^n$, with m, n, and H_0 the parameters, by the least-squares method, and compare the fit of equations in Problems 8.5 and 8.6 by their squares of deviations.

CHAPTER 9

SAMPLING THEORY

The sampling theory in the form of distributions of various sample characteristics, or statistics, is relevant for the description, analysis, and prediction of hydrologic random variables. Of particular significance are asymptotic normal, asymptotic non-normal, and exact sampling distributions of basic parameters currently used in hydrology. The experimental statistical (or Monte Carlo) method of derivations of sampling distributions have significant promise in hydrology. Confidence and tolerance intervals, limits, regions or limit curves are becoming more and more integral parts of the general investigation of hydrologic properties derived from available samples.

9.1. Basic Definitions and Approaches.

Concept of sampling. Basic principles of sampling theory are discussed in Chapter 4. Although the concept of sampling includes several subjects, it is mainly concerned with properties of sample statistics in the form of their distributions. Sometimes, it also includes the methods of carrying out sampling of random phenomena and stochastic processes to obtain a given amount of information with a minimum of cost and/or time. These methods are, however, more often encompassed under the topic of the design of the experiments. The sample statistic is any number that expresses a property of a sample, like various statistical moments, distribution parameters, order-statistics, correlation coeficients, characteristic frequencies, and similar. Therefore, the statistic should be selected according to the problem, and then and only then the study of how this statistic is distributed about its population value should be carried out. Placing confidence limits to a statistic, or drawing a confidence or tolerance interval, are subjects included in sampling theory. Estimation methods of statistics were given in the previous chapter, although they often may be treated under the sampling theory. Though testing hypotheses, including tests of goodness of fit, is sometimes included in the topic of sampling theory, it is given here as a separate subject, and dealt with in the next chapter. Thus, these two latter subjects are excluded from the narrower concept of sampling theory, as given in this chapter.

It is assumed that all hydrologic problems, which are successfully treated by methods of inferential statistics, can be conveniently classified in three categories of description, analysis, and prediction. Using this assumption, these three categories of sampling theory are discussed first.

Description. The total information available for a random univariate is usually given by its sample as the sequence of values $x_1, x_2, \ldots, x_N$. The question of condensation of information in the form of number descriptors, or mathematical models, arises for any random variable. The description is then defined as an extraction of information from a sample in the most condensed form. This extraction is called the reduction of the sample data by R. A. Fisher. If this reduction yields all or most of the information contained in a sample, this extraction is then complete. This information is measured by variances of sampling distributions of all statistics that represent the condensed information.

Many hydrologic variables are continuously observed. A large amount of data about them is already accumulated. Economy of data processing, handling, storage, and publishing, as well as their best use, require the reduction of information to the most condensed form of mathematical models and their estimated parameters, or simply in the form of a number of statistics. Sampling theory enables measurement of information thus extracted.

Exangle 9.1. Assume a time series of daily river flows with 100 years of observations to be available. This discrete series contains $N = 36{,}525$ values. The question arises whether these data can or cannot be described meaningfully by a small number of mathematical models, and by their parameters to be estimated from these data. Instead of publishing reports of dozens of pages, presenting data in graphical form, or keeping data on magnetic tapes or punched cards, these descriptive mathematical models and their parameters may contain information equivalent to all of these 36,525 numbers.

Assuming tests have shown that this series has periodic components in daily means, daily standard deviations, and daily first serial correlation coefficients of the first-order linear autoregressive model, as μ_τ-, σ_τ- and ρ_τ-periodic components, each with 365 values, $\tau = 1, 2, \ldots, 365$, then the mathematical model of time series composition is*)

$$x_i = \mu_\tau + \sigma_\tau (\rho_\tau \varepsilon_{i-1} + \xi_i),$$

with x_i the observed daily flows, $\varepsilon_i = (x_i - \mu_\tau)/\sigma_\tau$ a standardized variable, ξ_i the independent stochastic variable with $\xi_1 = \varepsilon_1$, and μ_τ, σ_τ, and ρ_τ the periodic components as defined above. Three equations of the type

$$\upsilon_\tau = \upsilon_0 + \sum_{j=1}^{m} (A_j \cos \lambda_j \tau + B_j \sin \lambda_j \tau)$$

describe these periodic components, with υ_0 the symbol for the means of μ_τ, σ_τ, and ρ_τ; υ_τ the symbol for the values of any of the three periodic components; A_j and B_j the Fourier coefficients for a given angular frequency $\lambda_j = 2\pi j / \omega$, with $\omega = 365$ days in this case, $j = 1, 2, \ldots, m$, and m the number of significant harmonics of λ_j. The remaining stochastic component ξ_i, after the three periodic components are removed, may be described by its probability distribution function $F(\xi_i)$. Let us assume further that each of the three periodic components may be described by three significant

*) The methods of structural analysis of hydrologic time series are dealt with in detail in the book by the same writer under the title Stochastic Processes in Hydrology 1972.

harmonics (m = 3). Each then needs an estimate of (2m + 1)-parametes, or 2m of A_j and B_j coefficients, plus υ_0 value, or a total of seven parameters per periodic component. If F (ξ_i) has 3 parameters only, then this description requires 5 equations and 24 parameters. The sampling theory and testing of hypotheses are effective techniques in deciding the goodness of fit of the above models, the significance of various harmonics in periodic components, and the reliability of estimated parameters.

Analysis. The application of analysis in sampling theory is mainly concerned with cause-effect relations, or relations of effect-effect when both are produced by the same or similar causative factors. It is often required in hydrology to find whether two or several variables — either conceived or not conceived as cause-effect — are related. Correlation analysis, or in general the multivariate analysis, provide techniques for estimating measures of association and of parameters of these relations. The sampling theory measures the reliability of these parameters, and supplies distributions for testing hypotheses in this analysis.

Hydrology is full of these types of problems, especially in the search for eventual cause-effect associations of variables. Similarly, hydrology has many problems in which various mathemtical models are analyzed to determine whether they fit well various relations. Selection of proper mathematical models, whether inferred simply by statistical inference or by a combination of physical investigation and statistical inference, is an important subject of the analysis of sample data.

Example 9.2. Various geometric parameters, which describe geomorphology of river basins, affect more or less characteristics of their unit hydrographs. In studying the relations between properties of a unit hydrograph and the river basin geometry, two aspects are often evident: (a) a very large number of geometric parameters have been used or proposed for the description of river basin geomorphology, and (b) nearly all these parameters are mutually dependent basin properties. To analyze their fundamental relations, multivariate analysis may reduce this large number of geometric parameters to a new set of significant parameters, much smaller in number and mutually independent, in the form of principal components, or independent factors. A successful physical explanation for these new parameters, of hydrologic relevance, represents an attractive solution in this analysis. The sampling theory is necessary to test the goodness of fit of various relations between the unit hydrograph characteristics and these new independent geometric parameters. This example has as an implication when the relations among many random variables are looked for.

Prediction. This part of sampling theory is concerned with prediction problems involving random variables, particularly with the reliability (precision, accuracy) of predicted values of a random variable from values of other random variables.

Two main aspects in hydrology are the prediction of any dependent variable by its causative factors, like flood characteristics predicted statistically from the rainfall and river basin causative variables, and the mutual transfer of information between variables, which is based on appropriate statistical transfer or prediction equations.

Example 9.3. The prediction of basic water quality parameters, such as temperature, oxygen content, concentration of dissolved solids, and similar, from parameters such as water discharge, water stage, climatic variables, and river basin characteristics in the form of a set of multiple regression equations is a need in water pollution control. Because data may be more readily available for a set of variables other than the set of variables needed in pollution control, these types of statistical prediction equations, or statistical transfer equations from one group of variables to another, often represent the most important statistical predictions in hydrology.

Two basic approaches for deriving sampling distributions. The two approaches available to a hydrologist for determining exactly or approximately the distribution of any sample statistic, or of parameters of a probability distribution, are: (a) the analytical approach, by means of mathematical statistical derivations, and (b) the experimental or Monte Carlo approach, by generating many samples of given size and characteristics, based on the property of the sample available. In this latter case, from each sample a statistic to be studied is estimated, and its experimental distribution is derived from all these estimates.

This latter method is of particular interest in hydrology. The availability of sophisticated digital computers permits generation of a large number m of samples of size N. For each statistic, there are m estimates from which its experimental sampling frequency distribution can be derived. This data generation method deserves a proper place in the application of statistics in hydrology. It is especially useful when the analytical method does not supply an accurate distribution of a given statistic or supplies none. This approach is the experimental statistical method of solving sampling distribution problems.

9.2. Parameters of Distributions of Basic Statistics.

Summary of equations of biased and unbiased estimates of basic sample parameters. Distributions of the sample parameters as estimates of population parameters are discussed in this text: the mean, $\bar{x}$; the variance, s^2; the standard deviation, s; the coefficient of variation, C_v; the skewness coefficient, C_s and the excess coefficient, E, for the univariate distributions; in addition, the covariance, s_{xy}, and the correlation coefficient, r, for bivariate distributions. The population parameters are estimated by the following expressions for sufficiently large sample sizes, say $N \geq 30$:

$$\bar{x} = \frac{1}{N}\sum_{i=1}^{N} x_1;\ s^2 = \frac{1}{N}\sum_{i=1}^{N}(x_i - \bar{x})^2 = m_2;\ s = \left[\frac{1}{N}\sum_{i=1}^{N}(x_i - \bar{x})\right]^{1/2} = m_2^{1/2} \quad 9.1$$

$$C_v = \frac{s}{\bar{x}};\ C_s = \frac{m_3}{m_2^{3/2}} = \frac{m_3}{s^3},\ \text{and}\ E = \frac{m_4}{m_2^2} - 3 = \frac{m_4}{s^4} - 3, \quad 9.2$$

in which m_2, m_3, and m_4 are sample central moments as estimates of population central moments, μ_2 μ_3, and μ_4.

For bivariate distributions, six parameters of Eqs. 9.1 and 9.2 are estimated for both x and y, while additional parameters, the covariance and the correlation coefficient, are estimated by

$$s_{xy} = \frac{1}{N}\sum_{i=1}^{N}(x_i - \bar{x})(y_i - \bar{y}),\ \text{and}\ r = \frac{s_{xy}}{s_x s_y}. \quad 9.3$$

Unbiased estimates, designated with a caret above the parameter symbol, of parameters of Eqs. 9.1 through 9.3 for small samples are

$$\hat{\bar{x}} = \bar{x};\ \hat{s}^2 = \frac{N}{N-1} s^2;\ \hat{s} = s\sqrt{\frac{N}{N-1}};\ \hat{C}_v = \frac{s}{\bar{x}}\sqrt{\frac{N}{N-1}}; \qquad 9.4$$

$$\hat{C}_s = C_s \frac{(N-1)N}{(N-2)^2};\ \hat{E} = \frac{m_4}{m_2^2}\frac{(N-1)(N^2-2N+3)}{N(N-2)(N-3)} -$$

$$3\frac{(N-1)(2N-3)}{N(N-2)(N-3)},\ \text{and}\ \hat{s}_{xy} = \frac{N}{N-2} s_{xy}, \qquad 9.5$$

and r by Eq. 9.3 with the unbiased values s_{xy}, s_x, and s_y. Some other approximations for unbiased values of C_s and E are given in Chapter 6. Population values of parameters in Eqs. 9.1 through 9.5 are designated, respectively, as $\mu, \sigma^2, \sigma, \eta, \gamma_1, \gamma_2 - 3, \sigma_{xy}$, and ρ. They are usually unknown constants in hydrology.

Parameters of sampling distributions. Each estimate $\hat{\alpha}$ of a parameter α is a function of sample values, or $\hat{\alpha} = g(x_1, x_2, \ldots, x_N)$. Each estimate is a known function of random variables. The distribution $F(\hat{\alpha})$ of $\hat{\alpha} = g(x_1, \ldots, x_N)$ is its sampling distribution, and $f(\hat{\alpha})$ is its sampling probability density function. Each parameter as a statistic estimated from a sample of a random variable is also a random variable. Its distribution is a function of the sample size, N, and of the distribution of the basic variable (s), F (x) or F (x, y). These conclusions apply also to multivariate distributions. For them, the total number of estimates of parameters is usually much greater than for univariate or bivariate cases.

Parameters of sampling distribution of the mean. The expected value of the sample mean $\bar{x}$ is the population mean μ; or the mean of the sample means is

$$E\bar{x} = \mu. \qquad 9.6$$

The sample mean converges with probability unity to the population mean as the sample size N goes to infinity.

The variance of the sample mean is [1, p. 345]

$$\text{var}\ \bar{x} = \frac{\sigma^2}{N} = \frac{\mu_2}{N}, \qquad 9.7$$

in which σ^2 is the population variance of x, an N is the sample size. It is assumed in Eq. 9.7 that sample means are independent, which can be nearly always considered as satisfied in hydrology for $N \geq 20$. Only for small values of N and highly dependent variables may this condition not be fulfilled. By using Eq. 9.7, the standard deviation of the sample mean is

$$\sigma(\bar{x}) = \frac{\sigma}{\sqrt{N}}. \qquad 9.8$$

The value of σ in Eqs. 9.7 and 9.8 is replaced by $\hat{s}$ as the unbiased estimate of σ. Then the coefficient of variation of $\bar{x}$ is $\sigma/\mu\sqrt{N}$, or its unbiased estimate is $\hat{s}/\bar{x}\sqrt{N}$. By using $(x_i - \mu)$, the third central moment [1, p. 345] of $\bar{x}$ is μ_3/N^2, so that the coefficient of skewness of $\bar{x}$ is

$$\gamma_1(\bar{x}) = \frac{\gamma_1}{\sqrt{N}}, \qquad 9.9$$

or for γ_1 estimated by the sample value C_s, $\hat{\gamma}_1(x) = C_s/\sqrt{N}$, where γ_1 and C_s are the population and the sample skewness coefficient of x, respectively.

The fourth moment [1, p. 345] of $\bar{x}$ is

$$\mu_4(\bar{x}) = \frac{3\mu_2^2}{N^2} + \frac{\mu_4 - 3\mu_2^2}{N^3}, \qquad 9.10$$

and the kurtosis coefficient of the sample mean is

$$\gamma_2(\bar{x}) = 3 + \frac{\gamma_2 - 3}{N}. \qquad 9.11$$

With the population kurtosis coefficient of x estimated by the sample excess coefficient, E, than $\hat{\gamma}_2 = E + 3$.

It follows from Eqs. 9.9 and 9.11 that $\gamma_1(\bar{x})$ is close to zero, and $\gamma_2(\bar{x})$ close to three, as soon as C_s and E parameters of x are small, and N is sufficiently large. For all practical purposes, the distribution of the sample mean $\bar{x}$ is normal, with μ and $\sigma/\sqrt{N}$ parameters; or because of unknown μ and σ, it has approximately $\bar{x}$ and $\hat{s}/\sqrt{N}$ as its parameters. For a random variable normally distributed, its sample mean is also normally distributed, $N(\mu, \sigma/\sqrt{N})$, or approximately $N(\bar{x}, \hat{s}/\sqrt{N})$, because $\gamma_1(\bar{x}) = \gamma_1/\sqrt{N} = 0$, and $\gamma_2(\bar{x}) = 3$.

The above distribution parameters of the sample mean are obtained under the assumption that x is an independent random variable. If x is serially correlated, it is possible to develop for every autocorrelation mathematical model of x-series, and for its every parameter, an effective sample size $N_e < N$, which takes into account this serial correlation. Because hydrologic variables are in most cases serially correlated, this reduction of the observed sample size N, to its effective sample size N_e, is an important technique for applying sampling theory in hydrology. This reduction of N to N_e is an expedient method in comparing hydrologic information in independent and dependent series of various sample sizes.

The variance of the sample mean for a linearly autocorrelated random variable is [2]

$$\text{var } \bar{x} = \sigma_x^2 \left[\frac{1}{N} + \frac{2}{N}\sum_{k=1}^{N-1}\left(1 - \frac{k}{N}\right)\rho_k\right], \qquad 9.12$$

in which the population autocorrelation coefficients, ρ_k, are defined by

$$\rho_k = \frac{\text{cov}\,(x_i,\ x_{i+k})}{(\text{var}\ x_i \cdot \text{var}\ x_{i+k})^{1/2}}, \tag{9.13}$$

which are estimated by the corresponding sample serial correlation coefficients r_k, and σ_x^2 by $\hat{s}_x^2$, so that Eq. 9.12 gives the effective sample size $N_e = \sigma_x^2/\text{var}\ \bar{x}$, or

$$N_e = \frac{N}{1 + 2 \sum_{k=1}^{N-1} \left(1 - \frac{k}{N}\right) r_k}. \tag{9.14}$$

Problems arise for high values of k, or close to $k = N - 1$, because r_k are unreliable estimates of ρ_k. For many hydrologic random series, r_k converges rapidly to $r = 0$ with an increase of k, or it starts to quickly fluctuate in narrow tolerence interval around $r = 0$, so that it is sufficient to select the first m values of r_k, which are significantly different from zero. Then Eq. 9.14 becomes

$$N_e = \frac{N}{1 + 2 \sum_{k=1}^{m} \left(1 - \frac{k}{N}\right) r_k}. \tag{9.15}$$

If the first serial correlation coefficient of the series of x, or ρ_1 of Eq. 9.13 for $k = 1$, is positive, and if the first-order autoregressive (or Markov) linear model, $x_i = \rho x_{i-1} + \varepsilon_i$, represents well the dependence of that series, with x_i a standardized random variable, then Eq. 9.12 gives a simple expression for the effective sample size [2] as

$$N_e = N\,\frac{1 - \rho_1}{1 + \rho_1}, \tag{9.16}$$

with ρ_1 estimated by the sample value r_1. So, for a series of $r_1 = 0.25$, $N_e = 3N/5$; for $r_1 = 0.50$, $N_e = N/3$, and for $r_1 = 0.75$, $N_e = N/7$. Because hydrologic stochastic variables (with eventual periodicities and trends removed) are often greatly serially correlated, Eq. 9.16, as an example, shows how important it is to reduce the observed sample size, N, to the effective sample size, N_e, before parameters of sampling distributions are computed. Equation 9.16 shows that autocorrelated hydrologic random variables require much longer periods of observation, $N = N_e\,(1 + \rho_1)/(1 - \rho_1)$, than an independent variable to obtain the same accuracy of the sample mean, or the same information in the sample mean.

An expression more accurate than Eq. 9.16 is [3,4]

$$N_e = \left[\frac{1}{N} \cdot \frac{1 + \rho_1}{1 - \rho_1} - \frac{2}{N}\,\frac{\rho_1\,(1 - \rho_1^N)}{(1 - \rho_1)^2}\right]^{-1}. \tag{9.17}$$

However, the second term on the right side of Eq. 9.17 is often small in comparison to the first term, and usually can be neglected for small ρ_1 and large N. For $\rho_1 = 0.10$ and $\rho_1 = 0.20$, and $N = 20$, the second term subtracts only 1.08 percent and 1.93 percent, respectively, from the first term. For $\rho_1 = 0.50$ and $N = 20$, this percentage is 6.67 percent. For small ρ_1 and/or large N (say $\rho_1 < 0.25$ and $N > 30$), Eq. 9.16 should be used. However, for large ρ_1 and/or small N (say $\rho_1 > 0.25$ and $N < 30$), Eq. 9.17 would give more accurate results than Eq. 9.16.

Distributions of sample means are often applied in hydrology to distributions of differences of subsample means or two sample means. The distribution of differences of two means of the two samples is discussed in Chapter 10. Usually, it is hypothesized whether these differences are significantly different from zero. Because many water resource problems depend greatly on the means, the distributions of sample means is an important subject of hydrology and of water resources planning and operation. Three examples are given for these distributions, all three taken from annual river flows, though conclusions are applicable to many other hydrologic random variables.

Example 9.4. To determine whether the series of annual flows [5] of the Rhine River at Basle, Switzerland, for 150 years of observations (1807 – 1957) is self-stationary (properties are independent of the absolute time), the series is divided into 15 periods of 10 years, and for each subseries the mean $\bar{x}$ is computed. This computation is done in modular coefficients, K_i, with $K_i = Q_i/\bar{Q}$, in which Q_i are annual flows and $\bar{Q} = 36.250$ cfs is the mean annual flow. Figures 7.7 and 7.8 show that K_i is approximately lognormally distributed, with $\bar{K}_i = 1.00$ and $s_K = 0.159$. The distribution of means $\bar{x}_{10}$ for $N = 10$ is required for this test. It was also found that the sample first serial correlation coefficient is $r_1 = 0.077$ for a 150 year long sample, while all other r_k values were not significantly different from zero. It is correct to reduce $N = 150$ to $N_e = 128.53$ by Eq. 9.16, and then to divide N_e into 15 subseries of $N_e = 8.57$. The values r_1 of subseries fluctuate in large limits, and the use of r_1 of each 10-year subseries is less reliable than the use of r_1 for $N = 150$ for the computation of N_e of $N = 10$. Sample central moments are: $m_2 = 0.02547$, $m_3 = 0.000437$, and $m_4 = 0.001791$. The skewness cofficient of K_i is $C_s = 0.108$, and its kurtosis coefficient is $k = 2.76$. However, Eqs. 7.24 and 7.25 of the lognormal distribution for $\eta = s_K$ give the skewness coefficient $\hat{\gamma}_1 = 0.480$, and the kurtosis coefficient $\hat{\gamma}_2 = 3.411$. Figure 7.7 shows that the fit of the lognormal curve on the right extreme is not the best because the extreme low values of K_i are not sufficiently large and representative for this lognormal function. This explains the difference between the computed C_s and k, and $\hat{\gamma}_1$ and $\hat{\gamma}_2$ of the fitted lognormal function. Unbiased parameters of the $\bar{K}$-distribution for $N_e = 8.57$ and for the above sample moments are: $E(\bar{K}) = 1$; $\text{var}\,\bar{K} = s_K^2/(N_e - 1) = 0.00295$; $\hat{s}(\bar{K}) = 0.0542$; $\hat{C}_s(\bar{K}) = C_s(N_e - 1)\sqrt{N_e}/(N_e - 2)^2 = 0.0556$, and $\hat{\gamma}_2(\bar{K}) = 2.935$ with γ_2 of Eq. 9.11 being 2.44. The normal distribution for $\bar{K}_{10}$ is applicable, with, N (1; 0.05421), for the test of the 10-year means being all from the same population. A direct computation of $s(\bar{K})$ of 15 means of $N = 10$ gives $s(\bar{K}) = 0.0383$ or $\hat{s}(\bar{K}) = 0.0394$.

Example 9.5. The average first serial correlation of annual river flows of 446 river gauging stations in western North America was found to be 0.197 [6, p. 25], with average flow records of 37 years. Equation 9.16 gives, then, an estimate of the average effective length of records of $N_e = 0{,}67N$, or $N_e = 24.6$ years, about 2/3 of the original sample size, provided the first-order Markov model is applicable. The average r_1 value of annual precipitation series of 1141 gauging stations [6] in the same region is 0.055 with 54 years the average length of observations. The estimate of average effective length of station records is then $N_K = 48.2$ years, a small reduction.

Example 9.6. There is a controversy about the mean annual flow and its distribution for the Colorado River at the Lee Ferry Station in Arizona, because of three factors: (1) flows were not measured between 1896 and 1922, but were estimated from data of surrounding stations; (2) there is significant water depletion, either by diversion out of the basin, or by consumption in the basin (reservoir evaporation, irrigation, small pond losses, and so forth); and (3) it seems that the period 1922–1960 had a sampling downward trend. Taking the reconstructed virgin flows [7] for the period of 1896–1959 (64 years of data), the mean is $\overline{Q} = 15.17 \cdot 10^6$ acre-ft, the standard deviation is $s_q = 4.24 \cdot 10^6$ acre–ft, and the first serial correlation coefficient is $r_1 = 0.184$. Equation 9.16 for this last value gives $N_e = 0.69 \cdot 64 = 44.2$, so that the standard deviation of the mean $\overline{Q}$ is $s(\overline{Q}) = s_q/\sqrt{N_e} = 0.64 \cdot 10^6$, and the distribution of $\overline{Q}$ is normal, $N(15.77 \cdot 10^6;\ 0.64 \cdot 10^6)$. Similarly, for the homogeneous sample reduced to depletion conditions of 1954–1957, the distribution of $\overline{Q}$ is normal, $N(12.89 \cdot 10^6,\ 3.96 \cdot 10^6)$, provided the first serial correlation of 0.184 is not predominately a sampling deviation in both cases of virgin and homogeneous 1954–1957 annual flows. As the average value r_1 for annual flows of rivers in western North America is 0.197, as stated in Example 9.5, the value $r_1 = 0.184$ for the Colorado River may be close to the true value, though the sampling downward trend would indicate that the true value is $\rho_1 < r_1$. If the interest is in determining the average 10-year virgin value of this river, of the probability that it will be exceeded 90 percent of 10-year periods, then the effective length, $N_e = 0.69 \cdot 10 = 6.9$ years, and the standard deviation of distribution of 10-year means is $s(\overline{Q}_{10}) = s_q/\sqrt{N_e} = 1.61 \cdot 10^6$ acre-ft. The value of the standardized normal variable for 90 percent exceedence is $t = -1.282$, so that the value required is $\overline{Q}_{10} = \overline{Q} + t_i\, s(\overline{Q}_{10}) = (15.17 - 1.282 \cdot 1.61)\, 10^6 = 13.11 \cdot 10^6$ acre-ft. Similarly, for the homogeneous sample reduced to depletion conditions of 1954–1957, $s(\overline{Q}_{10}) = 1.51 \cdot 10^6$ acre-ft, so that $\overline{Q}_{10} = (12.89 - 1.282 \cdot 1.51)\, 10^6 = 10.96 \cdot 10^6$ acre-ft. If the period 1922–1959 is used instead of 1896–1959, the above values will be smaller.

Parameters of sampling distribution of the variance. The expected value of the sample variance s^2 is asymptotically the population variance, or the mean of sample variances [1, p. 347] is

$$E(s^2) = \frac{N-1}{N}\sigma^2 = \left(1 - \frac{1}{N}\right)\sigma^2 \qquad 9.18$$

and the variance of the sample variance [1, p. 348] is

$$\mu_2(s^2) = \text{var } s^2 = \frac{\mu_4 - \mu_2^2}{N} - \frac{2(\mu_4 - 2\mu_2^2)}{N^2} + \frac{\mu_4 - 3\mu_2^2}{N^3}, \qquad 9.19$$

in which $\mu_2^2 = \sigma^4$. For a sufficiently large N, Eq. 9.19 is approximately

$$\text{var } s^2 = \frac{1}{N}\left(\mu_4 - \frac{N-3}{N-1}\sigma^4\right). \qquad 9.20$$

The population parameters σ, μ_2, and μ_4 in Eqs. 9.18 through 9.20 are estimated by the sample statistics, biased or unbiased: s, m_2, and m_4 (with m_2 an m_4 the sample central moments), and N is the sample size.

The third and fourth moments [1, p. 348] of the variance are

$$\mu_3(s^2) = \frac{\mu_6 - 3\mu_2\mu_4 - 6\mu_3^2 + 2\mu_2^3}{N^2} + 0\left(\frac{1}{N^3}\right), \qquad 9.21$$

and

$$\mu_4(s^2) = \frac{3(\mu_4 - \mu_2^2)^2}{N^2} + 0\left(\frac{1}{N^3}\right), \tag{9.22}$$

in which $0(1/N^3)$ are the second-order quantities in comparison with the main expressions of the right side of Eqs. 9.21 and 9.22. They can be neglected for $N \geq 20$. Equations 9.18, 9.19, 9.21, and 9.22 facilitate the computation of coefficients of variation, the skewness and the kurtosis coefficients for sampling distributions of the variance.

Example 9.4 has shown that in some problems, like testing hypotheses for the self-stationarity of a time series, a sample may be split into many small subsamples so that N may be very small, $N \ll 20$. In this case, $0(1/N^3)$ of Eqs. 9.21 and 9.22 may not be negligible. It may be found for Eq. 9.21 by developing further the following expected value of the third moment and retaining terms of the order $1/N^3$,

$$\mu_3(s^2) = E\left(m_2 - \frac{N-1}{N}\mu_2\right)^3; \tag{9.23}$$

for Eq. 9. 22 this can be done by further development of

$$\mu_4(s^2) = E\left(m_2 - \frac{N-1}{N}\mu_2\right)^4, \tag{9.24}$$

and retaining the additional term $0(1/N^3)$. So, both Eqs. 9.21 and 9.22 may be increased in accuracy in the case of small sample sizes.

The covariance between the sample mean and the sample variance is [1, p. 348]

$$\text{cov}(\bar{x}, s^2) = \frac{N-1}{N^2}\mu_3, \tag{9.25}$$

in which μ_3 is estimated by the sample third central moment, m_3. If the distribution of x-variable is symmetrical, or $E(m_3) = 0$, then for the normal and other symmetrical distributions the sample mean and the sample variance are linearly uncorrelated. However, since hydrologic variables are mostly skewed, the assumption currently used in most analyses of the independence of the sample mean and the sample variance is not valid for most variables in hydrology. The measure of correlation between $\bar{x}$ and s^2 is then

$$\rho(\bar{x}, s^2) = \frac{\text{cov}(\bar{x}, s^2)}{(\text{var } \bar{x} \cdot \text{var } s^2)^{1/2}}, \tag{9.26}$$

with var $\bar{x}$ given by Eq. 9.7, var s^2 given by Eq. 9.19, and cov $(\bar{x}, s^2)$ given by Eq. 9.25.

Equations 9.18 through 9.26 are valid for any distribution of x. For normal distributions of x, it can be shown that $\bar{x}$ and s^2 are not only linearly uncorrelated but also independent, so that s^2 is distributed with [1, p. 349]

$$E(s^2) = \frac{N-1}{N}\sigma^2, \text{ and var } s^2 = \frac{2(N-1)}{N^2}\sigma^4, \qquad 9.27$$

with σ^2 estimated by s^2, and σ^4 by s^4.

Similarly as it was done for the sample mean of autocorrelated variables the effective sample size $N_e(s^2) < N$ should replace N in Eqs. 9.18 through 9.27 with $N_e(s^2) \neq N_e(\bar{x})$. The effective sample size is different for each parameter, because variances of their sampling distributions are different.

The effective sample size in the case of sample variance for an autocorrelated variable is [2] approximately

$$N_e = \frac{N}{1 + 2\sum_{k=1}^{m} \rho_k^2}, \qquad 9.28$$

with the same objections raised as for Eqs. 9.12 through 9.15. In case the first autoregressive linear model is applicable, or $\rho_k = \rho_1^{|k|}$, for $|\rho_1| < 1$, then [2] approximately

$$N_e = N \cdot \frac{1-\rho_1^2}{1+\rho_1^2}, \qquad 9.29$$

with ρ_1 estimated by the sample value r_1. For $r_1 = 0.25$, $N_e = 0.872$ N, which is greater than the corresponding $N_e = 3N/5$ for the mean, or $r_1 = 0.50$ in the analysis of the sample variance is equivalent to $r_1 = 0.25$ in the analysis of the sample mean. For $r_1 = 0.75$, $N_e = 0.28$ N, or it is double the value of N_e for the mean.

Example 9.7. In the case of the series of annual river flows of the Rhine River [5] and the sampling distribution of the variance, $N_e = 9.87 \cong 10$, so that $N = 10$ can be used. Then, Eq. 9.18 gives $E(s_x^2) = 9s_x^2/10 = 0.0228$ for $s_x^2 = 0.02528$; for the sample moments m_2, m_3, and m_4 of Example 9.4, and for $m_6 = 0.000189$, Eq. 9.19 gives var $s_x^2 = 0.000104$. The standard deviation of the distribution of variance is $s(s_x^2) = 0.0102$. The skewness and kurtosis coefficients of this distribution are obtained from $m_3(s_x^2) = 0.738.10^{-6}$ and $m_4(s_x^2) = 0.393.10^{-7}$, computed by Eqs. 9.21 and 9.22, so that $C_s(s_x^2) = 0.695$ and $\hat{\gamma}_2(s_x^2) = 3.63$. They show that the distribution of variance is not close to a normal distribution. Because of a small $N_e = 9.87$, Eqs. 9.23 and 9.24 would give additional corrections for the two moments, so that now more accurate values of $m_3(s_x^2)$ and $m_4(s_x^2)$ would give the more accurate values of $C_s(s_x^2)$ and $\hat{\gamma}_2(s_x^2)$. As shown later by Eq. 9.64, the best distribution to use for the sample variance in this case is the gamma or χ^2-distribution, with two parameters.

Example 9.8. The sample mean and the sample variance of annual flows of the Rhine River are dependent, because these flows are lognormally or skewed distributed. Equation 9.26 for $N = 10$ and the data in Examples 9.4 and 9.7 give $\rho(\bar{x}, s^2) = 0.0734$. It is of the same order of magnitude as the autocorrelation coefficient of the series, $r_1 = 0.077$. The effective sample sizes should be used in determining the variances of basic sample statistics.

Parameters of sampling distribution of standard deviation. Similarly as in the case of variance, the sampling distribution of the standard deviation has [1, p. 353]

$$E(s) = \sqrt{\mu_2} + 0\left(\frac{1}{N}\right) = \sigma + 0\left(\frac{1}{N}\right), \qquad 9.30$$

and

$$\text{var } s = \frac{\mu_4 - \mu_2^2}{4\mu_2 N} + 0\left(\frac{1}{N^2}\right), \qquad 9.31$$

in which $0(1/N)$ and $0(1/N^2)$ are small quantities to be neglected for $N \geq 20$, and σ, μ_2, and μ_4 are estimated by the sample parameters s, $s^2 = m_2$, and m_4. For a normal distribution of x, the sufficiently accurate estimates are

$$E(s) = \sigma; \quad \text{var } s = \frac{\sigma^2}{2N} \cdot \qquad 9.32$$

The exact distribution of the standard deviation of a normal variable x, N (0, σ), is given by Eq. 9.66 for small values of N.

Example 9.9. The reservoir storage capacity necessary to regulate annual river flows on a given release regime, and during a given period n, is proportional to the standard deviation of annual flows, s_a, with all other conditions the same. Therefore, it is sufficient to determine the properties of these capacities for a standardized variable, and multiply them by s_a. The sampling distribution of s_a is required to determine various limits of this storage capacity. For a given water regulation pattern of the Colorado River, at the Lee Ferry Station, by all upstream reservoirs (assuming all sites have approximately the same values of s_a), the distribution of s_a has the following parameters for n = 10 years, [7], and for virgin flows of this river for the period 1896–1959 (64 years): (1) $E(s_a) = \sigma \approx s_a = 4.2 \cdot 10^6$ acre-ft, computed by Eq. 9.30; (2) For $m_2 = s_a^2 = 18.01 \cdot 10^{12}$ and $m_4 = 885.11 \cdot 10^{24}$ of 64 years, Eq. 9.31 gives var $s_a = 1.64 \cdot 10^{12}$, or $s(s_a) = 1.28 \cdot 10^6$ acre-ft. An appoximate value, $s(s_a) = 1.37 \cdot 10^6$, is obtained by Eq. 9.32, assuming the annual flows are normally distributed. The skewness coefficient of 64 values is $C_s = 1.60$, a rather unusually high value for annual flows. Therefore, $s(s_a) = 1.28 \cdot 10^6$ is a more accurate estimate. The exact distribution of s_a follows Eq. 9.66 with 6.9 degrees of freedom. The approximate distribution is normal, N $(4.24 \cdot 10^6;\ 1.28 \cdot 10^6)$. Similarly, for the homogeneous sample reduced on the depletion conditions of 1954–57, $E(s_a) = 3.96 \cdot 10^6$ acre-ft, and $s(s_a) = 1.07 \cdot 10^6$ acre-ft. This latter value is obtained by Eq. 9.31, because $C_s = 1.47$.

Parameters of sampling distributions of coefficients of variation, and skewness and kurtosis coefficients. These coefficients are functions of statistical moments. Expected values and variances of sampling distributions of the coefficient of variation C_v, of the skewness coefficient C_s, and of the kurtosis coefficient k, are given in this text.

The sample coefficient of variation has the expected value [1, p. 358]

$$E(C_v) = \eta = \frac{\sigma}{\mu}, \qquad 9.33$$

which is estimated by $\hat{s}/\bar{x}$ for small N, or by $C_v = s/\bar{x}$ for sufficiently large N, and the variance

$$\text{var } C_v = \frac{\mu^2(\mu_4 - \mu_2^2) - 4\mu\mu_2\mu_3 + 4\mu_2^3}{4\mu^4\mu_2 N}, \tag{9.34}$$

in which μ is the expected value of x, or its population mean, while the other parameters are central moments. They are estimated by the sample values $\bar{x}$, m_2, m_3, and m_4, respectively. For the normal distribution of x, truncated at $x = 0$, and for η fairly small, the variance of C_v is approximately [1, p. 358]

$$\text{var } C_v = \frac{\sigma^2}{2\mu^2 N}\left(1 + 2\frac{\sigma^2}{\mu^2}\right), \tag{9.35}$$

with μ estimated by $\bar{x}$ and σ^2 by $\hat{s}^2$.

The sample skewness coefficient has the expected value [1, p. 357]

$$E(C_s) = \gamma_1, \tag{9.36}$$

with γ_1 estimated by the sample skewness coefficient C_s, and the variance (multiplied by N) is

$$N \text{ var } C_s = \frac{\mu_6}{\mu_2^3} - 3\frac{\mu_3}{\mu_2^{3/2}}\frac{\mu_5}{\mu_2^{5/2}} - 6\frac{\mu_4}{\mu_3^2} + \frac{9\mu_3^2}{4\mu_2^3}\frac{\mu_4}{\mu_2^2} + \frac{35}{4}\frac{\mu_3^2}{\mu_2^3} + 9, \tag{9.37}$$

with μ_2, μ_3, μ_4, μ_5, and μ_6 estimated by the sample central moments m_2, m_3, m_4, m_5, and m_6. Equation 9.37 requires, therefore, the estimate of five i-th central moments, $i = 2, 3, 4, 5, 6$. Because of large errors in estimates of higher moments, Eq. 9.37 is applicable only for large samples, or when moments can be computed from the fitted probability function. For a two-parameter lognormal probability distribution, all moments are functions of the mean and the standard deviation of logarithms of a variable, x. Similarly, for the two-parameter gamma distribution, all moments are function of its two parameters. The application of Eq. 9.37 assumes that biases at the extremes (unrepresentative or high-in-error extreme values) are not such as to make the estimates of higher moments unreliable. For the normal population distribution of x [1, p. 357]

$$E(C_s) = 0, \text{ and var } C_s = \frac{6}{N}. \tag{9.38}$$

The sample kurtosis coefficient, k, has the expected value [1, p. 357]

$$E(k) = \gamma_2, \tag{9.39}$$

in which k is the sample kurtosis coefficient, γ_2 is the population kurtosis coefficient, and the variance is

$$\text{var } k = \frac{\mu_2^2\mu_8 - 4\mu_2\mu_4\mu_6 - 8\mu_2^2\mu_3\mu_5 - 4\mu_4^3 - \mu_2^2\mu_4^2 + 16\mu_2\mu_3^2\mu_4 + 16\mu_2^3\mu_3^2}{\mu_2^6 N}, \tag{9.40}$$

with all central moments estimated by the corresponding sample moments.

For the normal population distribution of x [1, p. 357],

$$E(k) = 3, \text{ or } E(E) = 0, \text{ and } \operatorname{var} k = \operatorname{var} E = \frac{24}{N}, \tag{9.41}$$

in which E is the excess coefficient, with $E = k - 3$.

The use of Eqs. 9.34, 9.37, and 9.40 would have been prohibitive from the computational point of view before the computer age. However, the current digital computers can now determine var C_v, var C_s, and var k, of these equations within a short span of computer time.

Example 9.10. The series of 13-day precipitation values is designated by $x_{p,\tau}$, at the gauging station of Durango, Colorado, for the period of n = 71 years (1895–1966), with $\tau = 1, 2, \ldots, 28$ being the sequence of 28 intervals of 13-days inside the year, and $p = 1, 2, \ldots, n = 71$ being the sequence of n years. It was found that 28 means of values of 13-day intervals, m_τ, and 28 standard deviations, s_τ, of these values about m_τ, follow the periodic movement of the year. Their 28 values of the coefficient of variation, $(C_v)_\tau = s_\tau / m_\tau$, either do not show a significant periodicity, or even very slight 12-month periodicity (depending on the test criteria). In other words, m_τ and s_τ are approximately proportional. To test whether the 28 values of $(C_v)_\tau$ do not significantly depart from a constant, $C_\tau = s/\bar{x}$, with $\bar{x}$ and s the mean and standard deviation of the entire series, these values are: $\bar{x} = 0.0523$ in., s = 0.0449 in., and $C_v = 0.8596$. Equation 9.33 states that $E(C_v) = C_v = 0.8596$, and Eq. 9.34 gives the variance, var $C_v = 0.000369$, with μ_2, μ_3, and μ_4 in Eq. 9.34 estimated by m_2, m_3, and m_4 of the entire series, with removed periodicities of m_τ and s_τ. This series is $\varepsilon_{p,\tau} = (x_{p,\tau} - m_\tau + \bar{x})\, s/s_x$ in which $\tau = 1, 2, \ldots, 28$, and $p = 1, 2, \ldots, 71$, with $m_2 = 4.3538$, $m_3 = 0.18251$, and $m_4 = 0.044474$. The estimates m_2, m_3, and m_4 of μ_2, μ_3, and μ_4 are reliable, because $N = 28\,n = 1988$ values, and it can be shown that the autocorrelation of $\varepsilon_{p,\tau}$ is negligible.

Parameters of sampling distributions of covariance and correlation coefficient. For the sample covariance, cov (x, y), of the bivariate (x, y), the expected value is [1, p. 358]

$$E[\operatorname{cov}(x, y)] = \frac{N-2}{N}\sigma_{xy}, \tag{9.42}$$

in which σ_{xy} is the population covariance estimated by the sample covariance, s_{xy}, and the variance

$$\operatorname{var} s_{xy} = \frac{\mu_{22} - \mu_{11}^2}{N} + 0\left(\frac{1}{N^2}\right), \tag{9.43}$$

in which

$$\mu_{22} = \frac{1}{N}\sum_{i=1}^{N}(x_i - \mu_x)^2 (y_i - \mu_y)^2, \text{ for } N \to \infty, \tag{9.44}$$

to be estimated by the sample value, m_{22}, with μ_x replaced by $\bar{x}$, μ_y by $\bar{y}$, and μ_{11}^2 in Eq. 9.43 replaced by s_{xy}^2.

For the sample correlation coefficient defined by

$$r = \frac{s_{xy}}{s_x s_y} = \frac{m_{11}}{\sqrt{m_{20} m_{02}}}, \tag{9.45}$$

with $|r| \leq 1$, the expected value is approximately

$$E(r) = \rho, \qquad 9.46$$

in which ρ is the population correlation coefficient to be estimated by the sample value, r, and the variance is [1, p. 359]

$$\text{var } r = \frac{\rho^2}{4N}\left(\frac{\mu_{40}}{\mu_{20}^2} + \frac{\mu_{04}}{\mu_{02}^2} + \frac{\mu_{22}}{\mu_{20}\,\mu_{02}} + \frac{4\mu_{22}}{\mu_{11}^2} - \frac{4\mu_{31}}{\mu_{11}\,\mu_{20}} - \frac{4\mu_{13}}{\mu_{11}\,\mu_{02}}\right), \qquad 9.47$$

in which ρ^2 is replaced by r^2. The population central moments μ_{ij} are

$$\mu_{ij} = \frac{1}{N}\sum_{k=1}^{N}(x_k - \mu_x)^i\,(y_k - \mu_y)^j, \text{ for } N = \infty, \qquad 9.48$$

with i = 1, 2, 3, 4, and j = 1, 2, 3, 4 in μ_{ij} of Eq. 9.48. All μ_{ij} moments are estimated by the sample central moments, m_{ij}. For a normal population distribution of x, Eq. 9.47 becomes [1, p. 359]

$$\text{var } r = \frac{(1 - \rho^2)^2}{N}, \qquad 9.49$$

and with ρ replaced by r, var $r \approx (1 - r^2)^2/N$. A more detailed discussion of the r-distribution is given in Chapter 11.

Example 9.11. The monthly precipitation values of two adjacent stations are often highly correlated, one reason being that variables from both stations have similar periodic components. The correlation coefficient between the monthly precipitation data at Fort Collins, Colorado, as the x variable, and ot Fort Morgan, Colorado, as the y variable, for the period of n = 54 years (1907 – 1963), or N = 12 n = 648, is $r_{x,y} = 0.637$. By removing the periodic components in monthly means and monthly standard deviations, m_τ and s_τ, the standardized variables $\varepsilon_{p,\tau} = (x_{p,\tau} - m_{\tau x})/s_{\tau x}$, and $\xi_{p,\tau} = (y_{p,\tau} - m_{\tau y})/s_{\tau y}$, have the correlation coefficient $r_{\varepsilon,\xi} = 0.552$, which measures the correlation between the stochastic independent components of x and y, or between the *noises*, by removing the periodic components as *signals*. The distribution of $r_{\varepsilon,\xi}$, has the mean, $E(r) = r_{\varepsilon,\xi} = 0.552$. The estimates of moments of Eq. 9.47 from N = 12 n = 648 values are: $m_{11} = r_{\varepsilon,\xi} = 0.552$; $m_{02} = 1$; $m_{22} = 4.224$; $m_{13} = 4.976$; $m_{31} = 4.053$, $m_{04} = 8.385$, and $m_{40} = 6.831$. Then Eq. 9.47 for ρ^2, estimated by $r^2_{\varepsilon,\xi}$, gives var $r_{\varepsilon,\xi} = 0.001607$. The test whether the regression prediction equations should be used in any transfer of information between monthly precipitation data of these two stations, can be performed.

9.3. Sampling Distribution Functions.

Two types of analytically derived sampling distributions. Sampling distributions analytically derived may be either *asymptotic* or *exact*. The asymptotic distribution requires N to be large. What this *largeness* is, is a relative concept in statistics.

Asymptotic normal sampling distributions. Distributions of sample moments and sample functions of moments (parameters described by moments) are asymptotically normal. Any sample moment is asymptotically normal for $N \to \infty$, with the expected values and the standard deviations of some of these moments given in the preceding subchapter.

The relevance in hydrology is mainly in functions of sample moments, such as the parameters $\bar{x}$, s_x, s_x^2, C_v, C_s, k, E, $\bar{y}$, s_y, s_y^2, s_{xy}, and r in the analysis of various properties of hydrologic random univariates or bivariates. The general theorem is that any sample property based on moments is also approximately normally distributed about the population characteristic, if N is sufficiently large. Some parameters converge faster than others to their population values as N increases. The mean of this sampling distribution is the population characteristic, nearly always estimated by the corresponding sample characteristic. The variance is of the form C/N, with C depending on the property investigated, under the condition that these sampling distribution parameters are finite.

Any sample central moment m_r, of univariate distributions, with r the order of the moment, has an asymptotic normal distribution. Its expected values and variances are [1, p. 350]

$$E(m_r) = \mu_r + O\left(\frac{1}{N}\right); \text{ and } N \text{ var } m_r = \mu_{2r} - 2r\mu_{r-1}\,\mu_{r+1} - \mu_r^2 + r^2\,\mu_2\,\mu_{r-1}^2 + + 0\left(\frac{1}{N^2}\right), \qquad 9.50$$

with the second and third order terms, $0\,(1/N)$ and $0\,(1/N^2)$, negligible for large N (say $N \geq 30$).

The sample order-statistic (percentile), x_p, is defined as the x-value for the sample probability: $P\,(X \leq x_p) = p$, and $q = 1 - p$. The sample percentile is asymptotically normal, with the expected value and the standard deviation [1, p. 369] given by

$$E\,(x_p) = \xi_p\,, \text{ and } s\,(x_p) = \frac{1}{f\,(\xi_p)}\sqrt{\frac{pq}{N}}\,, \qquad 9.51$$

in which the population value, ξ_p, is estimated by the sample value, x_p, and $f\,(\xi_p)$ by the sample frequency density at x_p.

In the case of the median value, M, or for $p = q = 0.50$, Eq. 9.51 becomes [1, p. 369]

$$E\,(M) = \xi_{0\cdot5}\,, \text{ and } s\,(M) = \frac{1}{2f\,(\xi_{0\cdot5})\,\sqrt{N}}\,. \qquad 9.52$$

For the normal variable with $f\,(\xi_{0\cdot5}) = 1/\sigma\sqrt{2\,\pi}$, Eq. 9.52 produces

$$E\,(M) = \mu\,, \text{ and } s\,(M) = \sigma\sqrt{\frac{\pi}{2N}} = \frac{1.2533\,\sigma}{\sqrt{N}}\,, \qquad 9.53$$

in which μ is the population median, and M is the sample median. Because $\bar{x}$, for the estimation of μ, has a smaller sampling standard deviation (error), $s\,(\bar{x}) = \sigma/\sqrt{N}$, than Eq. 9.53 gives for the median, the median should be estimated by the sample mean, $\bar{x}$, for the normal variable, and not by the sample median, M.

The joint distribution of two percentiles is also asymptotically normal. For $p_1 = 0.25$ and $p_2 = 0.75$, the lower and the upper quartiles, the semi-interquartile range, $Q = (x_{0 \cdot 75} - x_{0 \cdot 25})/2$, is asymptotically normal with the expected value, $E(Q) = (\xi_{0 \cdot 75} - \xi_{0 \cdot 25})/2$, and the standard deviation of Q [1, p. 370]

$$s(Q) = \frac{1}{8\sqrt{N}} \left[\frac{3}{f^2(\xi_{0 \cdot 75})} - \frac{2}{f(\xi_{0 \cdot 75})\, f(\xi_{0 \cdot 25})} + \frac{3}{f^2(\xi_{0 \cdot 25})} \right]^{1/2}, \qquad 9.54$$

which for a normal variable becomes $E(Q) = 0.6745\, \sigma$, and its standard deviation $s(Q) = 0.7867\, \sigma/\sqrt{N}$.

Asymptotic non-normal sampling distributions. The sample values, $x_1, x_2, .., x_N$, ordered from the smallest to the largest give the largest observed value, x_{max}, and the smallest observed value, x_{min}, as extremes. Their difference is the range of extremes, $R_e = x_{max} - x_{min}$. The m-th value from the top, and the m-th value from the bottom, as the order--statistics, are also random functions of sample values, or their sampling distributions are functions of F (x) and N, as well as of the order m, or $N - m + 1$, with $f(R_e) = \psi [F(x), N, m]$. The probability density function of the m-th value of x is [1, p. 370]

$$g_m(x) = N \binom{N-1}{m-1} [F(x)]^{N-m} [1 - F(x)]^{m-1} f(x), \qquad 9.55$$

in which $f(x) = F'(x)$. By substituting $y = N[1 - F(x)]$, Eq. 9.55 becomes

$$h_m(y) = \binom{N-1}{m-1} \left(1 - \frac{y}{N}\right)^{N-m} \left(\frac{y}{N}\right)^{m-1}, \qquad 9.56$$

for $0 \leq y \leq N$, and $h_m(y) = 0$ outside (0, N). For $N \to \infty$, and for any $y > 0$, Eq. 9.56 converges to

$$f_m(y) = \frac{y^{m-1}}{\Gamma(m)} e^{-y}, \qquad 9.57$$

which is one-parameter gamma probability density function, with $\alpha = m$. Similarly, for the m-th value from the bottom, or for the $(N - m + 1)$-th value from the top, the variable is $y = N\, F(x)$, and Eq. 9.57 is also valid for $f_{N-m+1}(y)$, with $\alpha = m$ the parameter. Therefore, the order-statistics are asymptotically distributed as one-parameter gamma distribution. For $m = 1$ or $m = N$, Eq. 9.57 becomes the simple exponential probability density function, $f(y) = f_N(y) = e^{-y}$.

For the normal distribution of x, and putting $z = -\ln y = -\ln N[1 - F(x)]$, the limit case of Eq. 9.56 for $N \to \infty$ becomes [1, p. 374]

$$f_m(z) = \frac{1}{\Gamma(m)} e^{-mz - e^{-z}}, \qquad 9.58$$

which is the double exponential probability density function, as the asymptotic distribution of the m-th and (N — m + 1)-th values of order--statistics. For m = 1, Eq. 9.58 reduces to Eq. 7.49. For N large, the m-th and (N — m + 1)-th values of x are related by simple linear transformations to variables having the limiting distribution given by Eq. 9.58. One of the simplest linear transformations is given by Eq. 7.48.

Applications in hydrology. Since the exponential and double exponential distributions are either extensively used or very much advocated for application to hydrologic extremes, the main question is how well distributions of Eqs. 9.57 and 9.58 approximate the exact distributions of Eq. 9.55 for various F (x), N, and m. This problem is not yet clear for many basic distributions applicable to hydrologic variables. Equation 9.55 is a bridge between the underlying distribution of x, and the asymptotic distributions of x_{max} and x_{min}. Present hydrologic practice is oriented to the study of the x_{max}-series, or the x_{min}-series as samples derived from the x-series but separated from the study of the x-distribution and the time dependence of the x-series. It is clear from Eq. 9.55 that $g_1(x_{max})$ and $g_2(x_{min})$, as p.d.f. of the first and the N-th values from the top, are highly dependent on F (x) as well as on N, for small N. Once F(x) is selected and fitted to x, for a given sample size, N, the distributions of x_{max} and x_{min} are exactly given by the p.d.f. of Eq. 9.55 for m = 1, or m = N. Difficulty arises because $g_1(x_{max})$ and $g_2(x_{max})$ change with N. Therefore, the limiting distributions for $N \rightarrow \infty$ should be used. There is a need for the study of extremes, not only for the derived samples of N maximum and N minimum values from a large sample of x by an empirical method but also by including all properties (distribution and time dependence) of the underlying variable, x.

Recently Todorovic [8] studied the maximum (supremum) of a random number of random variables in an interval. Under given conditions he obtained the distribution of this supremum composed of Poissonian distribution of the number of random variables and the underlying distribution of identically distributed random variables. Applied to flood peaks [9], as all flood peaks above a given basic value (when floods begin), and assuming all these peaks are distributed according to the simple exponential function, it has been shown that the distribution of annual (or any other interval) flood peaks are distributed as double exponential function of Eq. 7.49 or Eq. 9.58. Thus, the sampling distribution of interval supremum is an exact distribution and not an asymptotic distribution as derived by the previous authors, provided that the specified conditions are satisfied. The two advantages of this latter approach are the replacement of the asymptotic by the exact distribution and the more accurate estimates of two parameters by all flood peaks above the given level, instead of using for estimates only N flood peaks from N years of observations as in the Gumbel or other methods.

Equation 9.55 shows also that there is no theoretical justification for the use of Eqs. 9.57 and 9.58 for distributions of the maximum or minimum sums of n consecutive values of x, even as asymptotic sampling distributions.

Often, these equations are used for the maximum annual or seasonal flow of 10, 20, 30, 60, consecutive days or so, or for the smallest low flows for 7, 10, 15, 30, 60 days, and similar. As m changes for each of n values included in the sum, these values are then the sums of n variables, each differently distributed, and besides, they are mutually dependent. There is no proof that those sums under various conditions converge to distributions of Eqs. 9.57 and 9.58 as N goes to infinity, or N is sufficiently large.

Exact sampling distributions. Asymptotic sampling distributions are often grossly in error for small N. They depart sometimes significantly from the exact sampling distributions also for medium-size samples. Very often it is not known how significant these departures are. Exact sampling distributions are ideal cases for statistical inference. Since a parameter is a function of sample values, $g(x_1, x_2, \ldots, x_N)$, and since all x_i have the same distribution, $F(x)$, the problem is to find the distribution of a function of N identically distributed random variables. These g-functions are often complex and must be solved by approximate procedures. Their solution in the explicit form is not always either simple or possible. As soon as the distribution function, $F(x)$, of a random variable, x, is given, and a sample statistic defined, $\hat{a} = g(x_1, x_2, \ldots, x_N)$, knowledge of $F(x)$ and N determines the distribution of $\hat{a}$. The availability of digital computers has, however, decreased the need for the complex computations of exact distributions of statistics in explicit form, because the application of sampling theory is often satisfied with the numerical approximations of exact distributions.

Exact distributions can be obtained in explicit forms when x has a distribution to which the addition theorem applies. For these, the sums or the means of identically distributed variables have the same distribution function as individual variables. Such population distributions as the normal, gamma, binomial, and Poisson have this addition property.

The sums and means of independent variables, x_j, with various distributions of x_i, are

$$z_i = \sum_{j=1}^{m} x_j, \quad \text{and} \quad u_i = \frac{z_i}{m}. \tag{9.59}$$

If x_j is normal, $N(\mu, \sigma)$, the sampling distribution of z_i is also normal, $N(m\mu, \sigma\sqrt{m})$. The sampling distribution of u_i of the normal distributions is also normal, $N(\mu, \sigma/\sqrt{m})$. In the Cauchy distribution, u_i has the same distribution as x. Sums (z_i) or means (u_i) of independent gamma variables, $G(\alpha)$, are also gamma distributed, or the z_i-variable with $G(m\alpha)$ and the u_i-variable with $G(\alpha)$. If x_j is binomial with parameters n and p, or $B(n, p)$, then the sampling distribution of z_i is also binomial, $B(mn, p)$. If x_j is Poissonian, $P_0(\lambda)$, the sampling distribution of z_i is also Poissonian with the parameter $m\lambda$, or $P_0(m\lambda)$. There is no sense, however, in using u_i for the binomial and Poisson distributions, because they are not necessary integers like z_i.

Chi-square exact sampling distributions of parameters, drawn from the normal population distributions. It has been stated that the sample mean of a population normal distribution of x, N (μ, σ), is normally distributed, N $(\mu, \sigma/\sqrt{N})$. The quadratic sum of the sample of standardized variable ε is

$$\xi = \chi^2 = \sum_{i=1}^{N} \varepsilon_i^2 = \sum_{i=1}^{N} \left(\frac{x_i - \mu}{\sigma}\right)^2, \qquad 9.60$$

in which ε_i is the standardized independent random normal variable, N (0, 1), of x_i. This new variable, $\xi = \chi^2$, or chi-square, has an exact two-parameter gamma distribution of Eq. 7.39, with $\alpha = N/2$ and $\beta = 2$ (a constant) so that

$$f(\xi) = \frac{1}{2^{N/2}\,\Gamma(N/2)}\,\xi^{\frac{N}{2}-1}\,e^{-\xi/2}, \qquad 9.61$$

for $\xi \geq 0$, and $f(\xi) = 0$ for $\xi < 0$. Its mean, variance, and skewness coefficient are: $E(\chi^2) = N$, var $\chi^2 = 2N$, and $\gamma_1(\chi^2) = \sqrt{8/N}$. Equation 9.61 has, therefore, only one parameter, N. It is widely known as chi-square or χ^2-distribution. The sample size, N, is called by Fisher the number of degrees of freedom, designated in subsequent chapters by NDF, d.f., or simply as degrees of freedom.

In general, the number of degrees of freedom are sample sizes, N, decreased by υ, which is the number of constraints imposed either in fitting the probability function to empirical frequency distributions, or in defining parameters by using other parameters. For N values in the sample, NDF $= N - \upsilon$. For instance, the sample mean is defined by $\bar{x} = \Sigma x_i/N$, and when the sample variance is computed by $s^2 = \Sigma (x_i - \bar{x})^2/N$, the constraint imposed on $x_1, x_2, \ldots, x_N$ is already made by the mean of their sum being equal to $N\bar{x}$. Because NDF is defined as the difference of N and the number of constraints, the distribution of the sample variance has NDF $= N - 1$.

For $N = 1$, Eq. 9.61 becomes

$$f(\xi) = \frac{1}{\sqrt{2\pi}}\,\xi^{-\frac{1}{2}}\,e^{-\frac{1}{2}}, \qquad 9.62$$

with the parameters: $E(\chi^2) = 1$; var $\chi^2 = 2$ and $\gamma_1(\chi^2) = 2\sqrt{2}$.

The addition theorem is valied for the χ^2-distribution. If χ_1^2 is with NDF $= N_1$, and χ_2^2 with NDF $= N_2$, then the variable $\xi = \chi_1^2 + \chi_2^2$ is a χ^2-distribution with NDF $= N_1 + N_2$. Tables of χ^2-distribution with various NDF-values are currently available. It is often necessary to find the probability that χ^2 exceeds a given value χ_0^2 so that this tail probability is $P(\chi^2 > \chi_0^2) = 1 - F(\chi_0^2)$, which is usually available from the χ^2-tables. If χ_0^2 is defined by the probability $p\% = 100\,F(\chi_0^2)$, then χ_0^2 can be determined, and it is possible to compare the computed χ^2 with this χ_0^2.

As shown in the next chapter on testing statistical hypotheses, the application of the χ^2-distribution is associated with one of the best techniques for testing the goodness of fit of various functions to hydrologic data. In any attempt to replace the subjective tests of visual comparison of empirical curves with the fitted theoretical functions by an objective method the χ^2-distribution may best serve that purpose.

Various transformations of χ^2, or the functions of the random variable, χ^2, are associated with χ^2-distribution. For a normal variable x_i, N (0, σ), ε_i of Eq. 9.60 is $\varepsilon_i = x_i/\sigma$, or $x_i = \sigma\varepsilon_i$. Denoting $\xi_1 = \Sigma x_i^2 = \sigma^2\Sigma\varepsilon_i^2 = \sigma^2\xi$, then $\psi(\xi_1) \mid d\xi_1 \mid = f(\xi) \mid d\xi \mid$, or $\psi(\xi_1) = f(\xi)/\sigma^2$. Therefore, the following transformations of χ^2 have the corresponding distributions [1, p. 236]:

Variable, ξ_i	**Frequency function, $\psi(\xi_i)$, with $\xi_i > 0$**	
$\xi_1 = \sum_{i=1}^{N} x_i^2$	$\psi(\xi_1) = \dfrac{1}{2^{N/2}\,\sigma^N\,\Gamma(N/2)}\,\xi_1^{\frac{N}{2}-1}\,e^{-\xi_1/2\sigma^2}$	9.63
$\xi_2 = \dfrac{1}{N}\sum_{i=1}^{N} x_i^2$	$\psi(\xi_2) = \dfrac{(N/2)^{N/2}}{\sigma^N\,\Gamma(N/2)}\,\xi_2^{\frac{N}{2}-1}\,e^{-N\xi_2/2\sigma^2}$	9.64
$\xi_3 = \sqrt{\sum_{i=1}^{N} x_i^2}$	$\psi(\xi_3) = \dfrac{2}{2^{N/2}\,\sigma^N\,\Gamma(N/2)}\,\xi_3^{N-1}\,e^{-\xi_3^2/2\sigma^2}$	9.65
$\xi_4 = \sqrt{\dfrac{1}{N}\sum_{i=1}^{N} x_i^2}$	$\psi(\xi_4) = \dfrac{2\,(N/2)^{N/2}}{\sigma^N\,\Gamma(N/2)}\,\xi_4^{N-1}\,e^{-N\xi_4^2/2\sigma^2}$	9.66

In other words, the variables ξ_1/σ^2, $N\xi_2/\sigma^2$, ξ_3^2/σ^2, and $N\xi_4^2/\sigma^2$ all follow the χ^2-distribution of Eq. 9.61.

Example 9.12. For Example 9.7 of the Rhine River, the distribution of the sample variance of K_i, with $s_K^2 = 0.02528$, should follow the χ^2-distribution of Eq. 9.64 with two parameters, $\alpha = N/2$ and $\beta = 2\sigma^2/N$. Replacing N by $N_e = 9.87$, and σ^2 by s_K^2 then the first estimates are $a_1 = 4.94$ and $b_1 = 0.00522$. For the computed parameters of Example 9.7, $E(s_x^2) = 0.0228$ and var $s_x^2 = 0.000104$. Then Eq. 7.40 gives the second estimates $a_2 = 4.67$ and $b_2 = 0.00488$. Considering that the a_2 and b_2 estimates are obtained by the method of moments of Eqs. 7.40, rather than by the method of maximum likelihood (which is the method that should be used in case $\alpha < 10$), the agreement between the first estimates a_1 and b_1, and the second estimates a_2 and b_2, is very good.

Example 9.13. If deviations u_x, u_y, and u_z of instantaneous velocity components from their respective means at a point of a turbulent flow are all normal, N (0, σ), then the instantaneous variation of the velocity scalar, $u = \sqrt{u_x^2 + u_y^2 + u_z^2}$, around the mean velocity is distributed according to Eq. 9.65, with NDF = 3 and $\xi_3 = u/\sigma$, or

$$f(u) = \sqrt{\frac{2}{\pi}}\,\frac{u^2}{\sigma^2}\,e^{-u^2/2\sigma^2},$$

which is Maxwell distribution of Eq. 7.75 with $\alpha = \sigma\sqrt{2}$.

Exact distributions of some statistics related to the variance of the population normal distributions. It was stated earlier that the sample mean, $\bar{x}$, and the sample variance, s^2, of a normal distribution,

N (μ, σ), are both uncorrelated and independent. For the symmetrical non-normal distributions, they are linerarly uncorrelated but may be dependent. For skewed distributions, they are correlated, therefore dependent.

The statistic $\sqrt{N}(\bar{x} - \mu)/\sigma$ is normally distributed, N (0, 1) with $\bar{x}$ the sample mean, and μ and σ the population mean and standard deviation. The statistic Ns^2/σ^2 has χ^2-distribution with NDF = N − 1, with s^2 the sample variance. Then the unbiased estimate of the variance,

$$\xi_s = \frac{Ns^2}{N-1} = \frac{1}{N-1} \sum_{i=1}^{N} (x_i - \bar{x})^2$$

is distributed as

$$\psi(\xi_s) = \left(\frac{N-1}{2\sigma^2}\right)^{\frac{N-1}{2}} \frac{1}{\Gamma\left(\frac{N-1}{2}\right)} \xi_s^{\frac{N-3}{2}} e^{-(N-1)\xi_s/2\sigma^2}. \qquad 9.67$$

This is the same as saying that the statistic $\xi = (N-1)\hat{s}^2/\sigma^2$ has the χ^2-distribution with NDF = N − 1.

Student-distribution of sample parameters of normal population distributions. For N + 1 variables, $\xi, \xi_1, \xi_2, \ldots, \xi_N$, all independent and normal, N $(0, \sigma)$, the new variable is formed as

$$\eta = \sqrt{\frac{1}{N} \sum_{i=1}^{N} \xi_i^2}, \qquad 9.68$$

with η taken as always positive. Consider the ratio

$$t = \frac{\xi}{\eta} = \frac{\xi}{\sqrt{\frac{1}{N} \sum_{i=1}^{N} \xi_i^2}}, \qquad 9.69$$

in which ξ and η are two independent variables. The ratio t follows the Student distribution [1, p. 238]

$$\psi(t) = \frac{1}{\sqrt{N\pi}} \frac{\Gamma\left(\frac{N+1}{2}\right)}{\Gamma\left(\frac{N}{2}\right)} \left(1 + \frac{t^2}{N}\right)^{-\frac{N+1}{2}} \qquad 9.70$$

It is also often called the t-distribution. The parameter N in Eq. 9.70 is the number of degrees of freedom, or NDF = N. Thus the t-distribution is independent of the standard deviation σ of the basic variables ξ and η, and it is unimodal, symetrical, E (t) = 0, the r-th moments are finite for $r < N$, the mean is finite for $N > 0$, the variance is finite for $N > 2$, and var t equals $N/(N-2)$.

The probability for $|t|$ to exceed a value t_0 in any direction is the integral of the two tails, or $P(|t| > t_0) = 2[1 - \Theta(t_0)]$, where $\Theta(t_0)$

is the integral of $\varphi(t)$ of Eq. 9.70 from zero to t_0. These values $\varphi(t_0)$ of the t-statistic are often tabulated in books on statistics. For large N, the statistic t is asymptotically normal, N (0, 1). For small N, the t-distribution deviates significantly from the normal distribution. The probability of a large deviation from the mean, for small N, is much greater for the t-distribution than for the normal distribution.

The statistic, called the Student's ratio,

$$t = \frac{\sqrt{N}\,(\bar{x} - \mu)}{\hat{s}} = \frac{\sqrt{N-1}\,(\bar{x} - \mu)}{s}, \qquad 9.71$$

which is a different statistic from $\sqrt{N}\,(\bar{x} - \mu)/\sigma$, follows the Student t-distribution with NDF = N — 1, or

$$\varphi(t) = \frac{\Gamma\left(\frac{N}{2}\right)}{\Gamma\left(\frac{N-1}{2}\right)\sqrt{\pi(N-1)}}\left(1 + \frac{t^2}{N-1}\right)^{-\frac{N}{2}}. \qquad 9.72$$

Equation 9.72 may serve the purpose of testing hypotheses of how significant deviations of sample means, $\bar{x}_i$, are from some hypothetical value of the population mean, μ, or for testing the significance of the difference between the two sample means. This latter case is discussed in detail in Chapter 10.

Example 9.15. A precipitation station is situated in a seemingly homogeneous climatic region. Its mean is $K_1 = 0.90$ in modular coefficients, in comparison with the regional mean of $K = 1.00$, while the coefficient of variation of 26 annual values at this station is $C_v = 0.50$. What is the distribution of the difference? Equation 9.71 gives $t = -1.00$, and it is distributed according to Eq. 9.72, with NDF = 25.

Fisher F- and z-distributions of sample parameters of normal population distributions. Assume m + n independent random normal variables, N (0, σ), given as: $\xi_1, \xi_2, \ldots, \xi_m$ and $\eta_1, \eta_2, \ldots, \eta_n$, and combine them to form

$$\xi = \sum_{i=1}^{m} \xi_i^2 \text{ and } \eta = \sum_{j=1}^{n} \eta_j^2, \qquad 9.73$$

in which ξ and η are now two positively valued variables. The ratio

$$F = \frac{\xi}{\eta} = \frac{\sum_{j=1}^{m} \xi_i^2}{\sum_{i=1}^{n} \eta_j^2}, \qquad 9.74$$

follows the Fischer's F-distribution [1, p. 242] given for F = x as

$$f_{mn}(x) = \frac{\Gamma\left(\frac{m+n}{2}\right)}{\Gamma\left(\frac{m}{2}\right)\Gamma\left(\frac{n}{2}\right)} \cdot \frac{x^{\frac{m}{2}-1}}{(x+1)^{(m+n)/2}}, \qquad 9.75$$

for $F > 0$. Like the t-distribution, this F-distribution is independent of σ. For $m = 1$, the variable nx has the expression of the same form as the square of t of Eq. 9.70.

Fisher's z-variable is defined by

$$e^{2z} = \frac{n}{m} x = \frac{\frac{1}{m} \sum_{i=1}^{m} \xi_i^2}{\frac{1}{n} \sum_{j=1}^{n} \eta_i^2}, \qquad 9.76$$

with $E(e^{2z}) = n/(n-2)$, and $\text{var}(e^{2z}) = 2n^2 (m+n-2)/m(n-2)^2 \cdot (n-4)$, for $n > 4$. The z-variable has the distribution [1, p. 243]

$$\psi(z) = 2m^{\frac{m}{2}} n^{\frac{n}{2}} \frac{\Gamma\left(\frac{m+n}{2}\right)}{\Gamma\left(\frac{m}{2}\right)\Gamma\left(\frac{n}{2}\right)} \frac{e^{mz}}{(me^{2z}+n)^{(m+n)/2}}. \qquad 9.77$$

Application of the χ^2-, t-, F-, and z-distributions in hydrology. It is difficult to apply Eqs. 9.61, 9.70, 9.75, and 9.77 in hydrology because not too many basic hydrologic variables are normally distributed. These distributions are used in hydrology as the best approximations to distributions of various parameters, which meet the above conditions, as the only exact distributions available. The smaller the absolute values of the skewness and excess coefficients of a variable x, the more accurate the application of the above four distributions to hydrologic variables.

Experimental determination of sampling distributions. Contrary to opinions held by some statisticians, data generation or the Monte Carlo method of determining sampling distributions of various statistics has a future in applied sciences, particularly in hydrology. Once the technique of generating new long samples from the properties of available samples is well mastered, m new samples, each of a given size, may be generated with m as large as necessary. The values of a statistic α, which are estimated from m generated samples of size N, represent a new sample of size m giving the experimental distribution or sometimes even the experimental time dependence of this statistic. This is an approximate but sufficiently accurate way of determining sampling distributions of statistics for hydrology objectives.

Many hydrologic time series are a composite of periodic components (daily or monthly means and standard deviations, or in some cases, also other parameters) and a stochastic component. Because of diurnal and annual periodicities in several basic parameters, it is difficult to find theoretically the sampling distribution of various statistics of such complex hydrologic variables. Even if a variable follows a stationary (time-invariant) process, it is often difficult to analytically obtain either an exact, an aproximate, or even an asymptotic distribution of some statistics relevant to hydrology. The use of the experimental, Monte Carlo statistical method, and digital computer simulation of new samples, is an expedient way of solving the problem of obtaining approximate sampling distributions. The better the simulation of properties of a complex time series, and the larger the number of simulated series by the Monte Carlo method, the closer the approximation to the exact sampling distribution.

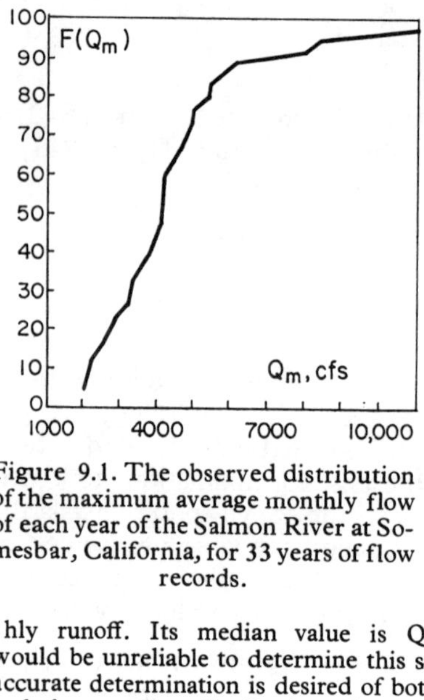

Figure 9.1. The observed distribution of the maximum average monthly flow of each year of the Salmon River at Somesbar, California, for 33 years of flow records.

Example 9.16. Figure 9.1 shows the observed frequency distribution curve of the largest average monthly flow in each year of the Salmon River at Somesbar, California, for 33 years of observation. Any extrapolation of distribution in Fig. 9.1 to obtain the maximum monthly value of a 100-year return period, or of Q_m for $P(Q \geq Q_m) = 0.01$, gives uncertain results. To find the distribution of this maximum value by theoretical consideration is even more uncertain because of periodicities in monthly means and standard deviations. After the series of monthly flow is decomposed, and all parameters estimated for various mathematical models, the use of the Monte Carlo simulation method or data generation method produced 100 samples each 100 years long (or each sample 1200 months long). By selecting the highest monthly value for each sample 100 years long, a new discrete series of 100 extreme values is obtained. Figure 9.2 gives this experimental statistical frequency distribution of the 100-year extremes of the annual average monthly runoff. Its median value is $Q = 43{,}000$ cfs. But, using this example, it would be unreliable to determine this statistic and its limits from Fig. 9.1. If a more accurate determination is desired of both the mean of the 100-year return period value, and the sampling distribution, than given by Fig. 9.2, the number of generated samples may be made much larger, so that the relative frequency of a 100-year value is better estimated than if only 100 samples are generated. These estimates are limited by the information contained in the original sample of 396 average monthly river flows, and by the economy of simulating data on a digital computer.

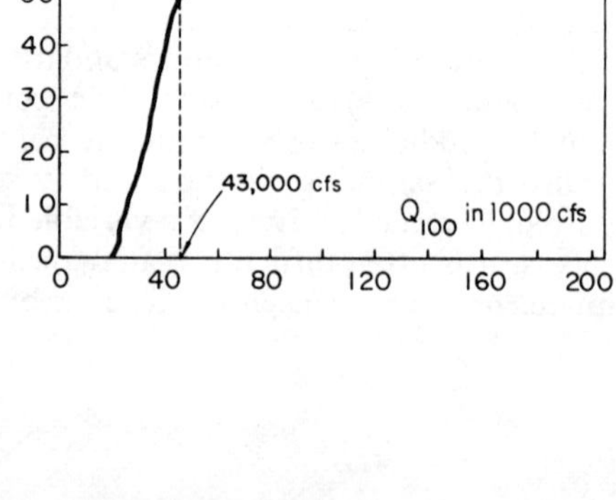

Figure 9.2. The experimental sampling distribution of the maximum average monthly flow in each year, of a 100-year return period, or $P(Q_{max} \geq Q_{100}) = 0.01$ of the Salmon River at Somesbar, California. One hundred generated samples of monthly flows, each sample 100-years long, produce 100 values of the 100-year maximum value.

9.4. Confidence and Tolerance Intervals and Limits.

Definitions and properties of confidence intervals and limits. The confidence interval is defined as a range around an estimated sample statistic, $\hat{\alpha}$. If a statement is made that the probability of this interval to include the population value α is $1 - \varepsilon$, where $0 < \varepsilon < 1$, this statement is expected to be correct $(1 - \varepsilon)$ n times and incorrect εn times, out of n cases. This probability, $P_c = 1 - \varepsilon$, is called the confidence level. The confidence interval at the 95 percent level ($P_c = 0.95$, or 95 percent, and $\varepsilon = 0.05$) means that out of a very large number n of samples of equal size, N, it is expected that 95 percent of the time the statement about the population value α being inside the confidence interval will be correct, while 5 percent of the time it will not. Because each sample of size N has a different estimate, α_i, the confidence interval changes from one sample to another because the interval is a function of the corresponding α_i, and the sampling distribution has usually $E(\hat{\alpha}) = \alpha$, with $\hat{\alpha}$ the sample estimate in each case. The confidence limits are numerical boundaries of the confidence interval for a given estimate $\hat{\alpha}$. The confidence interval and the confidence limits are based on a fundamental restriction, namely that the population value α is not known, and only an estimate of it in the form of the sample statistic $\hat{\alpha}$ is available. If α_1 and α_2 designate the changing confidence limits of sample statistics $\hat{\alpha}$, computed for a large number of samples of size N, then for all these samples the statement

$$P(\alpha_1 \leq \alpha \leq \alpha_2) = P_c = 1 - \varepsilon \qquad 9.78$$

represents the probability definition of the confidence interval and its limits. The additional condition is that $\hat{\alpha}$ divides the interval between α_1 and α_2 of the sampling distribution in two subranges, each with equal probability P/2 for symmetrical sampling distributions. This can be expressed then by

$$P(\alpha_1 \leq \alpha_i \leq \hat{\alpha}) = P(\hat{\alpha} \leq \alpha_i \leq \alpha_2) = P_c/2, \qquad 9.79$$

with α_1 and α_2 the confidence limits, $\hat{\alpha}$ the estimate of the population value α, and α_i any possible value of α.

Because the estimate $\hat{\alpha}$ is a function, $g(x_1, x_2, \ldots, x_N)$, of sample values, the confidence limits are also functions of sample values, or $\alpha_1 = g_1(x_1, x_2, \ldots, x_N)$, and $\alpha_2 = g_2(x_1, x_2, \ldots, x_N)$. As $x_1, x_2, \ldots, x_N$ change from sample to sample, all three values $\hat{\alpha}$, α_1, and α_2 also change from one sample to another. The confidence interval and its limits depend also on the estimation method used in obtaining $\hat{\alpha}$. The efficient estimates have the narrowest confidence intervals.

For arbitrary values $0 < \varepsilon_1 < 1$, and $0 < \varepsilon_2 < 1$, with $\varepsilon_1 + \varepsilon_2 = \varepsilon$, and $\varepsilon_1 \geq \varepsilon_2$, the probability

$$1 - P_c = \int_{-\infty}^{\alpha_1} g(\hat{\alpha})\, d\alpha + \int_{\alpha_2}^{+\infty} g(\hat{\alpha})\, d\alpha = \varepsilon_1 + \varepsilon_2 = \varepsilon \qquad 9.80$$

satisfies Eq. 9.78. To satisfy Eq. 9.79, the condition is $\varepsilon_1 = \varepsilon_2 = \varepsilon/2$, or each tail of $g(\hat{\alpha})$ has the same probability $\varepsilon/2 = (1 - P_c)/2$. Either the analytical or the experimental sampling distribution of $\hat{\alpha}$ is required to compute the limits that satisfy Eqs. 9.78 and 9.79.

Definitions and properties of tolerance intervals and limits. If the population value α is known, or if it is assumed to be a given value, and all other population parameters that define the sampling distribution of the statistic $\hat{\alpha}$ are also known, then an interval can be put around α. The distribution of a sample estimate, $\hat{\alpha}$, in the form $g(\hat{\alpha}; \alpha)$ depends now on the sample size and the population parameters including α, and $\hat{\alpha}$ is the random variable. In this case, an interval or its limits are drawn about α, that provide a specified proportion of all population values of $\hat{\alpha}$-statistic that may be expected to lie inside the interval with a given degree of tolerence, P_t. Such intervals and limits are called tolerance intervals and tolerance limits. Even when some of the population parameters in $g(\hat{\alpha}; \alpha, \beta, \ldots)$ must be estimated by the sample values, except α, one still may speak in terms of tolerance intervals or tolerance limits. Equations 9.78 through 9.80 are also valid, except that the sampling distribution of $\hat{\alpha}$ is about the known or assumed α, as the expected value of $g(\hat{\alpha}; \alpha, \beta, \ldots)$, and the tolerance limits are mainly a function of the sample size, N. Designating the tolerance limits by α_1 and α_2, the expected percentage of all the $\hat{\alpha}$-population falling within the tolerance interval is $P_t = 1 - \varepsilon$; the expected percentage for it to fall outside the interval is $1 - P_t = \varepsilon$. This latter case is applicable to many practical problems when α is assumed to be given or known.

The application of tolerance intervals and limits is a useful procedure in hydrology for testing various hypotheses. An example is the testing of the linear independence of two time series, or testing whether a given series is or is not serially correlated. Then, the population correlation coefficient, or the population serial correlation coefficients, are assumed to be zeros, and the tolerance limits are drawn about them in order to investigate whether the sample estimates fall inside or outside the tolerance interval.

Determining confidence or tolerance limits. Selecting the probability P_c or P_t of Eqs. 9.78 through 9.80 is a decision made a priori. Usually, it is a convention agreed upon prior to analysis of the sampling distribution, and prior to any decision making based on the statistical inference. In hydrology, the probabilities P_c or P_t are conventionally 0.90 or 0.95 (90 percent, or 95 percent, but values of 0.80 or 0.99 also have been used). Another definition of the probability level refers to the two tails as

$$1 - P_c = P(\alpha < \alpha_1; \alpha > \alpha_2); \text{ or } 1 - P_t = P(\hat{\alpha} < \alpha_1; \hat{\alpha} > \alpha_2). \qquad 9.81$$

The value $1 - P_c$ is the probability that the population value α in a very large number of samples lies outside the confidence limits. The value $1 - P_t$ is the probability that the statistic $\hat{\alpha}$ lies outside the tolerance limits. These are probabilities of a value being at one of the two tails of a sampling distribution. In this case, $1 - P_c = 1 - P_t = 0.10$ or 0.05, or in some cases 0.20 or 0.01.

The frequency of any class interval (or the frequency density of any variable value) of an empirical distribution may be considered as a statistic. Its two confidence limits can be determined in each particular case. Because all frequencies represent a set of statistics, the confidence limits of a continuous frequency curve are two limiting continuous curves. The sampling variance of the sample frequency density is smallest for the expected value of a variable. It increases as the deviations of x values from the expected value increase. The confidence interval of a statistic for a given sample size is related to the sampling variance, and increases with an increase of this variance. Therefore, the confidence interval around an empirical distribution curve, called the confidence region, expands as the variable values depart from its expected value. Similarly, for a bivariate distribution, represented as a frequency surface, the confidence limits are two surfaces about this frequency distribution. The accuracy of determining confidence limits is directly related to the reliability of the sampling distribution determined for any statistic. The limiting curves of sampling distributions of relative frequencies about the assumed or known population distribution of a random variable x are conceived as tolerance limit curves, and between them is the tolerance region.

It is often important to draw confidence limit curves to any hydrologic empirical curve that is used in various applications. Examples are flood peaks and low flow probability distributions, or any other hydrologic curve which involves sampling errors. A water resource planner, or a designer of waterworks, can then check the risk or effect involved, if the population curve departs significantly from the sample curve. Practically, no hydrologic probability curve or any other line, which has statistical sampling properties, should be drawn without a confidence region. Often the approximate confidence limits are drawn when equations for computation of exact limits either do not exist or are time consuming to determine.

Example 9.17. A river flow is regulated by a given reservoir capacity to supply water to several users. The average annual water available for delivery in a period of $n = 10$ years is proportional to the average annual inflow, Q_i, of each n-year period into the reservoir, but inversely proportional to the standard deviation, s_i, of annual flows in a given regulating period of n years, or $W_i = a\,Q_i/s_i$. This supply is then $W_i = a/\eta_i$, with $\eta_i = s_i/Q_i$ the coefficient of variation of annual inflows of each 10-year period, and a the proportionality constant. The expected value of η_i is approximately $E(\eta_i) = s(Q)/Q$ in which s (Q) is the overall standard deviation of annual river inflows, and Q is the annual mean. Assume that Q_i and s_i are independent parameters. To find the confidence limits of the average annual water supply during each 10-year period, at the probability level $P_c = 0.80$, the distribution for $n = 10$ of the coefficient of variation, C_v, as an estimate od η_i is necessary. In this example, $Q = 900$ cfs and $s(Q) = 450$ cfs. The mean coefficient of variation is $\overline{C}_v = \overline{\eta} = 0.50$, which gives $W = 2a$. According to Eq. 9.35 and assuming C_v is approximately normally distributed, then $\text{var}\,C_v = \overline{\eta}^2 (1 + 2\overline{\eta}^2)/2n = 0.01875$. For $P_c = 0.80$, $t = \pm 1.28$ of the standard normal distribution, $\sqrt{\text{var}\,C_v} = 0.137$, and the limits of C_v are $0.50 \pm 1.28 \cdot 0.137$, or 0.325 and 0.675. The confidence limits of the average annual water supply for a regulation period of 10 years are $W_1 = a/0.675 = 1.48$ a, and $W_2 = 3.07$ a, with the mean $W = 2a$.

If the values Q = 900 cfs and s (Q) = 450 cfs are very close to population parameters, the average annual water supply will be 80 percent of the time, out of a large number of 10-year period, in the limits 1.48a to 3.07a.

Joint confidence limits of two or more parameters. There is a current hydrologic need for applying confidence limits to two or more parameters simultaneously. A typical example is the water yield of large reservoirs with overyear regulation capacity. The annual inflows are characterized by the mean annual flow, μ, their standard deviation, σ, and the first serial correlation, ρ_1, for a first-order linear autoregression dependence of annual flow series. These three parameters are estimated by sample statistics, $\bar{x}$, s, and r_1. They may or may not be dependent parameters. The storage capacity for a given regulation rule is a function of s and r_1. The water supply, W, is primarily a function of $\bar{x}$, so that $W = f(\bar{x}, s, r_1)$. To place confidence limits on a computed benefit for n years of operation, limits must be placed on all three parameters $\bar{x}$, s, and r_1. This is a simple task if a population distribution is normal because $\bar{x}$ and s are independent; they may also be independent of r_1. However, for a skewed distribution, $\bar{x}$ and s are dependent, and they may be also dependent with r_1. A logarithmic transformation may normalize the case and justify making the assumption that $\bar{x}_n$, s_n, and r_1 are mutually independent, at least approximately.

For two parameters $\bar{x}$ and s, the confidence region is approximately an ellipse, while for the three parameters, $\bar{x}$, s, and r_1, the confidence space is approximately an ellipsoid. Assume that the probability is P_c of all values $\bar{x}$, s, and r_1 inside the ellipsoid for a very large number of samples of size n. From this, the computation of confidence limits of $W = f(\bar{x}, s, r_1)$ would be relatively difficult. Two practical approaches are feasible. First, if the W-function is determined, and the marginal distributions of $\bar{x}$, s, and r_1 are known, and these parameters are mutually independent, the distribution of W, or of the corresponding benefit, can be computed and from it the confidence limits. Second, the function of W may be difficult to obtain, which makes it necessary to investigate the problem experimentally. Besides, $\bar{x}$, s, and r_1 may be mutually dependent. In this case, the Monte Carlo method of data simulation may be the only practical way of applying reliable confidence limits to water supply and the resulting benefits. In this method, the joint distribution of $\bar{x}$, s, and r_1 for the dependent case, or marginal distributions for the independent case, are divided in such a way that a sufficiently large number of ($\bar{x}_i$, s_i, and $r_{1,i}$)-values, with equal frequencies of class-regions around them, may be singled out. By simulating samples of annual flows of n years, and for each ($\bar{x}_i$, s_i, $r_{1,i}$) -value, n estimates of W are obtained, and from them the sampling distribution of W. This case underlines importance of the experimental statistical inference method in hydrology.

Example 9.18. A simple hydrologic example is given as an approximate graphical estimate of the confidence region of a distribution. Assume that the distribution of a hydrologic variable, x, plots approximately as a straight line in cartesian-probability scales, or x is normally distributed, as N ($\bar{x}$, s). By knowing distributions of $\bar{x}$ and s, the limits of $\bar{x}$, as x_1 and x_2, and of s, as s_1 and s_2, for a given probability level can be determined. By plotting the values x_1 and x_2 at 50 percent probability, and by drawing through them lines with slopes corresponding to s_1 and s_2 in such a way that all four lines diverge from the straight line fitted to the distribution, the upper and lower confidence limit lines are obtained. They form the confidence region, which is an approximate but often accurate estimate for current hydrologic applications. If the distribution is lognormal, the same procedure may be used for logarithms of the variable as outlined for the normal distribution. This case is shown in Fig. 9.3 for the distribution of the

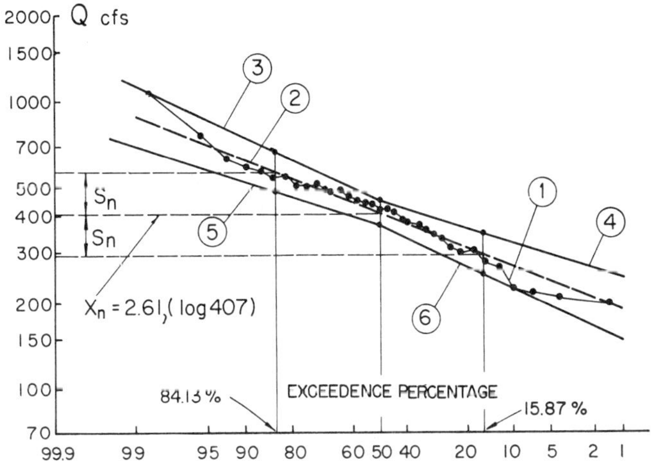

Figure 9.3. Applying graphical estimates of the confidence curves of the confidence region for the lognormal distribution of annual minimum one-day flows of the Meramec River at Eureka, Missouri, 1922–1956 ($N = 35$): (1) plotted points by $P(x_m) = m/(N + 1)$; (2) lognormal straight line fit; (3) upper-left straight-line limit obtained by using the upper limits of $\bar{x}_n$ and s_n; (4) upper-right straight-line limit obtained by using the upper limit of $\bar{x}_n$ and the lower limit of s_n; (5) lower-left straight-line limit obtained by using the lower limits of $\bar{x}_n$ and s_n, and (6) lower-right straight-line limit obtained by using the lower limit of $\bar{x}_n$ and the upper limit of s_n.

minimum annual one-day low flow of the Meramec River at Eureka, Missouri, for the period 1922–1956 (35 years). Graphical estimates of the mean, $\bar{x}_n$, and the standard deviation, s_n, of logarithms in Fig. 9.3 are: $\bar{x}_n = 2.61$, or it is log 407; $s_n = 0.143$, or it is (log 567 – log 294)/2. For a confidence level of 90 percent the standard normal variable gives $t = \pm 1.64$. For the normal distribution of the mean $\bar{x}_n$ of logarithms for $N = 35$, using Eq. 9.8, the confidence limits are $\bar{x}_n \pm t\, s_n/\sqrt{N} = 2.61 \pm 0.04$, or 2.65 and 2.57. Assuming the standard deviation s_n of logarithms is approximately normal, and using Eq. 9.32, $E(s_n) = s_n = 0.143$, $s(s_n) = s_n/\sqrt{2N} = 0.0174$, and its confidence limits on the 90 percent level are $0.143 \pm 1.64 \cdot 0.017$, or 0.171 and 0.115, respectively. The four straight lines of the confidence region are plotted in Fig. 9.3 as described above.

9.5. Content of Information in Various Parameters.

Concept of information. Following the thoughts of R. A. Fisher, the concept of information in any statistic, $\hat{\alpha}$, estimated from a sample is defined as the inverse of the variance of that statistic, or

$$I = \frac{1}{\text{var } \hat{\alpha}} = \frac{1}{\sigma^2(\hat{\alpha})}, \tag{9.82}$$

in which $\sigma^2(\hat{\alpha})$ is determined either analytically or experimentally. Since the variance is always positive, the information, I, varies from zero to infinity. This is valid if a population distribution is normal .

Information content of the mean. In the case of an independent variable, Eq. 9.8 gives the variance of the sample mean so that the information content of the mean, provided the variable is not serially dependent, is

$$I = \frac{N}{\sigma^2}. \tag{9.83}$$

For serially dependent variables and the same sample variance, the information content of the sample mean is smaller than given by Eq. 9.83 [4]. To use Eq. 9.83, the effective sample size N_e, as given by Eqs. 9.14 through 9.17, should be used instead of N.

Information content of other parameters. Similarly as for the mean, Eq. 9.82 is applicable to other parameters. For the sample variance, with the approximate variance of the sample variance given by Eq. 9.20, the information contained in the sample variance for sufficiently large N is

$$I(s^2) = \frac{1}{\text{var } s^2} = \frac{N(N-1)}{(N-1)\, m_4 - (N-3)s^4}. \tag{9.84}$$

Example 9.19. The information content in the mean of various hydrologic time series, x_t, is often of interest. The time interval Δt into which a series is divided (such as daily, weekly, monthly, seasonal, annual precipitation or runoff) is called here the time series measure. Every hydrologic series x_t for a given Δt has a random independent component ε, with the sample mean $\bar{\varepsilon}$ and the population variance σ_ε^2. The periodic components in the mean and standard deviation, or the linear dependence in the stochastic component, are considered as being deterministic properties of the series. The sample mean $\bar{\varepsilon}$ has the variance

$$\text{var } \bar{\varepsilon} = \frac{\text{var } \varepsilon}{N} = \frac{\sigma_\varepsilon^2}{N} = \frac{\sigma_\varepsilon^2 \Delta t}{T}.$$

As Δt increases, say from 1-day to 1-month values, σ_ε^2 decreases but not as rapidly as Δt increases. The product $\sigma_\varepsilon^2\, \Delta t$, therefore, increases with an increase of Δt, or the information content in the sample mean $\bar{\varepsilon}$ decreases with an increase of Δt for T = = constant. Though the mean river discharge is the same whether computed from daily or annual flows, the variance of the mean of independent stochastic component, ε, is smaller if determined from daily discharges rather than from annual discharges [10].

9.6. Problems for Discussion and Solutions.

9.1. For Problems 6.1 through 6.3 find the variances of the computed sample mean, variance, standard deviation, coefficient of variation, skewness, and the excess coefficients, assuming the basic variable is normally distributed.

9.2. For Problem 6.8 find the variance of the computed sample correlation coefficient.

9.3. Parameters of central tendency, dispersion, skewness, and flatness are of different reliabilities. The variance of the sample parameter about the population value is a measure of the accuracy of each parameter. Using this criterion, and assuming the basic variable normally distributed, sort parameters, whose sampling distributions are discussed in this chapter, in order of accuracy from the most to the least accurate parameter.

9.4. For chi-square, Student-t, and Fisher-F distributions define the type of variables that follow these distributions, describe the conditions necessary for their application, and give two to three examples of application for each of these three functions in hydrology.

9.5. For the curves of straight-line fits in the form of normal and lognormal distributions of Problem 7.7, draw graphically the approximate confidence lines on the 95 percent probability level, following the procedure given in this chapter.

9.6. For $N = 10$, find the confidence limits on the 95 percent level for the mean and the standard deviation of annual river flows of the Rhine River, with its basic data taken from Example 9.4, and assume the mean is normally distributed and the standard deviation is distributed according to Eq. 9.66. Draw the confidence limit curves similarly as it was done in Example 9.18.

9.7. For Example 9.11, find the confidence limits on the 95 percent level of $|r_{\varepsilon,\xi}|$, assuming the correlation coefficient is Beta-distributed with the boundaries at $+1$ and -1.

9.8. The average 25-year benefit of a water resources project with a large reservoir is $B = A_0 + A_1 \bar{x} + A_2 s^{-1} r_1^{-n}$, in which $\bar{x}$ is the mean annual flow, s is the standard deviation of annual flows, and r_1 is the first serial correlation coefficient of annual flows. The parameters $\bar{x}$, s, and r_1 are independent, and A_0, A_1, A_2, and n are constants. Describe practical methods for finding the 90 percent limits to the benefit B.

CHAPTER 10

TESTING HYPOTHESES AND GOODNESS OF FIT

Because hydrologic information is primarily derived from sample data many practical hypotheses about the populations of hydrologic random variables must be postulated and tested. Available techniques in mathematical statistics for testing various types of hypotheses are relevant to hydrology. Because hydrologic practice is filled with fitting functions to the empirical relations, the goodness of fit tests of these functions are necessary steps in drawing from data the most reliable hydrologic information.

10.1. Definitions of Hypotheses and Their Testing.

Definition of a hypothesis. A statistical hypothesis is an assumption having a logical basis in a problem about the properties of distributions of random variables. Presently in hydrologic practice this assumption may refer to a distribution function, to some parameters of this function, or to any property associated with the distribution of a random variable. In the case of hydrologic series, this assumption may refer to any parameter or any other characteristic of their sequential dependence.

Example 10.1. Observed times, x, between successive rainfall storms in a thunderstorm region and for a particular time interval of the year are assumed to be exponentially distributed as

$$f(x; \Theta) = \Theta e^{-\Theta x}. \tag{10.1}$$

This is the gamma distribution of Eq. 7.39 with parameters $\alpha = 1$ and $\beta = 1/\Theta$. Several hypotheses about this assumed distribution may be constructed. First, a hypothesis may be advanced that the probability density function of Eq. 10.1 fits well the frequency density curve of observed times, x_i, $i = 1, 2, \ldots, N$, between the successive storms. This hypothesis is designated as H_0 or the null hypothesis. An alternative hypothesis to this first one is that Eq. 10.1 does not fit well the empirical distribution, which is designated as H_1. A parameter is necessary to measure this goodness of fit. These two alternative hypotheses are decided on the basis of the sampling distribution of this parameter, with regions defining what is an acceptable and what is an unacceptable fit. In this case, testing hypotheses becomes equivalent to testing the goodness of fit of a probability function to the empirical distribution.

Another H_0 hypothesis, related to Θ in Eq. 10.1 for the precipitation series at a given station, may be that Θ does not differ significantly from the mean regional value Θ_0, obtained as the arithmetic mean of the Θ_i values at several precipitation stations. The alternative H_1 hypothesis is that Θ differs significantly from Θ_0. What this significant difference is must be clearly stated, but it can be done only in probability terms.

A further H_0 hypothesis can be that the empirical distribution of n regional Θ_i values for n precipitation stations does not differ from the sampling distribution of Θ parameter of Eq. 10.1, with the mean value of Θ_i being the sampling average regional value Θ_0. The alternative H_1 hypothesis is that the regional variation in Θ_i is larger than the expected sampling variation. If H_0 is accepted, Θ_0 as a more accurate estimate may replace Θ_i of any station, but for H_0 rejected, this replacement cannot be justified.

Testing of hypotheses. The concept of testing hypotheses is based on methods used for accepting or rejecting a hypothesis. Because it is possible to conceive of very large numbers of hypotheses for the distributions of random variables, the procedures used must be such that they render the decision in each case of a particular hypothesis. This leaves much freedom for choosing various decisions. The proper way to use hypotheses in hydrology is by postulating them as required in a problem, or by constituting them with all relevant alternatives needed in an exhaustive investigation.

Regardless of the method used for testing hypotheses, the sampling distribution of the parameter used for the test must be obtained. Then, the total range of potential variations of this parameter is divided into an acceptance region and a rejection region. The rejection region is called the critical region, and it may be either only on one tail or both tails of a sampling distribution. The acceptance region may be composed only of the central part of this range, which is between the two tails of critical regions, or the acceptance region is located between a boundary at one side of the sampling distribution (say, a finite boundary, or even the infinite value), and a value defining the one-tail critical region at the other side of the sampling distribution. In conclusion, a test of H_0 hypothesis is reduced to the problem of choosing its critical region or regions.

If there are only two hypotheses, null hypothesis H_0 and alternative hypothesis H_1, four cases may result from tests: (1) the H_0 hypothesis is true and the test accepts it, which is a correct decision; (2) the H_0 hypothesis is true but the test rejects it, which means the decision is incorrect; this is called the *type I error*; (3) the H_1 hypothesis is true and the test accepts it, which is a correct decision, and (4) the H_1 hypothesis is true but the test rejects it, which means the decision is incorrect, which is called the *type II error*. Table 10.1 shows these four cases.

Table 10.1
Definition of Types of Errors Associated with Testing Hypotheses

Decision	True Situation	
	Hypothesis H_0 is true	Hypothesis H_1 is true
Accept hypothesis H_0	Correct decision	Type II error
Reject hypothesis H_0	Type I error	Correct decision

The size of the type I error is the probability of rejecting the H_0 hypothesis when it is true. The size of the type II error is the probability of accepting the H_0 hypothesis when H_1 is true. The words size and probability are synonyms in this case. The size of the type I error is designated by α, and the size of the type II error is designated by β. The general principle of testing hypotheses is in choosing the size β of the type II error as small as possible for the same selected size α of the type I error. The tolerance level, P_t, which was discussed in the previous chapter, determines readily the size of the type I error, which is $\alpha = \varepsilon = 1 - P_t$, and it is either 0.05 or 0.10 (it is also sometimes 0.20 or 0.01, as the extremes used in current hydrology applications). Then the size β of the type II error may be greater than, equal to, or smaller than α. Most applied statisticians, and most hydrologists using inferential statistics in their work, pay attention primarily to α, or to the size of the type I error. From the various tests, which all have the same value α of the type I error, the test with the minimum value β of the type II error should be selected. These sizes α and β of the two types of errors are schematically shown in Fig. 10.1. The power of a test of H_0 hypothesis is measured by the function $1 - \beta(\Theta)$ of the true parameter Θ. The size of the error $\beta(\Theta)$ is determined for various realistic hypotheses for the true values of Θ, that is, the probability for the computed sample statistic, $\hat{\Theta}$, to fall in the noncritical region when Θ is the true parameter value. It is usually easier to work with $1 - \beta(\Theta)$ as the power function of Θ giving the probability of the sample statistic falling within the rejection region of the test when Θ is the true parameter value, than to work with $\beta(\Theta)$.

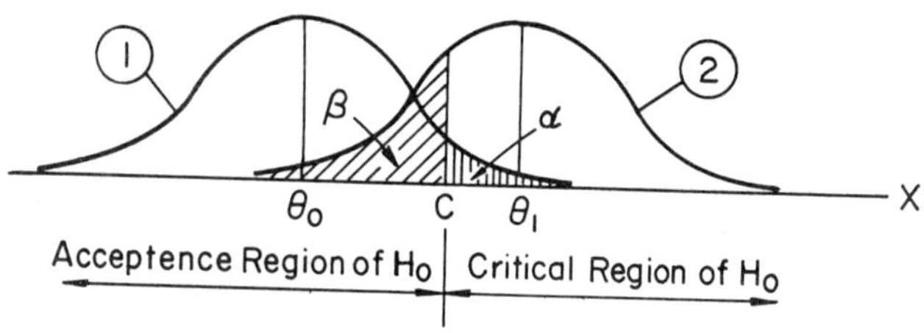

Figure 10.1. Schematic representation of the acceptance and rejection (critical) regions in the test of hypotheses: (1) distribution of $x = \Theta$ assuming H_0 ($\Theta = \Theta_0$) is true, and (2) distribution of $x = \Theta$ assuming H_1 ($\Theta = \Theta_1$) is true.

For a detailed analysis of tests of hypotheses, refer to the general exposition of the theory by J. Neyman [1], and to books by P. G. Hoel [2, p. 46—56, and p. 212—228], H. Cramer [3, p. 525—536], and Ostle Bernard [4].

Example 10.2. In the case of testing the H_0 hypothesis for a time series not linearly autocorrelated, the alternatives may be all hypotheses that do involve time series linearly autocorrelated. Therefore, the case of null hypothesis and one alternative hypothesis may be broadly understood as the first being a particular and well defined hypothesis, while the second may encompass either some particular or all other hypotheses which are opposite or adverse to the first one. This case is very current in hydrology. If the test refers to the fit of the normal function to an empirical distribution, as the H_0 hypothesis, the alternative H_1 hypothesis may include all non-normal probability functions.

Testing of hypotheses related to a sample statistic. For any statistic, $\hat{\Theta}$, computed from a sample and related to the distribution of a random variable, a population value Θ may be postulated. The H_0 hypothesis is then that the difference $\hat{\Theta} - \Theta$ is not significant from zero, while the H_1 hypothesis is that this difference is significant, with prescribed rejection regions in both cases. The tolerance interval about the postulated value Θ is often applied, in this case, as the acceptance region in testing hypotheses if $\hat{\Theta}$ can be both, greater or smaller than Θ. If the sample statistic is inside the tolerance interval, H_0 is accepted; otherwise H_0 is rejected. This case is for the two-tail test because the value $\hat{\Theta}$ at either of the tails of the $F(\hat{\Theta}; \Theta)$-distribution, as critical regions, means a rejection of H_0. However, if the alternative hypothesis is that $\Theta_1 < \hat{\Theta}$, or the converse $\Theta_1 > \hat{\Theta}$, the one-tail tests should be used, which are shown later.

Practical tests of this two-tail type are performed in hydrology by computing the sampling distributions of various statistics, as described in Chapter 9, by postulating their various population values with the corresponding construction of relevant hypotheses and by drawing the tolerance intervals about the postulated values. Cases often arise in hydrology in testing whether two means, two variances, or two values of any other statistic of two subseries of a given series, or of series of two neighboring stations, are or are not statistically significantly different. These types of hypotheses are often related to testing the presence of non-homogeneity and inconsistency in hydrologic data. Whether a hydrologic series is independent or dependent is another application of testing hypotheses.

Example 10.3. A subsample of size N_1, of a sample of size N, is given as the variable x_1 and another of size $N_2 = N - N_1$ as the variable x_2. Their expected values and population variances are: μ_1, μ_2, σ_1^2, and σ_2^2. Subsample means and variances are: $\bar{x}_1$, $\bar{x}_2$, s_1^2, and s_2^2. For small subsamples unbiased variances are used. The H_0 hypothesis is as follows, two subsamples have means from the same normal population, $N(\mu, \sigma)$. The H_1 hypothesis is as follows, two subsamples have unequal population means, which are from two different normal populations $N(\mu_1, \sigma_1)$ and $N(\mu_2, \sigma_2)$. The case may be a test of non-homogeneity in a series. Two cases may be investigated in practice: (1) the variances are considered constants, σ, σ_1, or σ_2, while the means vary from subsample to subsample, and (2) the variances as well as the means vary from one subsample to another. This discussion refers only to the case of variances being considered as constants. The two hypotheses may be redefined in terms of differences of means $\Delta x = \bar{x}_1 - \bar{x}_2$, with $E(\Delta x) = \mu_1 - \mu_2 = \varepsilon$. If Δx is not significantly different from zero, for the constant variance estimated from the sample of size $N_1 + N_2 = N$, it represents the H_0 hypothesis. If Δx is significantly different from zero, and the variances σ_1^2 and σ_2^2 are estimated from each of the two subsamples, it represents the H_1 hypothesis.

For H_0, $E(\Delta x) = \mu_1 - \mu_2 = \varepsilon = 0$, and

$$\text{var}(\Delta x) = \frac{\sigma^2}{N_1} + \frac{\sigma^2}{N_2} = \frac{N\sigma^2}{N_1 N_2}, \qquad 10.2$$

and for H_1, $E(\Delta x) = \mu_1 - \mu_2 = \varepsilon \neq 0$, and

$$\text{var}(\Delta x) = \frac{\sigma_1^2}{N_1} + \frac{\sigma_2^2}{N_2}. \qquad 10.3$$

By using the estimates $\bar{x}_1$, $\bar{x}_2$, s_1^2, and s_2^2, and the variance σ^2 is estimated by s^2 from the sample $N = N_1 + N_2$ for H_0, then its mean $\bar{x}$ is

$$\bar{x} = \frac{N_1\bar{x}_1 + N_2\bar{x}_2}{N}, \qquad 10.4$$

and its sample variance is

$$s^2 = \frac{N_1 s_1^2 + N_2 s_2^2 + N_1(\bar{x}_1^2 - \bar{x}^2) + N_2(\bar{x}_2^2 - \bar{x}^2)}{N}, \qquad 10.5$$

with $\bar{x}$ given by Eq. 10.4.

For H_0, the distribution of Δx is approximately normal, N $(0, s_0)$, or the expected value is zero, and the sample standard deviation is

$$s_0 = \left(\frac{N s^2}{N_1 N_2}\right)^{1/2}, \qquad 10.6$$

in which s^2 is given by Eq. 10.5. For H_1, the distribution of Δx is normal, N (ε, s_a), with the expected value ε and the sample standard deviation

$$s_a = \left(\frac{s_1^2}{N_1} + \frac{s_2^2}{N_2}\right)^{1/2}, \qquad 10.7$$

in which s_1^2 and s_2^2 are variances of variables x_1 and x_2.

For very small N_1 and N_2 the Student t-distribution with N-2 degrees of freedom is applicable for the distribution of Δx. For sufficiently large N_1 and N_2, Δx is normal, N $(0, s_0)$, with s_0 given by Eq. 10.6 in testing the H_0 hypothesis. If the alternative hypothesis H_1 is $\mu_1 \neq \mu_2$, the two-tail test should be used. However, if the alternative hypothesis H_1 is $\mu_1 > \mu_2$ a right-tail test should be used, while for the hypothesis $\mu_1 < \mu_2$, a left-tail test should be used as the best tests. The two-tail test and the tolerance interval approach are given here as an example. For a given tolerance level, $P_t = 1 - \alpha$, the standard normal variable gives $t_{\alpha/2} = t$, so that the tolerance limits are $\pm ts_0$. If $|\Delta x| < ts_0$, the H_0 hypothesis is accepted, othervise it is rejected. In this latter case, another population value $\varepsilon = \mu_1 - \mu_2$ is postulated, with Δx normally distributed, N (ε, s_a), and s_a given by Eq. 10.7 for testing this H_1 hypothesis, which is now postulated as a new null hypothesis. In most cases, the estimate s_a of Eq. 10.7 is smaller than s_0 of Eq. 10.6, so that the tolerance interval of H_1 is narrower about ε than the tolerance interval of H_0 is about $\varepsilon = 0$. This relation of s_0 and s_a assumes that a rejection of H_0, with $\varepsilon = 0$, automatically rejects H_1 with $\varepsilon = 0$, or ε must be significantly larger or smaller than zero for H_1 to be accepted. If the new null hypothesis has the alternative of $\mu_1 - \mu_2 < \varepsilon$, the left-tail test should be used, and this will increase the rejection region on the left side. A similar test would be for $\mu_1 - \mu_2 > \varepsilon$ with the right-tail test used.

Fixed-size sample tests versus sequential sampling tests. The previous discussion refers to the samples of fixed sizes. All sample data are used in making decisions whether to accept or reject a hypothesis. Another approach is by adding one sample value at a time in the sequential sampling. An experiment or an observational procedure is set up, and every time a new value is obtained the test is performed. These are sequential sampling tests with three alternatives: to accept a hypothesis, to reject a hypothesis, or to continue sampling. As soon as a hypothesis is accepted or rejected, observations or experiments are stopped. This approach requires less data than the sample of fixed size, and is useful in observations designed only for statistically proving or disproving a hypothesis. The procedure of testing hypotheses of fixed-size samples does not allow decisions to be made before the entire sample is obtained, nor does it permit additional sampling without reorganizing the entire procedure.

Example 10.4. Detecting changes with time of various parameters are examples of sequential sampling tests in hydrology. If the objective for seeding clouds is to ascertain that particular types of storms produce a precipitation increase, while other types decrease the precipitation by cloud seeding, the sequential sampling test is relevant. There is no point in conducting experiments beyond the point a hypothesis is accepted or rejected. Similarly, the study of the effects of some new river basin measures on the mean river flow, its variance, its various water quality parameters, and so forth, has sense until a decision is made about the type and amount of these effects. Sequential sampling tests are appropriate in cases of detecting or proving non-homogeneity, inconsistency, expected changes, and any other influences produced in hydrologic time series.

10.2. Testing Simple and Composite Hypotheses.

Simple and composite hypotheses. If a hypothesis includes the postulated population values of all parameters of a distribution, it is a simple hypothesis; otherwise, it is a composite hypothesis. In fitting the three-parameter lognormal distribution, a postulate of all three parameters, μ_n, σ_n, and γ, as three population constants, makes a simple hypothesis. If σ_n parameter is postulated as being equal to s_0, and μ_n and γ are given as smaller or greater than some given values, x_0 and γ_0, then it represents a composite hypothesis. The hypothesis H_0 is simple in Example 10.3 because the population difference of two means being zero, and the variance a constant estimated from the sample data for the normal distribution, is specified.

Tests of simple hypotheses. If the simple hypotheses approach is used, the size α of the type I error is selected in advance, and the best test of these simple hypotheses is defined when the size β of the type II error is a minimum. The Neyman-Pearson theorem is used to find this best test [1, p. 214—216]. In case of a one-parameter distribution function, $f(x;\Theta)$, two simple hypotheses are: H_0 with $\Theta = \Theta_0$, and H_1 with $\Theta = \Theta_1$, but with $\Theta_1 < \Theta_0$ assumed. A parameter is defined as the ratio of two likelihood functions, L_1 and L_0, as

$$\frac{L_1}{L_0} = \frac{\prod_{i=1}^{N} f(x_i;\Theta_1)}{\prod_{i=1}^{N} f(x_i;\Theta_0)}, \qquad 10.8$$

in which L_1 is the likelihood product function when H_1 is true, and L_0 when H_0 is true.

The critical region of size α is defined by a constant υ. If $L_1/L_0 \geq \upsilon$, the ratio is inside the critical region; otherwise the ratio is outside the critical region. If this is true, the critical region of α gives the best test. As L_1/L_0 goes from zero to infinity, the size α of critical region goes from unity to zero, permiting the computation of the proper value υ for a given α.

Example 10.5. In the case of fitting Eq. 10.1 to observed times, x, between the precipitation storms in a given interval of the year, two hypotheses are: H_0 for $\Theta = \Theta_0$ and H_1 for $\Theta = \Theta_1$, with $\Theta_1 < \Theta_0$ (which is different from $\Theta_1 > \Theta_0$ in Fig. 10.1). Then Eq. 10.8 gives

$$\frac{L_1}{L_0} = \left(\frac{\Theta_1}{\Theta_0}\right)^N e^{-N\bar{x}(\Theta_1 - \Theta_0)}, \qquad 10.9$$

in which $N\bar{x} = \Sigma x_i$, $i = 1, 2, \ldots, N$, and N is the sample size. For H_0 to be rejected, L_1/L_0 should be greater than υ, while H_1 requires $\Theta_1 < \Theta_0$. Because a division of an inequality by $\Theta_1 - \Theta_0$ reverses the inequality sign, Eq. 10.9 for $L_1/L_0 \geq \upsilon$ becomes

$$\bar{x} \geq \frac{N(\ln \Theta_1 - \ln \Theta_0) - \ln \upsilon}{N(\Theta_1 - \Theta_0)} = x_0, \qquad 10.10$$

with υ defined by the size α of the selected critical region of H_0. Equation 10.10 gives the condition under which the selected size α of the type I error will produce the best test in the same time.

For particular numerical values, $\Theta_0 = 2$, $\Theta_1 = 1$, and $N = 25$, Eq. 10.10 gives $x_0 = \ln 2 + \ln \upsilon/25$. Because x_0 is a monotonic function of υ and $\upsilon > 0$, the critical region of $\bar{x}$ is determined by α for $P(\bar{x} \geq x_0) = \alpha$. This is the right one-tail test. If $\alpha = 0.05$, $\bar{x}$ should be greater than x_0 determined for this α at the right tail for rejecting H_0. Because $\bar{x}$ is approximately normally distributed for N large, the critical region of $\bar{x} \geq x_0$ can be determined from the approximate normal distribution of sample means of Eq. 10.1. For $\Theta_0 = 2$, Eq. 10.1 becomes $f(x; 2) = 2e^{-2x}$, and it is the gamma probability function of Eq. 7.39 with the parameters 1 and 0.5, or $E\,x = 0.5$, and $\text{var}\,x = 0.25$. Therefore, the means $\bar{x}$ for the sample size $N = 25$ have $E\,\bar{x} = 0.50$, and $s(\bar{x}) = 0.10$, or $\bar{x}$ is normal, $N(0.5, 0.10)$. The value x_0 of Eq. 10.10 is $x_0 = E\,\bar{x} + t_\alpha \sigma(\bar{x}) = 0.50 + + 1.28 \cdot 0.10 = 0.628$, for $\alpha = 0.05$. If $\bar{x} \geq 0.628$ in the sample of times x between storms, with $N = 25$, H_0 will be rejected; this will ensure the best test or the size β of the type II error will be a minimum. If $\bar{x} < 0.628$, the H_0 hypothesis should be accepted, or $\Theta = 2$.

Though Eq. 10.10 is referred to as testing of simple hypotheses ($\Theta = \Theta_0$ versus $\Theta = \Theta_1$, and $\Theta_1 < \Theta_0$ in this case), it can be used also for testing composite hypotheses (say $\Theta = \Theta_0$, and $\Theta < \Theta_0$, or $\Theta > \Theta_0$), provided that a range of Θ for the second hypothesis is specified, either smaller or greater than Θ_0 of the H_0 hypothesis.

Example 10.6. For the problem of Example 10.3 a known constant variance, σ^2, is assumed, and the test of the H_0 hypothesis, $E(\Delta x) = \mu_1 - \mu_2 = \varepsilon = \varepsilon_0 = 0$, is performed, while the alternative hypothesis is $\varepsilon = \varepsilon_1$, with $\varepsilon_1 < 0$. If Δx is approximately normal, $N(\varepsilon, \sigma)$, then its probability density function is

$$f(\Delta x; \varepsilon) = \frac{e^{-(\Delta x - \varepsilon)^2/2\sigma^2}}{\sigma\sqrt{2\pi}}. \qquad 10.11$$

The ratio L_1/L_0 of Eq. 10.8 has only one term x in each likelihood product function, so that

$$\frac{L_1}{L_0} = e^{\,\varepsilon_1(2\Delta x - \varepsilon_1)/2\sigma^2} \geq \upsilon. \qquad 10.12$$

By taking logarithms, and by reversing the inequality sign when the inequality is divided by a negative ε_1, Eq. 10.12 becomes

$$\Delta x \leq \frac{2\sigma^2 \ln \upsilon + \varepsilon_1^2}{2\varepsilon_1} = (\Delta x)_0. \qquad 10.13$$

Because $\sigma^2 > 0$, $\upsilon > 0$, $\varepsilon_1^2 > 0$, and $\varepsilon_1 < 0$, the right side of the inequality of Eq. 10.13 gives a negative value of $(\Delta x)_0$. Because $(\Delta x)_0$ is a value of Δx-distribution on the left tail, determined by the size α of the critical region of the type I error in Eq. 10.11, for $\varepsilon_0 = 0$ and for the given σ^2, then $\Delta x \leq (\Delta x)_0$ means a rejection of H_0, while the hypothesis $\varepsilon_0 = 0$ is accepted for $\Delta x > (\Delta x)_0$. This left-tail test is the best test, or the probability of $\Delta x \leq (\Delta x)_0$ is α when H_0 is true, and the size β of the type II error is a minimum. This example also shows that the left-tail critical region of the Δx-distribution should be used for any alternative hypothesis $\varepsilon_1 < 0$, and it is easy to show that the right-tail critical region of the Δx-distribution should be used for rejecting H_0 when the H_1 hypothesis is $\varepsilon_1 > 0$, for $\varepsilon_0 = 0$. For $\varepsilon_1 \neq \varepsilon_0$, the two-tail test should be used, but there is no best critical region or regions for all possible alternatives of ε_1 being either greater or smaller than ε_0.

Likelihood ratio tests. When the Neyman-Pearson criterion is not applicable, and in the case of composite hypotheses, the principle of the likelihood ratio test may produce a good test. If there are several parameters in a distribution function $f(x; \Theta_1, \Theta_2, \ldots, \Theta_m)$, and the H_0 hypothesis is $\Theta_1 = \Theta_{1,0}$, $\Theta_2 = \Theta_{2,0}, \ldots, \Theta_m = \Theta_{m,0}$, while the H_1 hypothesis has unspecified values of these parameters, then a good test may be obtained by the likelihood ratio test. The best estimates $\hat{\Theta}_1, \hat{\Theta}_2, \ldots, \hat{\Theta}_m$ of Θ_1, $\Theta_2, \ldots, \Theta_m$ are obtained by the maximum likelihood method. Then their likelihood function is $L(\hat{\Theta}) = \prod_{i=1}^{N} f(x; \hat{\Theta}_1, \hat{\Theta}_2, \ldots, \hat{\Theta}_m)$. If H_0 is true, then the likelihood function of the postulated parameters is $L(\Theta_0) = \prod_{i=1}^{N} f(x_i; \Theta_{1,0}, \Theta_{2,0}, \ldots, \Theta_{m,0})$. The ratio of the two numbers $L(\Theta_0)$ and $L(\hat{\Theta})$ is

$$\lambda = \frac{L(\Theta_0)}{L(\hat{\Theta})}. \tag{10.14}$$

Because $L(\Theta_0)$ and $L(\hat{\Theta})$ are both functions of sample values, x_1, $x_2, \ldots, x_N$, the parameter λ is a random variable having a sampling distribution, $g(\lambda)$. If H_0 specifies only some values Θ_i, but not all, then the unspecified Θ_j parameters, with $i + j = m$, are estimated also by the maximum likelihood method, keeping all Θ_i values as specified constants. The denominator is a maximum of $L(\Theta)$ for $\hat{\Theta}_i$ values, while the numerator is a maximum of $\hat{\Theta}_j$ parameters, or $L(\Theta_0) \leq L(\hat{\Theta})$ so that $0 < \lambda < 1$. If λ is close to unity, the H_0 hypothesis is reasonable, otherwise it is not. The distribution of this parameter serves for defining the acceptance and the rejection regions of the H_0 hypothesis. The critical value λ_0 determines the size (probability) α of the type I error by

$$\alpha = \int_0^{\lambda_0} g(\lambda)\, d\lambda, \tag{10.15}$$

so that for $0 \leq \lambda \leq \lambda_0$, the H_0 hypothesis is rejected, and for $\lambda_0 < \lambda \leq 1$ it is accepted.

The likelihood ratio test is a compromise when the alternative hypothesis has no restrictions. In several cases a monotonic relation between λ and a sample parameter, whose distribution is known, may be established, so that there is no need to obtain the $g(\lambda)$-density function. It is then sufficient to use $\alpha/2$ at each of the two tails of the known distribution of that sample parameter to define the critical regions. When it is not possible to find such a simple relation between λ and a sample parameter, two ways are open to an investigator, the asymptotic approximation to the distribution of a function of λ for large N, and the experimental Monte Carlo method.

In the first case, it can be shown that under particular conditions the function $(-2 \ln \lambda)$ of the parameter λ (where ln is the logarithm for the natural base e) is asymptotically chi-square distributed, with the number of degrees of freedom (NDF) being the number of parameters that are specified by the H_0 hypothesis. Because a small value of λ gives a large value of $(-2 \ln \lambda)$, the critical value λ_0 becomes now $-2 \ln \lambda_0 = \chi_0^2$, where $P(\chi^2 > \chi_0^2) = \alpha$ and NDF = m. This is a simple chi-square one--tail test. If H_0 specifies only one parameter, like in the case of the simple hypotheses discussed in the previous examples, NDF = 1, and Eq. 9.62 is applicable to $(-2 \ln \lambda_0)$ as the asymptotic approximation of the likelihood ratio test.

For small N the above approximation by the function $(-2 \ln \lambda)$ may be in error. The Monte Carlo simulation method can be used to find the experimental distribution of λ, or of $(-2 \ln \lambda)$. In this latter case it also can be determined how much this experimental distribution deviates from the asymptotic chi-square distribution.

Example 10.7. For the problem in Example 10.6, for the simple H_0 hypothesis $\varepsilon = \varepsilon_0 = 0$, and the alternative H_1 hypothesis $\varepsilon = \varepsilon_1 \neq 0$, Eq. 10.11 gives the likelihood function

$$L(\varepsilon) = (2\pi)^{-N/2} \sigma^{-N} \exp\left[-\frac{1}{2\sigma^2} \sum_{i=1}^{N} (\Delta x - \varepsilon)^2\right], \qquad 10.16$$

and $\partial \ln L(\varepsilon)/\partial \varepsilon$ gives

$$\frac{\partial \ln L(\varepsilon)}{\partial \varepsilon} = \frac{1}{\sigma^2} \cdot \sum_{1}^{N} (\Delta x - \varepsilon) = 0. \qquad 10.17$$

For the maximum of $\ln L(\varepsilon)$, Eq. 10.17 gives $\Delta x = \varepsilon$. Because there is only one value of Δx, Eq. 10.16 becomes

$$L(\varepsilon = 0) = (2\pi)^{-N/2} \sigma^{-N}. \qquad 10.18$$

For $\varepsilon \neq 0$, the ratio of Eq. 10.14, taking into account Eqs. 10.16 and 10.18, gives

$$\lambda = \exp\left[-\frac{1}{2\sigma^2} (\Delta x - \varepsilon_1)^2\right], \qquad 10.19$$

or

$$-2 \ln \lambda = \left(\frac{\Delta x - \varepsilon_1}{\sigma}\right)^2. \qquad 10.20$$

According to Eqs. 9.60 through 9.62, $(-2 \ln \lambda)$ is distributed as χ^2 of Eq. 9.62, because NDF = 1, while Eq. 10.19 implies that the distribution of λ may be replaced by the normal distribution of Δx, so that either the test of the normal distribution of Δx, or the right-tail test of the chi-square distribution may be used in this problem.

Equality of variances. Examples 10.3, 10.6, and 10.7 are related to the problem of equality of population means of subsamples of a given sample, or to the equality of population means of samples of two different variables. This type of investigation is relevant to hydrology. Tests are discussed in these three examples under the condition of the population variances being specified constants for the variables of compared means. An important application is the testing of self-stationarity of hydrologic time series. For self-stationarity tests of hydrologic time series a sample of size N is divided into m subsamples, each of size $n = N/m$, and various parameters — particularly the mean and the variance — are estimated from the subsamples. Then comparison is made for the equality of population values of each parameter across the subsamples. The equality of variances is usually one of the most required tests.

Let $x_1, x_2, \ldots, x_m$ represent m variables, given either by each subsample or sample data, and assume that they are normal independent variables with population means $\mu_1, \mu_2, \ldots, \mu_m$, and population variances $\sigma_1^2, \sigma_2^2, \ldots, \sigma_m^2$, with the sample sizes $n_1, n_2, \ldots, n_m$, for $\Sigma n_i = N$, $i = 1, 2, \ldots, m$ in the case of m subsamples or samples. The null hypothesis is $\sigma_1^2 = \sigma_2^2 = \ldots = \sigma_m^2$. It can be shown that the likelihood ratio of Eq. 10.4 in this case [2, p. 225—226] is

$$\lambda = \frac{N^{N/2} \prod_{i=1}^{m} s_i^{n_i}}{\left(\sum_{i=1}^{m} n_i s_i^2 \right)^{N/2}}, \tag{10.21}$$

in which s_i is the maximum likelihood estimate of σ_i. Usually the distribution $g(\lambda)$ is not available, and must be approximated by the distribution of the function $(-2 \ln \lambda)$, or $g_1(-2 \ln \lambda)$. In the above case, $(-2 \ln \lambda)$ is chi-square distributed with $NDF = m - 1$ instead of m because the variances are only assumed to be equal in value, but that value is not specified.

For very small values of n_i, a more accurate test is given by the parameter υ [2, p. 227]:

$$\upsilon = \frac{-2 \ln k}{1 + \dfrac{1}{3(m-1)} \left(\sum_{i=1}^{m} \dfrac{1}{n_i - 1} - \dfrac{1}{N - m} \right)}, \tag{10.22}$$

which is distributed as chi-square with $NDF = m - 1$, and k is estimated by

$$k = \frac{\prod_{i=1}^{m} \left(\dfrac{n_i s_i^2}{n_i - 1} \right)^{(n_i - 1)/2} \left[\sum_{i=1}^{m} (n_i - 1) \right]^{\sum_{i=1}^{m} (n_i - 1)/2}}{\left(\sum_{i=1}^{m} n_i s_i^2 \right)^{\sum_{i=1}^{m} (n_i - 1)/2}}. \tag{10.23}$$

If n_i's are equal, or $n = N/m$, then Eq. 10.22 becomes

$$\upsilon = -\frac{3(N-m)}{3N-2m+1}\left[(N-m)\ln m + (N-m)\sum_{i=1}^{m} \ln (s_i^2) - \right.$$

$$\left. -2\ln \sum_{i=1}^{m} (s_i^2)^{(N-m)/2}\right], \qquad 10.24$$

and υ is chi-square distributed with NDF $= m - 1$.

In the case of only two variances, s_1^2 and s_2^2 with N_1 and N_2 the sample sizes, a simpler test of the equality of their population variances is the Fisher test using the ratio $F = s_1^2/s_2^2$, for which Eq. 9.75 is applicable with $N_1/2$ and $N_2/2$ degrees of freedom. Instead of F, the Fisher z statistic of Eq. 9.76 may be computed, and a test performed whether this z is not significantly different from $z_0 = 0$ by using Eq. 9.77. This distribution is often tabulated in books and is called the Fisher z-tables.

Example 10.8. Two small samples have the same sizes: $N_1 = N_2 = 10$, with $\bar{x}_1 = 6.0$ and $\bar{x}_2 = 5.7$, and variances $s_1^2 = 0.064$ and $s_2^2 = 0.024$. The testing of the equality of population means and of population variances is made under the small sample conditions. The difference $\Delta x = \bar{x}_1 - \bar{x}_2 = 0.3$ gives the t-statistic of Eq. 9.71, under the H_0 hypothesis of $\varepsilon_0 = \mu_1 - \mu_2 = 0$, as

$$t = \frac{(N-1)^{1/2}(\Delta x - \varepsilon_0)}{(s_1^2 + s_2^2)^{1/2}} = \left(\frac{9}{0.088}\right)^{1/2}(6.0 - 5.7) = 3.034.$$

The Student t-distribution table gives the probability $P = 0.0072$ of $t = 3.034$ being exceeded for NDF $= 2N - 2 = 18$, so that for $\alpha = 0.05$, the null hypothesis of μ_1 and μ_2 being equivalent must be rejected.

If the equivalence of population variances is tested, the F ratio of unbiased variances is equal here ($N_1 = N_2 = 10$) to the ratio of biased variances, or

$$F = \frac{s_1^2}{s_2^2} = \frac{0.064}{0.024} = 2.663,$$

so that Eq. 9.76 for $N_1 = N_2$ gives $z = 0.5 \ln F = 0.49$. For $N_1 - 1 = 9$ and $N_2 - 1 = 9$ degrees of freedom, the Fisher z-table gives for $\alpha = 0.05$ the critical value $z_0 = 0.58$. Because $z < z_0$, the null hypothesis of $\sigma_1^2 = \sigma_2^2$ is accepted.

10.3. Testing the Goodness of Fit of Probability Functions to Empirical Distributions.

Goodness of fit. As defined in the previous text, goodness of fit tests require a measuring parameter for fitting the discrete and continuous distribution function to empirical distributions. Two such parameters are discussed here, the chi-square and the Smirnov-Kolmogorov parameters.

The chi-square test. Let a discrete variable have k mutually exclusive random events, or a continuous variable be divided into k class intervals, also mutually exclusive events. Let f_i, $i = 1, 2, \ldots, k$, represent the relative frequencies of these random events or of class intervals for a sample of size N, and let the discrete or continuous distribution functions have the probability mass of random events or the probability of class intervals as p_i, $i = 1, 2, \ldots, k$. For the observed sample size N, the sample absolute frequencies are $N_i = Nf_i$, in which f_i are the relative frequencies, and the expected values of absolute frequencies are Np_i.

It has been shown that the chi-square parameter,

$$\chi^2 = \sum_{i=1}^{k} \frac{(N_i - Np_i)^2}{Np_i}, \qquad 10.25$$

will have chi-square distribution with the number of degrees of freedom equal to $k - 1$, for a sufficiently large N. It can be shown that the distribution of the function $(-2 \ln \lambda)$, of λ, is approximated well by the χ^2-distribution. It can be also shown that the likelihood ratio test is equivvalent to the test by the parameter χ^2 of Eq. 10.25 [2, p. 376—378]. The chi-square test prescribes the critical value χ_0^2 for a given α, so that for $\chi^2 < \chi_0^2$, the null hypothesis of a good fit is accepted, otherwise for $\chi^2 \geq \chi_0^2$ it is rejected.

In general, the number of class intervals should be $k \geq 5$ and the expected values of absolute frequencies should be $Np_i \geq 5$ for the distribution with NDF $= k - 1$ to well approximate the distribution of the parameter of Eq. 10.25. If $k < 5$, the expected absolute frequency should be $Np_i > 5$. The p_i value represents the integral of the probability density function between the lower and upper class limits of the i-th class interval for continuous variables. The length of class intervals may be equal or unequal. In the latter case, the class lengths are selected in such a way that all have equal probabilities, $p_i = 1/k$, or equal expected absolute frequencies, Np_i. Then, the distribution of these Np_i expected values over the unequal class intervals is uniform.

Two cases currently prevail in practice: parameters of distribution functions are assumed known, and parameters of distribution functions must be estimated from the sample data. In the first case, the number of degrees of freedom is $k - 1$, or a degree of freedom is lost because of the constraint $\Sigma p_i = 1$, $i = 1, 2, \ldots, k$. In the second case, the number h of parameters in the distribution function, estimated by the maximum likelihood method, represents h additional constraints so that NDF $= k - h - 1$.

The only condition for continuous variables and the selection of class intervals with equal probabilities of the fitted function is $p_i = 1/k$. Several functions or empirical distribution curves may satisfy this condition, for samples small and small k; but for large N the number k of class intervals may be also large, which makes a better discrimination of fits among the similar distribution functions. This conclusion is intuitive.

The binomial distribution with estimated parameters p and n, or with $h = 2$, has NDF $= k - 3$. If n is specified by the problem, NDF $= k - 2$; if p is assumed, NDF $= k - 1$. For Poisson distribution, if its parameter λ is known, NDF $= k - 1$, otherwise NDF $= k - 2$.

As the hydrologic samples are often of the order $N = 30 - 60$, six or seven class intervals are attractive numbers for k. Because the fitted probability functions usually have two or three parameters, NDF often equals four or three in these cases. Table 10.2 gives the values of $F(\chi^2) = P(\xi \leq \chi^2)$ for NDF = 3 and NDF = 4.

Table 10.2
Chi-Square Distribution Values for NDF = 3 and NDF = 4.

$F(\chi^2)$	NDF = 3	NDF = 4	$F(\chi^2)$	NDF = 3	NDF = 4
0.001	0.019	0.074	0.550	2.650	3.680
0.005	0.072	0.207	0.600	2.950	4.040
0.010	0.115	0.297	0.650	3.290	4.430
0.020	0.185	0.429	0.700	3.665	4.878
0.025	0.216	0.484	0.750	4.108	5.385
0.050	0.352	0.711	0.800	4.642	5.989
0.075	0.468	0.890	0.850	5.296	6.725
0.100	0.584	1.064	0.900	6.251	7.779
0.150	0.808	1.360	0.925	6.920	8.480
0.200	1.005	1.649	0.950	7.815	9.488
0.250	1.213	1.923	0.975	9.348	11.143
0.300	1.424	2.195	0.980	9.837	11.668
0.350	1.640	2.460	0.990	11.345	13.277
0.400	1.875	2.740	0.995	12.838	14.860
0.450	2.110	3.040	0.999	16.268	18.465
0.500	2.366	3.357			

To compute the class limits, with the expected frequencies being an equal $p_i = 1/k$, the normal distribution tables can be used in fitting both the normal and the lognormal distribution functions. For the gamma distribution the incomplete gamma function values of Eq. 7.36, with $p = \alpha - 1$ and $u\sqrt{p+1}$ for the upper integration limit, must be obtained from the incomplete gamma tables for given $\alpha = p + 1$. These limits of class intervals, obtained from I (u, p) of Eq. 7.36, for the one-parameter gamma function are then easily transformed to class limits of the two-parameter and three-parameter gamma functions, once their parameters are estimated by the maximum likelihood method. In the absence of I (u, p)-tables, a simple method is to compute F (x) values of the gamma function for a set of arbitrary values of x, plot the F (x) function, and divide the range of unity of F (x) into k equal intervals, and then determine graphically the values x_i for the limits of class intervals with equal probabilities 1/k. Computer programs exist for estimating these class limits which can be determined by using an iterative computational procedure.

Example 10.9. Figure 10.2 gives an example of fitting the two-parameter lognormal function to the empirical distribution of annual river flows (given in modular coefficients. $K_i = Q_i/\overline{Q}$) of the Weldon River at Mill Grove, Missouri, for the years 1930–1960. The maximum likelihood estimates of parameters of the lognormal function for K_i-

variable are $\hat{x}_n = 0.317$, and $\hat{s}_n = 0.840$. This function is divided into $k = 7$ unequal class intervals with equal probabilities, $p_i = 1/7$, as shown in Fig. 10.2. The expected absolute frequencies of unequal class intervals are $Np_i = 31/7 = 4.43$, and the observed absolute frequencies N_i are plotted in Fig. 10.2. Because $h = 2$ the number of degrees of freedom is NDF = 4. For NDF = 4 and $\alpha = 0.05$ the χ^2-distribution table gives the critical value $\chi_0^2 = 9.488$, while Eq. 10.25 gives $\chi^2 = 2.222$ for this example. Because $\chi^2 < \chi_0^2$ in this case, and because $F(\chi^2) = 0.307$, and the critical value χ_0^2 has $F(\chi_0^2) = 0.950$, the null hypothesis of a good fit of the two-parameter lognormal distribution function is accepted.

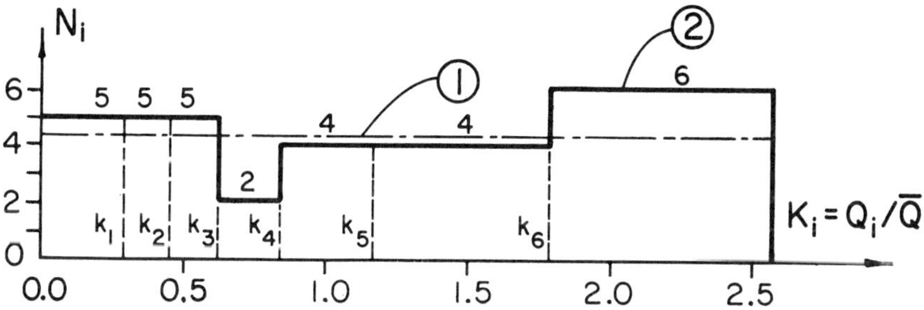

Figure 10.2. Fitting the two-parameter lognormal function to the empirical distribution of annual flows of the Weldon River at Mill Grove, Missouri: (1) expected absolute frequencies of equal probability class intervals, and (2) histogram of observed absolute frequencies of the same class intervals.

For this same river the fit of the normal distribution gives $\chi^2 = 11.267$ for NDF = 4, and the fit of the two-parameter gamma distribution has $\chi^2 = 2.674$ for NDF = 4 Comparing these three two-parameter distribution functions, the best fit is by the two-parameter lognormal distribution, while the fit by the normal distribution must be rejected ($\chi^2 = 11.267 > \chi_0^2 = 9.488$). In the case of the three-parameter distributions, NDF = 3, and for $\alpha = 0.05$ the critical χ_0^2 value is 7.815. The fit of the three-parameter lognormal function gives $\chi^2 = 4.030$ and $F(\chi^2) = 0.742$, while the fit of the three-parameter gamma function gives $\chi^2 = 9.474$ with $F(\chi^2) = 0.976$. Only the three-parameter lognormal distribution has an acceptable fit ($\chi^2 = 4.030 < \chi_0^2 = 7.815$); however, it is not as good a fit as the fit for the two parameter lognormal or gamma functions.

Tests using contingency tables. A contingency table is defined as the classification of two variable observations in a two-way table, like Table 5.1 given in Example 5.5. Marginal and class-field frequencies are produced by this two-way classification. Let one variable have m class intervals (or m states, or m random events), and the other variable n intervals. The corresponding marginal absolute frequencies are N_i and N_j, while the class-field absolute frequencies are N_{ij}, for the sample of size N. The expected absolute marginal and class-field frequencies are Np_i, Np_j, and Np_{ij}, and the chi-square parameter is defined as

$$\chi^2 = \sum_{i=1}^{m} \sum_{j=1}^{n} \frac{(N_{ij} - Np_{ij})^2}{Np_{ij}} . \qquad 10.26$$

Assuming the two variables independent, then $p_{ij} = p_{i.} p_{j}$, so that Eq. 10.26 becomes

$$\chi^2 = \sum_{i=1}^{m} \sum_{j=1}^{n} \frac{(N_{ij} - Np_i\, p_j)^2}{N\, p_i p_j}. \tag{10.27}$$

Because p_i and p_j are unknown, they are estimated by the marginal frequencies $f_i = N_i/N$ and $f_j = N_j/N$, which can be shown to be the maximum likelihood estimates of p_i and p_j [2, p. 253—254]. In this case, Eq. 10.27 becomes

$$\chi^2 = \sum_{i=1}^{m} \sum_{j=1}^{n} \frac{(N\, N_{ij} - N_i\, N_j)^2}{N\, N_i\, N_j}. \tag{10.28}$$

Taking into account that $\Sigma\, p_i = 1$, $i = 1, 2, \ldots, m$, $\Sigma p_j = 1$, $j = 1, 2, \ldots, n$, and m and n being the numbers of statistics of f_i and f_j, respectively, then $m + n - 2$ is the total number of estimated parameters or the number of constraints introduced. The number of class fields is mn, so that NDF $= mn - 1 - (m + n - 2) = (m - 1)(n - 1)$. Therefore, χ^2 of Eq. 10.28 has chi-square distribution with NDF $= (m - 1)(n - 1)$, provided that the H_0 hypothesis is true and N is sufficiently large. This chi-square test by using the contingency table is another method of investigating whether the two variables are independent or not.

Tests by Smirnov-Kolmogorov statistic. Previous tests have all been designed on the basis that a distribution function is known or assumed, its parameters are estimated from the sample data, and the tests are made by using the statistic measuring the goodness of fit to an empirical distribution. These methods are usually called parametric methods. If no distribution function of a given type is specified, and no estimating or testing of its parameters is performed, and the only condition specified in testing statistics of an empirical distribution is that the distribution is continuous, then the tests involved are called nonparametric. Sometimes they are also called distribution-free tests. One such test is performed by the Smirnov-Kolmogorov statistic as a nonparametric test.

Let the sample data $x_1, x_2, \ldots, x_N$ be in ascending or descending order $x_1', x_2', \ldots, x_N'$, and the plotting positions $P(x_1') = m/(N + 1)$ be determined for these values, and the empirical distribution plotted, as described in Chapter 5. Also let the true population value for $P(x_i')$ be $F(x)$, and their maximum difference be defined as

$$\Delta = \max | F(x) - P(x_i')| . \tag{10.29}$$

This statistic, Δ, has a sampling distribution. Let Δ_0 be the critical value for the size α of the type I error of this distribution defined by

$$P\{\max | F(x) - P(x_i') | \geq \Delta_0\} = \alpha. \tag{10.30}$$

On the basis of this equation a confidence interval about each value $P(x_i')$ is defined by

$$P\{F(x) \leq [P(x_i') - \Delta_0];\ \text{and}\ F(x) \geq [P(x_i') + \Delta_0]\} = \alpha \tag{10.31}$$

for all values x'_i. The distribution of Δ as a nonparametric statistic does not depend on F (x), so neither does Δ_0. Equation 10.31 makes it possible to draw two confidence lines, at the confidence level $P_c = 1 - \alpha$, about the empirical frequency distribution for the unknown distribution function F (x). This is defined as the confidence region or band. Here the distribution F (x), rather than the probability density function f (x), is involved. Table 10.3 gives Δ_0 for various values of the sample size N and several values of the size α of type I error. Comparing Δ-test to χ^2-test, there is no condition in a Δ-test that the absolute class frequencies should be greater than five, and no sorting of data in class intervals is required.

Table 10.3

Critical Value Δ_0 of the Smirnov-Kolmogorov Statistic Δ, for Various Values of N, and Values α Currently Used in Hydrology

N	α			
	0.20	0.10	0.05	0.01
5	0.45	0.51	0.56	0.67
10	0.32	0.37	0.41	0.49
15	0.27	0.30	0.34	0.40
20	0.23	0.26	0.29	0.36
25	0.21	0.24	0.27	0.32
30	0.19	0.22	0.24	0.29
35	0.18	0.20	0.23	0.27
40	0.17	0.19	0.21	0.25
45	0.16	0.18	0.20	0.24
50	0.15	0.17	0.19	0.23
N > 50	$\frac{1.07}{\sqrt{N}}$	$\frac{1.22}{\sqrt{N}}$	$\frac{1.36}{\sqrt{N}}$	$\frac{1.63}{\sqrt{N}}$

The statistic Δ may be used also as the measure in tests of goodness of fit. Because the parameters of a distribution function F (x), fitted to an empirical distribution, must be estimated from sample data, a Δ-test is not an exact test, but an approximation. Because Δ is obtained from the absolute maximum difference between F (x) and P (x'_i), under the condition the parameters of F (x) are estimated from the sample data, Δ is likely to be smaller than the true Δ_t if the population parameters of F (x) are known. This bias of $\Delta - \Delta_t$ is not known, however. Regardless of this bias, a Δ-test is useful and quick, if the critical value Δ_0 is selected somewhat smaller than Δ_0 of Table 10.3. In conclusion, for the application of a Smirnov-Kolmogorov Δ-test to the graphical fitting of straight lines of F (x) in transformed coordinates, as discussed in Chapter 7, the right practice would be to draw the straight lines in such a way that the maximum positive and the minimum negative values of [F (x) — P (x'_i)] are approximately equal.

Example 10.10. Figure 10.3 shows the same curves used in Example 7.11 and Fig. 7.13 for fitting the three-parameter gamma function, by the maximum likelihood method of estimating the three parameters, to the empirical distribution of annual 9-month low flow volumes of the Little Arkansas River. This figure shows that the largest Δ-value is $\Delta_1 = 0.080$. For $N = 33$, and a selected $\alpha = 0.05$, Table 10.3 gives $\Delta_0 = 0.236$. Because $\Delta_1 = 0.080 < \Delta_0 = 0.236$, the fit in Fig. 7.13 is acceptable by the Smirnov-Kolmogorov test.

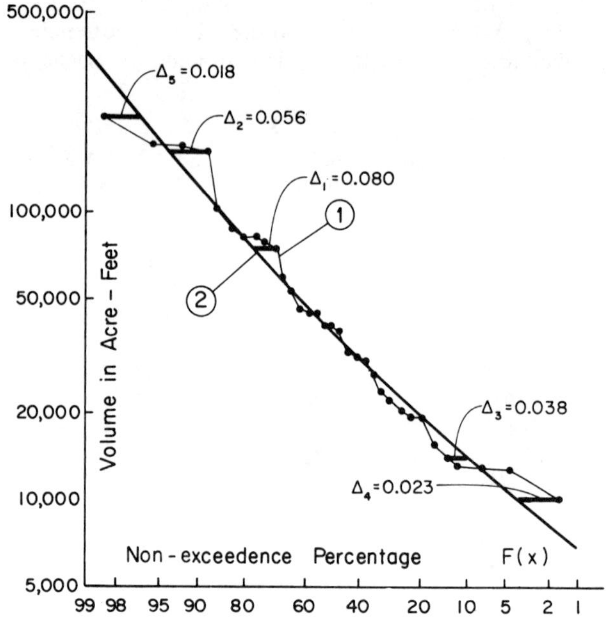

Figure 10.3. Smirnov-Kolmogorov Δ-test of fitting the three-parameter gamma function (Pearson Type III) to the empirical distribution of the annual 9-month, low-flow water volume of the Little Arkansas River at Valley Center, Kansas. (1) empirical distribution, and (2) fitted distribution function.

10.4. **Problems for Discussion and Solutions.**

10.1. For the annual river flow series, x, with x normally distributed, and the sample size $N = 60$ years, it is suspected that inconsistency (systematic errors) in data are present in the second 35 years of observation. In terms of modular coefficients, $K_i = x_i/\bar{x}$, the following is given: $\bar{K} = 1$ for $N = 60$, $\bar{K}_1 = 1.09$ for the first 25 years, or $N_1 = 25$; the unbiased variance, $\hat{s}^2 = 0.036$ for $N = 60$; the unbiased variance $\hat{s}_1^2 = 0.045$ for $N_1 = 25$, and the time dependences of K_i of separate periods $N_1 = 25$ and $N_2 = 35$ are negligible. Find $\bar{K}_2$ and $\hat{s}_2^2$, and test the equivalence of population means and population variances of these two subsamples, N_1 and N_2.

10.2. For the data of Example 1.9 and Fig. 1.1, test whether the population means of the two subsamples, $N_1 = 29$ years (1870–1878) and $N_2 = 57$ years (1899–1955), are equivalent for the size $\alpha = 0.05$ of the type I error, assuming that both subsamples have the same unbiased estimate 0.155 of the standard deviation of K_i, or for the annual flows given in modular coefficients. What is the probability that the difference of the two subsample means would be exceeded by the sampling variation for population means assumed to be equal?

10.3. For examples (a), (b), (c), and (d) in Problem 7.7 perform chi-square test of the goodness of fit, and determine whether the fit should be accepted or rejected by using the size $\alpha = 0.05$ of the type I error. If the sample size is $N \geq 35$ use seven class intervals for grouping the data; if $N < 35$, use either five or six class intervals. The length of class intervals should be such that the probabilities of a given function are the same in each class interval, (i. e., 1/7, 1/6, or 1/5). Use the straight line fits when appropriate to determine the limits for the class intervals that satisfy the condition of equal probability of class intervals. Show the histogram of observed absolute frequencies in order to compare them with the uniform distribution of expected absolute frequencies of equal-probability class intervals.

10.4. For the examples (a), (b), (c), and (d) in Problem 7.7, for which straight-line fits are obtained graphically, perform the simple Smirnov-Kolmogorov test by using Δ_0 values of Table 10.3 for $\alpha = 0.05$. Show graphically how the straight-line should be changed to satisfy the condition that the maximum positive and minimum negative deviations between the observed cumulative frequencies and the fitted function by the straight line are equal.

10.5. In Table 5.1 of Example 5.5 combine the 10 class intervals of x variable into 5, and 10 class intervals of y variable into 5 class intervals, and compute the corresponding marginal absolute class frequencies and absolute class-field frequencies. Perform chi-square test by using this contingency table to prove that the variables x and y are dependent.

CHAPTER 11

CORRELATION AND REGRESSION

Both simple and multiple correlation and regression analyses are among the oldest statistical techniques used in hydrology. The main objectives of this analysis were the transfer of information between points at which the same variable was observed, or between two among several variables observed simultaneously. This includes the completion of missing data in hydrologic series, and the prediction of a variable from the observed one or several other variables.*)

11.1. **Definitions of Correlation and Regression.**

Correlation. The correlative associations rather than being pure functional relations are basic relations encountered in hydrologic practice. Correlation is defined as the association of two or more random variables, that only partly explains the total variation of a variable by the variation of other random variables involved in the association equation. The effect of unaccounted or neglected random variables and errors is responsible for the remaining or unexplained part of the variation. This concept of correlative association requires numbers or statistics to measure the degree of association in any particular case. These statistics must have two extremes: (a) limits that represent the total correlation, equivalent to functional relations, and (b) values that represent the absence of any correlation. Therefore, these two extremes are: functional relation and no correlative relation. Various degrees of correlative association then fall between the two cases.

Example 11.1. Typical examples of correlative associations of random variables in hydrology are: runoff-rainfall, sediment discharge-water runoff, oxygen content--water temperature, and similar, for the two-variable correlation, and runoff as a function of rainfall characteristics, river basin geometric properties, soil and vegetation factors, antecedent moisture conditions of the basin, and similar factors, for the correlative association of many random variables.

---

*) Some material in this chapter is taken from the writer's contribution to HANDBOOK OF APPLIED HYDROLOGY, Section 8-II, Statistical and Probability Analysis of Hydrologic Data, Part II, Regression and Correlation Analysis. Handbook is edited by Ven Te Chow. Use of text is permitted by McGraw-Hill, Inc., copyright C 1964.

Regression. Regression (an unfortunate but widely used term) represents a mathematical equation expressing one random variable as being correlatively related to another random variable, or to several random variables. All variables on the right side of this equation do not need to be random variables. The regression equation may be any function that can be fitted to a set of points of observed variables. The selection of the function to be fitted to points determines the type and the degree of correlative association. Determining mathematical models of correlative association of two or more variables, so that the best prediction of one variable can be obtained from the other variable or variables, is referred to as regression analysis, and the models are called regression functions.

If two variables, given as series of concurrent values (x_i, y_i), show a concentration around an imaginary curve when plotted on a graph, like in Fig. 11.1, then for a large sample there will always be a distribution

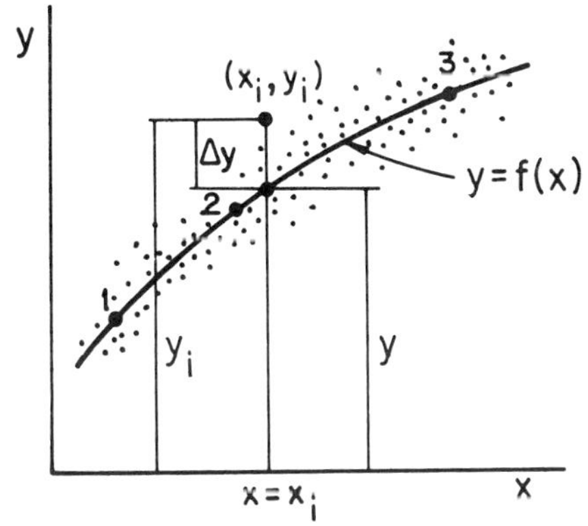

Figure 11.1. Schematic representation of a bivariate regression and correlation analysis, with 1, 2, 3, the means of y for given $x = x_i$.

of y values for a given value of x_i, or more precisely, a distribution of y values for an interval, Δx, around a given x_i. The mean value, y_0, of all y values for this given interval Δx around x_i is the expected value of y for the given $x = x_i$. A curve fitted to all mean values, y_0, is called the regression line of y versus x. On the other hand, the curve fitted to all expected values, x_0, for the given $y = y_i$, defines the regression line of x versus y. These two lines do not coincide, and have different parameters, showing the regressional relations between the two variables.

The functional relation between two variables assumes that all points lie closely on a curve. Inasmuch as the spread of points around the regression lines may actually be great or small, the degree of association of the two variables involved is the correlation, and is defined by parameters of correlation. Correlation is greater when the points are closer to the lines. In short, a two-variable regression problem considers the frequency distribution of one variable when another is held fixed at each of several levels.

Classification of correlations and regressions. The case of two random variables is referred to as bivariate (or bivariable) distribution, and the relation between these two variables is called the simple or bivariate correlative association, and the simple or bivariate regression. The case of a random variable related to several (more than one) random variables is referred to as the multiple correlative association and the multiple regression. The linear equation of relations represents the *simple linear and the multiple linear correlation and regression,* which depends on whether two or more variables are involved. The opposite cases are the *simple nonlinear and the multiple nonlinear correlation and regression.*

The random variables may be normally or non-normally (usually skewed) distributed. Their series (sequences) can be independent or dependent. Therefore, four categories of random variables may be studied: normal-independent, normal-dependent, skewed-independent, and skewed-dependent. Multiple correlation and regression may involve all of these four types of random variables. The association of serially independent normal random variables is first analyzed in this text, with the effects of skewness and serial dependence then discussed.

Steps in the regression analysis. Whether a simple or a multiple correlation or regression analysis is performed, the five steps are necessary for complete information about this association:

(1) Selecting a function of correlative relation, simple or multiple, of linear or any nonlinear type;

(2) estimating parameters (statistics) that measure the degree of correlative association;

(3) testing the significance of statistics that measure the correlative association;

(4) estimating parameters of regression equation, and

(5) testing the significance of regression parameters, or drawing the confidence limits about the fitted regression functions.

Two types of correlative association in hydrology. The following two types of regressions are current in hydrology.

(a) Cause-effect based relations, where a random variable, y, is correlatively related to causal factors, x_i, which produce or affect the outcomes of y. Typical examples are a runoff-rainfall relation, because the rain is the basic causal factor of runoff, with river basin geometric, soil, moisture, and climatic factors affecting the basic cause-effect relation.

(b) Relations of random variables, which have the same causative factors, such as the correlative association of the runoff of a river to the runoff of an adjacent river, or the association of rainfall variables at the adjacent precipitation stations, and similar. Both cases are similarly treated in correlative association.

11.2. Linear Bivariate Correlation.

Definition of the measure of correlation. The statistics that measure the correlative association should be independent of the choice of the coordinate origin and of the scales of measurement of variables x, y, z, ... This means that measures are dimensionless values. The ratios $\Delta x/\sigma_x$, $\Delta y/\sigma_y$, $\Delta z/\sigma_z$, ... satisfy these conditions. They are part of any measure of correlative association, where Δx, Δy, Δz, ... are deviations from the corresponding population means, μ_x, μ_y, μ_z, ..., and σ_x, σ_y, σ_z, ... are population standard deviations. Therefore, for two variables, x and y, a measure giving the average sum $\frac{1}{N}\sum_1^N \frac{\Delta x}{\sigma_x}\frac{\Delta y}{\sigma_y}$, when $N \to \infty$ (or is independent of N), satisfies the above conditions.

If two variables, x and y, are independent, their covariance is zero, as shown in Fig. 11.2 (a). The converse is not true. If $x^2 + y^2 = c^2$, it

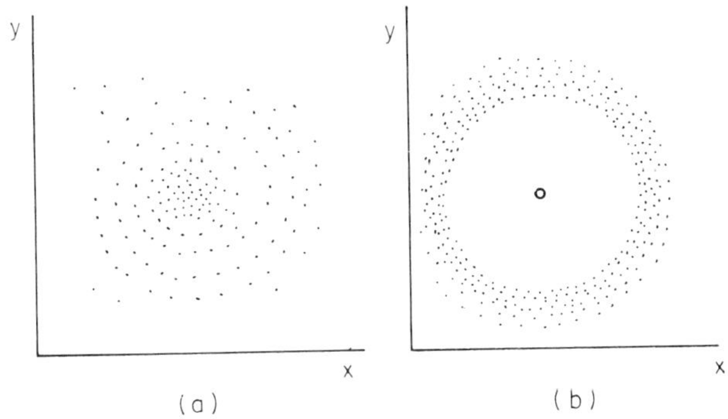

Figure 11.2. Linearly uncorrelated associations of two variables: (a) linearly uncorrelated but also independent relation of x and y, (b) linearly uncorrelated but dependent (circularly correlated) relation of x and y.

can be shown that cov (x, y) = 0 in this case, though the two components, x and y, of the bivariate distribution, F (x, y), are dependent, as shown by Fig. 11.2 (b). In the case of investigating linear dependence between x and y, it is a sufficient condition that cov (x, y) = 0, for x and y to be linearly independent.

The dimensionless measure of correlative association between x and y is the population (product-moment) correlation coefficient, ρ (x, y), defined as

$$\rho(x, y) = \frac{\text{cov}(x, y)}{(\text{var } x \text{ var } y)^{1/2}} = \frac{\sigma_{xy}}{\sigma_x \sigma_y} = \frac{E[(x-\mu_x)(y-\mu_y)]}{[E(x-\mu_x)^2 E(y-\mu_y)^2]^{1/2}}, \qquad 11.1$$

with ρ (x, y) being written in the subsequent text as ρ or ρ_{xy}. If x and y are independent, $\rho = 0$. If $\rho = 0$, x and y are linearly uncorrelated,though they may be dependent. It is easy to prove that $-1 \leq \rho \leq 1$, because $|\sigma_{xy}| \leq \sigma_x \sigma_y$. The two variables are properly functionally linearly depedent, or $c_1 x + c_2 y = c_3$ if $\rho = +1$ or $\rho = -1$, which depends whether $-c_1/c_2$ is positive or negative. The condition for the functional linear dependence is $\rho^2 = 1$. The equivalent and necessary condition stated in the form of the correlation matrix is

$$\begin{vmatrix} \sigma_{xx} & \sigma_{xy} \\ \sigma_{yx} & \sigma_{yy} \end{vmatrix} = 0, \qquad 11.2$$

or $\sigma_x \sigma_y = \sigma_{xy}$, with $\sigma_{xx} = \sigma_x^2$, $\sigma_{yy} = \sigma_y^2$, and $\sigma_{xy}^2 = \sigma_{xy}\,\sigma_{yx}$. This condition may be extended to random multivariates, namely that their components have a proper (functional) linear dependence if the determinant $|\sigma_{ij}| = 0$. The measure ρ is estimated from the sample data by the coefficient r.

The distinction should be clear between the concepts of correlation and dependence. As stated above, for $\rho = 0$ the variables may be dependent; they are linearly independent but may be dependent in other ways, such as points being around a circle, like in Fig. 11.2 (b). If points are narrowly spread around a circle, any straight line does not fit well the circle points. In this case the variables x and y are circularly correlated and dependent, but are linearly uncorrelated. For points spread very narrowly around a straight line, the case is opposite. The linear correlation is high but the circular correlation (non-linear correlation by fitting of the circular function) is very small, or the variables x and y are linearly correlated and dependent but are circularly uncorrelated. Therefore, distinction should be made in hydrology for two random variables between the terms *correlated or uncorrelated*, and *dependent and independent*. The correlated variables are always dependent, but uncorrelated variables may be either independent or dependent.

The magnitude of ρ determines the strength of the linear association. The sign of ρ indicates whether y increases (ρ positive) or decreases (ρ negative) as x increases. A scatter graph of plotted points, such as in Fig. 11.2 (a), indicates a hypothesis of no linear correlation between x and y. By a geometrical interpretation, the more the isolines of a bivariate frequency density surface concentrate along a straight line, the greater the absolute value of correlation coefficient (see Fig. 11.6). Only when all isolines converge to, and coincide with, the straight line does $|\rho| = 1$.

The value ρ^2 is called the determination coefficient. It measures that part of variation of a variable that is explained by the correlative linear association with the other variable.

Estimation of the correlation coefficient. Assume that N simultaneous observations of the random variables x and y are available. The i-th pair of observations is denoted by (x_i, y_i), $i = 1, \ldots, N$. The graphical approach is the simplest way to determine from the sample of size N whether the two variables are or are not associated. By plotting y_i against x_i a graph consisting of N points is obtained, usually as a scatter graph. It reveals any pronounced association, linear or clearly nonlinear, but it may fail to reveal a more tenuous association. Ideally, a scatter graph of points having a shot-gun appearance indicates that there may be no association between x and y, as shown in Fig. 11.2 (a).

The sample estimate of ρ is

$$r = \frac{s_{xy}}{s_x s_y} = \frac{\sum_{i=1}^{N} (x_i - \bar{x})(y_i - \bar{y})}{N s_x s_y} = \frac{\sum_{i=1}^{N} x_i y_i - N \bar{x} \bar{y}}{N s_x s_y}, \quad 11.3$$

in which s_{xy} is the estimate of σ_{xy}, $\bar{x}$ and $\bar{y}$ are the estimates of μ_x and μ_y, and s_x and s_y are the estimates of σ_x and σ_y, of Eq. 11.1. The unbiased estimates should be used for small N.

In the case of grouped data in class intervals and class fields (see Table 11.4),

$$r = \frac{\sum_{i=1}^{m} \sum_{j=1}^{n} f_{ij} (\Delta x_i \Delta y_j)}{\left[\sum_{i=1}^{m} f_i \, (\Delta x_i)^2\right]^{1/2} \left[\sum_{j=1}^{n} f_j \, (\Delta y_j)^2\right]^{1/2}}, \quad 11.4$$

in which m and n are the numbers of class intervals for x and y, respectively, f_i and f_j are their marginal class frequencies, Δx_i and Δy_j are the distances of class marks from $\bar{x}$ and $\bar{y}$, respectively, and f_{ij} are class field frequencies.

Statistical inference about the correlation coefficient. Although r differs from zero, x and y may be linearly uncorrelated (or even independent), because a non-zero value of r may be entirely attributed to sampling errors. In this case, the population correlation coefficient $\rho = 0$.

Under the assumption of a bivariate normal distribution of x and y, Fisher has derived the distribution of r as an asymmetrical probability function, bounded on both sides ($r = 1$, $r = -1$) and with only two parameters, ρ and N. For $\rho = 0$, it is a symmetrical bounded distribution

$$f(r) = \frac{(1 - r^2)^{(N-4)/2}}{B\,[1/2, (N-2)/2]}, \quad 11.5$$

in which B [ ] is the beta function with the parameters 1/2 and (N − 2)/2. The absolute skewness coefficient (positive for $+\rho$, negative for $-\rho$) of Eq. 11.5 increases as ρ^2 increases. Four probability density functions of r are given as examples in Fig. 11.3.

For ρ close to zero and for large N, the central part of f(r)-function may be approximated by a normal function in the limits $\rho - 3\sigma_r$ and $\rho + 3\sigma_r$, with σ_r the standard deviation of r. The estimate of σ_r is

$$s_r = \frac{1 - r^2}{\sqrt{N}}. \quad 11.6$$

If the two limits about r, given as $r \pm 3 s_r$ (or in a stronger test, $r \pm 4 s_r$), are of the same sign as r, it is considered that ρ estimated by r is significantly different from zero. If they are of the different signs, the converse is accepted. This test may be often performed in hydrology when N is large and r small, at least as a first approximate test.

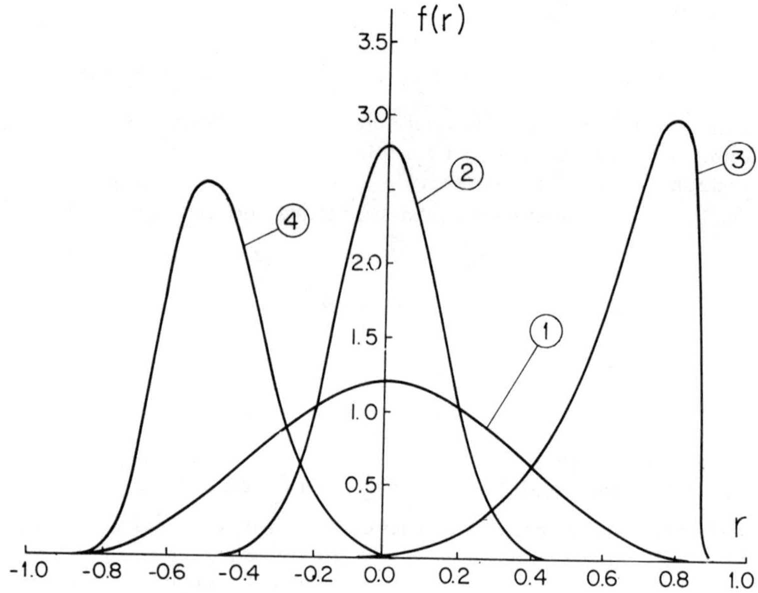

Figure 11.3. Distribution of the sample correlation coefficient r for four cases: (1) $\rho = 0$, $N = 10$; (2) $\rho = 0$, $N = 50$; (3) $\rho = 0.80$, $N = 10$, and (4) $\rho = -0.50$, $N = 30$.

The z-transformation, introduced by Fisher, as

$$z = \frac{1}{2} \ln \frac{1 + r}{1 - r} \tag{11.7}$$

is approximately normal, with the mean μ_z,

$$\mu_z = \frac{1}{2} \ln \frac{1 + \rho}{1 - \rho}, \tag{11.8}$$

and the standard deviation

$$\sigma_z = \frac{1}{\sqrt{N - 3}}. \tag{11.9}$$

The standardized value t of z may be assumed first as approximately normal for t in the neighborhood of zero, N (0, 1), so that

$$t = \frac{z - \mu_z}{\sigma_z} = \frac{\sqrt{N - 3}}{2} \ln \left[\frac{1 + r}{1 - r} \cdot \frac{1 - \rho}{1 + \rho}\right]. \tag{11.10}$$

Because ρ is never known, it must be assumed in any test. Two tests using Eqs. 11.7 through 11.10 are often performed in hydrology:

(a) Whether r is not significantly different from zero on a given probability level is determined. If the level is 95 percent, then $|t|$ of Eq. 11.10 for $\rho = 0$ should be smaller than 1.96.

(b) Whether r_1 and r_2 are not significantly different is determined. If r_1 is from a sample of size N_1, and r_2 of size N_2, then $\varepsilon = z_1 - z_2$, with $E(\varepsilon) = 0$, and the standard deviation of ε is

$$\sigma_\varepsilon = \sqrt{\frac{1}{N_1 - 3} + \frac{1}{N_2 - 3}}, \qquad 11.11$$

so that the standardized variable $t = \varepsilon/\sigma$ is approximately normal, N (0,1). This makes it possible to test the hypothesis that r_1 and r_2 are from the same population. Figure 11.4 gives z as a function of r, to facilitate the transformation of Eq. 11.7.

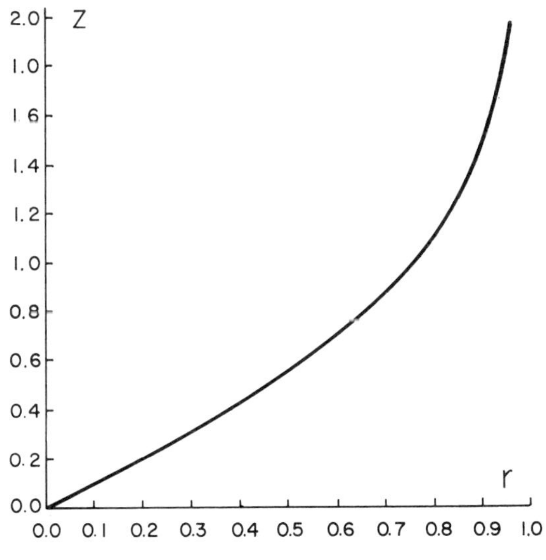

Figure 11.4. Relation of the transformed variable z to the variable r, given by Eq. 11.7.

A more exact test of the hypothesis that r is not significantly different from zero is given by

$$t = \frac{r\sqrt{N - 2}}{\sqrt{1 - r^2}}, \qquad 11.12$$

in which t follows Student distribution with (N − 2)-degress of freedom. Figure 11.5 gives the solutions of Eq. 11.12 for the values of r that are significantly different from zero at 10, 5, 2 and 1 percent probability levels, as a function of N. For comparative purposes, the curves for 20 and 50 percent levels are also plotted in Fig. 11.5.

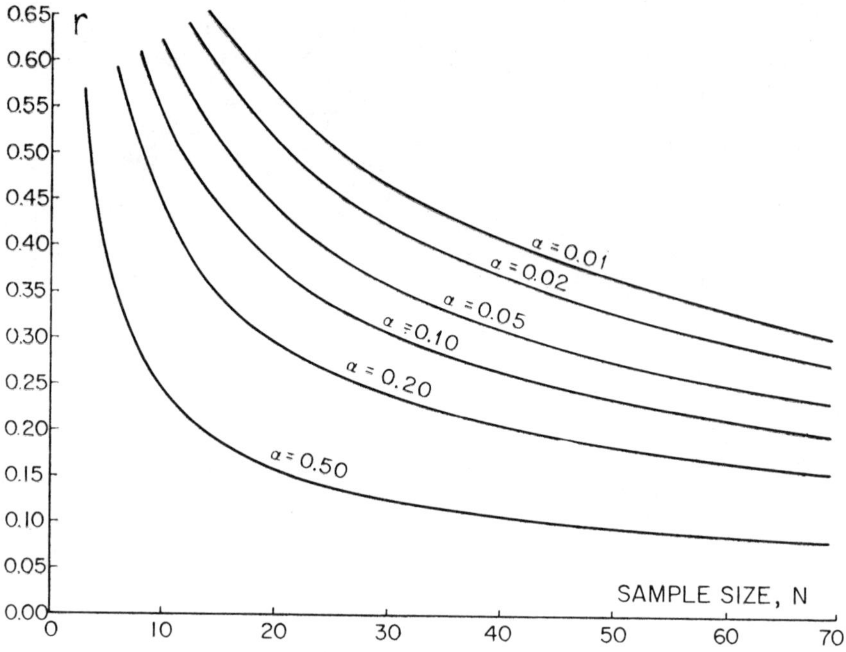

Figure 11.5. Relation r = f (N, α) for testing by Eq. 11.12 that r is not significantly different from ρ = 0, for α the probability of 0.01, 0.02, 0.05, 0.10, 0.20 or 0.50 that r will be exceeded by sampling variations.

For the most rigorous test, Fisher z -distribution should be used for the transformed variable z for any ρ or N.

The sample product-moment correlation coefficient, r, can be computed for any set of pairs of observations. If the bivariate probability distribution is skewed, r is not necessarily the best estimate of the linear correlation of the two random variables. A rigorous test of significance for r can be made only if the bivariate probability distribution is normal. Unfortunately, most hydrologic variables are not normally distributed. The skewness in data is often sufficiently large so that a normal approximation of probability distribution is poor. Transformations such as the square root, the cube root, the logarithm, and others are employed to normalize the data. In the hydrologic literature, one will find that the computed values of r refer often to transformed variables.

There is no precise evaluation of the effect of asymmetry on the estimate and distribution of r, and on the testing of r for significance by using the normal distribution function as the base. Various sampling experiments have been performed to study the effects of non-normality. They have indicated that the effect of asymmetry in small samples is not too large. Interest in hydrologic studies is often in determining whether two random variables are or are not associated, rather than in determining the exact degree of association. In this respect, the asymmetry of variable distributions does not seriously affect the use of the assumption of a normal bivariate distribution. A detailed evaluation of correlation methods in hydrology supports this notion.

Table 11.1

Seasonal and Annual Total Precipitation in Inches at Flagstaff (F.) and Fort Valley Valley (Ft. V.), Arizona

Year	Summer Apr-Sept F.	Summer Apr-Sept Ft. V.	Winter Oct.-Mar F.	Winter Oct-Mar Ft. V.	Annual Jan-Dec F.	Annual Jan-Dec Ft. V.
1920	8.16	9.44	11.17	14.47	19.33	23.91
21	14.46	16.12	7.97	11.30	22.43	27.42
22	11.18	9.26	13.89	15.96	25.07	24.72
23	7.95	11.87	13.12	15.22	21.07	27.09
24	8.15	8.51	8.59	10.49	19.74	19.00
1925	12.98	17.14	6.10	6.61	19.08	23.75
26	5.55	7.65	11.03	12.77	16.58	20.42
27	14.07	14.51	9.76	13.30	23.83	27.81
28	7.65	12.46	7.23	8.69	14.88	21.15
29	9.72	15.24	5.80	8.89	15.52	24.13
1930	11.06	13.18	10.49	12.47	21.55	25.65
31	11.07	13.54	9.27	11.36	20.34	24.90
32	11.84	10.11	10.13	12.89	21.97	23.00
33	10.03	12.33	5.57	7.77	15.60	20.10
34	8.24	10.13	6.56	7.97	14.80	18.10
1935	6.96	10.87	9.76	13.56	16.72	24.43
36	11.19	16.83	8.11	11.66	19.30	28.49
37	8.41	11.71	11.00	16.53	19.41	28.24
38	7.56	9.52	13.02	15.91	20.58	25.43
39	7.24	8.91	5.67	8.17	12.91	17.08
1940	11.19	13.24	10.03	15.60	21.22	28.84
41	13.71	13.59	11.31	17.67	25.02	37.26
42	4.80	5.20	5.10	10.07	9.90	15.27
43	7.74	10.90	10.04	12.51	17.78	23.41
44	5.18	6.59	12.32	15.19	17.50	21.78
1945	8.05	7.16	9.56	13.20	17.61	20.36
46	12.33	10.46	*n*9.41	12.97	21.74	23.43
47	8.47	8.73	4.66	4.66	13.13	13.39
48	6.34	9.21	9.01	8.28	15.35	17.49
49	14.53	15.91	11.96	12.92	26.49	28.83

Example 11.2. Precipitation in the Southwest of the United States is associated with two storm patterns. Summer precipitation occurs as thunderstorms of a random spatial distribution. The winter precipitation is from frontal storms of broader coverage. At lower elevations, the total precipitation is generally larger in the summer than in the winter. At higher elevations, the summer and winter precipitation are nearly equal. The winter precipitation at two locations is likely to be more correlated than the summer precipitation. The summer, winter, and annual precipitation in inches for Flagstaff and Fort Valley, Arizona, are given in Table 11.1. The correlation parameters are summarized in Table 11.2, with Flagstaff as variable y and Fort Valley as variable x; the means and standard deviations are given in inches.

Table 11.2
Summary of Correlation Analysis of Flagstaff versus Fort Valley Precipitation

	$\bar{x}$	s_x	$\bar{y}$	s_y	s_{xy}	r
Summer	9.53	2.75	11.34	3.09	6.58	0.77
Winter	9.25	2.51	11.95	3.18	6.98	0.87
Annual	18.78	3.87	23.30	4.30	13.90	0.84

Equation 11.12 is used to test the hypothesis of the computed r values not being significantly different from zero. For $N = 30$, the value of r, which is significant at the 5 percent level, is 0.36. Since the value $r = 0.36$ is less than three values of r in Table 11.2, the hypothesis that $\rho = 0$ is rejected.

The correlation coefficient for winter precipitation is larger than for summer precipitation. To test the significance of their difference, it is assumed that both samples are from the same population, and that the difference between the corresponding z-values is normally distributed with the standard deviation given by Eq. 11.11. From Fig. 11.4 the values of z are 1.02 and 1.32, so that $\varepsilon = 0.30$, and $s_\varepsilon = 0.272$ for $N = 30$. The standardized value is $t = \varepsilon/s_\varepsilon = 1.11$. This is less than $t = 1.96$ for the 5 percent exceedence level. The hypothesis that the difference between the two correlation coefficients is not significantly different from zero is accepted.

Assume that $\rho = 0.8$. The hypothesis is tested that the three r-values of Table 11.2 do not differ significantly from $\rho = 0.8$. By using Eq. 11.10 the t-values are -0.42, 1.21, and 0.63, for summer, winter, and annual correlation, respectively. Since each absolute value is less than 1.96, the above hypothesis is accepted for the 5 percent level.

Assuming that the three correlation coefficients, 0.77, 0.87, and 0.84, are estimates of three different ρ values, taking a level of significance at 5 percent, and using Eqs. 11.7 through 11.9, and Fig. 11.3, the confidence limits r_1 and r_2 are: (a) for $r = 0.77$, $r_1 = 0.89$ and $r_2 = 0.57$; (b) for $r = 0.87$, $r_1 = 0.94$ and $r_2 = 0.74$, and (c) for $r = 0.84$, $r_1 = 0.92$ and $r_2 = 0.69$. The confidence limits are such that each of the three ρ-values would be inside of each of the three confidence intervals at the 5 percent level. The three r-values thus may be assumed not to be significantly different.

11.3. Spurious Correlation and Interstation Correlation.

Spurious correlation. The data used in correlation analysis should be reasonably homogeneous. If the scatter graph shows a clustering of points in two or more groups, a seemingly large correlation may result, due merely to the heterogeneity in data. Such correlations are called spurious. Suppose the correlation between the monthly flows for two rivers is investigated. Moreover, assume that the dry and wet seasons are concurrent for both rivers. The scatter graph will exhibit the two clustering of flows, one for dry and one for wet months. If the heterogeneity of the data is disregarded, the computed value of r will be large, particularly if the high values are much greater than the low values. The correlation for each of the clusters may be close to zero. However, as shown below the main reasons for spurious correlation in hydrology and hydraulics are mathematical.

Assume that each of the four random variables x, y, z, and w is normally distributed. The mean, standard deviation, and coefficient of variation of x are denoted by μ_x, σ_x, and η_x, respectively, and similarly for y, z, and w. The correlation coefficient between x and y is ρ_{xy}, and similarly is denoted for the correlation coefficients between any two of the four random variables.

Pearson [1] has given the correlation coefficient, ρ, between ratios x/z and y/w as

$$\rho = \frac{\rho_{xy}\,\eta_x\,\eta_y - \rho_{xw}\,\eta_x\,\eta_w - \rho_{yz}\,\eta_y\,\eta_z + \rho_{zw}\,\eta_z\,\eta_w}{(\eta_x^2 + \eta_z^2 - 2\rho_{xz}\,\eta_x\,\eta_z)^{1/2}\,(\eta_y^2 + \eta_w^2 - 2\rho_{yw}\,\eta_y\,\eta_w)^{1/2}}. \qquad 11.13$$

Thus the correlation measure between the ratios is in terms of the above six correlation coefficients and four coefficients of variation. The correlation coefficient ρ is zero when the four random variables are mutually uncorrelated. That is, if $\rho_{xy} = \rho_{xw} = \rho_{yz} = \rho_{zw} = 0$, then $\rho = 0$.

Assume that z = w, so that $\rho_{zw} = 1$ and $\eta_z = \eta_w$. In this case, ρ reduces to

$$\rho = \frac{\rho_{xy}\,\eta_x\,\eta_y - \rho_{xz}\,\eta_x\,\eta_z - \rho_{yz}\,\eta_y\,\eta_z + \eta_z^2}{(\eta_x^2 + \eta_z^2 - 2\rho_{xz}\,\eta_x\,\eta_z)^{1/2}\,(\eta_y^2 + \eta_z^2 - 2\rho_{yz}\,\eta_y\,\eta_z)^{1/2}}. \qquad 11.14$$

This is the correlation coefficient for two ratios with the same denominator, z, or x/z and y/z. If $\rho_{xy} = \rho_{xz} = \rho_{yz} = 0$ in Eq. 11.14, ρ does not become zero but

$$\rho = \frac{\eta_z^2}{(\eta_x^2 + \eta_z^2)^{1/2}\,(\eta_y^2 + \eta_z^2)^{1/2}}. \qquad 11.15$$

Equation 11.15 gives the measure of spurious correlation among ratios having a common denominator. Assuming further that $\eta_x = \eta_y = \eta_z$, then $\rho = 0.50$.

Suppose for the three random variables, x, y, and z, the ratios x/z and y/z are formed, and that $\eta_x = \eta_y = \eta_z$. Let ρ_{xy} be different from ρ_{xz} and ρ_{yz}, but $\rho_{xz} = \rho_{yz}$. By substituting these values into Eq. 11.14, it becomes

$$\rho = \frac{1}{2}\left(1 + \frac{\rho_{xy} - \rho_{xz}}{1 - \rho_{xz}}\right). \qquad 11.16$$

Equation 11.16 shows how the correlation coefficient of ratios diverges from the spurious value $\rho = 0.50$ as ρ_{xy} and ρ_{xz} differ from zero. For $\rho_{xy} > \rho_{xz}$ the correlation between the ratios will be greater than the spurious value 0.50. But for $\rho_{xy} < \rho_{xz}$ the correlation coefficient is less than 0.50.

The matter of correlation between ratios occupied the attention of statisticians at the turn of the Century. In more recent statistical texts, this matter has been largely neglected. The correlation between ratios, however, may be fairly common in hydrology and hydraulics.

Pearson [1], and Benson [2] in a summarized form, have shown that the spurious correlation appears under various conditions, including products of random variables.

The dimensional analysis in fluid mechanics, and investigation of ratios and products in hydrology, may often involve one or more random variables which appear in dimensionless numbers or in various other ratios. The correlation is in most cases spurious. Even if variables are not stochastic but are measured with significant random errors, these errors may produce spurious correlation. Every design for investigating relations in hydrology should first look at whether the set-up would or would not produce spurious correlation.

Example 11.3. Table 11.2 gives parameters for this example. Spurious correlation is investigated between ratios of summer versus winter precipitation for the two precipitation stations. Furthermore, the correlation coefficient of summer versus winter precipitation is 0.10 for Flagstaff, and 0.14 for Fort Valley. The coefficients of variation for summer and winter precipitation for the two stations are, respectively, 0.290 and 0.271, and 0.272 and 0.266. The corresponding r-values for summer-summer and winter-winter precipitation are 0.77 and 0.87. The correlation coefficients of summer precipitation of Flagstaff versus winter precipitation of Fort Valley is 0.24, and for winter precipitation of Flagstaff versus summer precipitation of Fort Valley is -0.19. By using Eq. 11.13, the correlation coefficient of ratios is $\rho = 0.52$. It is smaller than either r values of summer-summer and winter-winter precipitations for the two stations. Ratios of summer to winter precipitations for Flagstaff and Fort Valley have means of 1.11 and 1.03, and standard deviations of 0.55 and 0.50, respectively. The covariance is $s_{xy} = 0.165$, and $r = 0.60$ for these ratios. This value is greater than the computed value by Eq. 11.13.

Application of bivariate correlation to interstation correlation. A hydrologic variable is gauged at n points in a region. Their lengths of record are usually different, but equal sample size N will be used in the following discussion. If N_i differs from station to station, either all station data should be reduced to the same sample size N, or the expressions given below should be adjusted to take care of unequal sample series. The estimates of population mean (μ_j) and variance (σ_j^2) of the j-th station from the annual values x_{ij} (i is for the year, j is for the station) are respectively

$$\bar{x}_j = \frac{1}{N} \sum_{i=1}^{N} x_{ij}; \quad s_j^2 = \frac{1}{N} \sum_{i=1}^{N} (x_{ij} - \bar{x}_j)^2. \qquad 11.17$$

If a variable parameter of particular interest of these stations is relatively homogeneous, especially for the reduced variables with the mean of zero and the variance of unity, and if the n variables, $j = 1, 2, \ldots, n$, are mutually uncorrelated (each pair of variables has an insignificant correlation), then n-point records of sample size N can be pulled together to form a simple record of nN-observations. This method of analysis is called the station-year method, and its objective is to extract the most reliable information possible about a statistic of interest.

The variables of these n points in a region are likely to be mutually correlated. The pulling together of n sets of N observations into a single sample size nN can yield only as much information as some lesser number $n_e N$, where $n_e < n$, with n_e called here the effective number of points (stations) in the region.

The estimates of the regional mean and its regional variance are

$$\bar{x} = \frac{1}{n} \sum_{j=1}^{n} \bar{x}_j, \qquad 11.18$$

and

$$\operatorname{var} \bar{x} = s^2(\bar{x}) = \frac{1}{n} \sum_{j=1}^{n} s_j^2 + \frac{2}{n^2} \sum_{j=1}^{n-1} \sum_{i=j+1}^{n} r_{ij} s_i s_j, \qquad 11.19$$

in which $\bar{x}_j$ and s_j^2 are estimated by Eq. 11.17, s_i and s_j are standard deviations of any pair of point variables, and r_{ij} is the sample product-moment correlation coefficient between variables of points i and j. For $i = j$, $r_{ij} = 1$. If all point variables have the sample mean zero ($\bar{x}_j = 0$, s_j^2), then $\bar{x} = 0$, and for s_j^2 reduced to a common variance s_0^2 the mean correlation coefficient is

$$\bar{r} = \frac{2 \sum_{j=1}^{n-1} \sum_{i=j+1}^{n} r_{ij}}{n(n-1)}, \qquad 11.20$$

and Eq. 11.19 becomes

$$s^2 = \frac{s_0^2}{n} [1 + \bar{r}(n-1)]. \qquad 11.21$$

Therefore, for $\bar{r} = 0$, $s^2 = s_0^2/n$, and for $\bar{r} = 1$, $s^2 = s_0^2$.

For n_e the effective number of mutually uncorrelated series at n_e points (n_e is not necessarily an integer), the variance of the regional mean is $s^2 = s_0^2/n_e$. By equating it with Eq. 11.21, then

$$n_e = \frac{n}{1 + \bar{r}(n-1)}. \qquad 11.22$$

If $\bar{r} = 0$, $n_e = n$, and if $\bar{r} = 1$, then $n_e = 1$. The effective number of stations is highly sensitive to $\bar{r}$ but much less sensitive to n. In the limit, if n tends to infinity, n_e tends to $1/\bar{r}$. Even for small $\bar{r}$, values n_e cannot increase appreciably by a very large increase in n. Although n sequences can be pulled together to form a sequence of nN observations, they yield only as much information about the mean as a sequence of $n_e N$ observations formed by n_e mutually uncorrelated series, provided that series at n points are not also serially correlated (time dependent).

Example 11.4. Benson [3] employed 160 stations having an average sample size of $N = 30$ in studying a regional flood frequency equation for New England. The estimated $\bar{r}$ was about 0.3. This gave $n_e = 3.3$, showing that n_e is very sensitive to $\bar{r}$. The large sample of $nN = 4800$ values is reduced to $n_e N = 100$ independent values for the study of floods. From this discussion, one may conclude that the interstation correlation severely limits regional studies in hydrology. The discussion in the forthcoming text on the application of regional regression analyses, however, will show that the interstation correlation may be beneficial.

Correlation of two serially dependent variables. The serial correlation is used here to account for the effect of serial dependence in each of the two variables x and y on the statistical inference for the correlation coefficient. An approximate method of inference is obtained by the computation of reduced size of a sample. Thus, the computed standard deviation s_r, of distribution of correlation coefficient, r, and the confidence intervals associated with it become greater than in the case of serially uncorrelated variables.

The serial correlation coefficients are defined as cross-product moment correlation coefficients (linear association) between the members of a series that are k-units apart. The k-th serial correlation coefficient, r_k, is the r-value of two variables, x_i and x_{i+k}, with x_i composed of the first $N - k$ members and x_{i+k} of the last $N - k$ members of a series. According to Bartlett [4], the reduced or effective sample size, N_e, for the test of significance of r for the dependent stochastic series is approximately

$$N_e = \frac{N}{1 + 2r_1r_1' + 2r_2r_2' + \ldots + 2\,r_k r_k' + \ldots}, \qquad 11.23$$

in which r_k and r_k' are serial correlation coefficients for the variables x and y.

Equation 11.23 is not valid for series of x and y with the cyclic movement, or trends and jumps. If monthly flows of two rivers are investigated for linear correlation, and if both rivers follow an annual cycle of wet and dry seasons, the sequence of serial correlation coefficients gives a correlogram that may be very close to the cosine function for both series. The positive and negative values of r_k and r_k' would coincide in two series so that the denominator in Eq. 11.23 would be infinite. Similar objections are for trends and jumps. Therefore, the cyclic movement and trends and jumps in the series should be first removed. Equation 11.23 is valid only if both variables are dependent stochastic variables. If one is independent, the dependence of the second variable does not affect N.

If interstation correlation is used for pulling together data of n stations in a regional investigation of the association of two hydrologic variables, and if all n series of these two variables are time dependent but not periodic, then the effective size of the sample of the pulled-together data may be estimated as an approximation by

$$N_e\, n_e = \frac{Nn}{[1 + \bar{r}(n-1)]\,(1 + 2\overline{r_1r_1'} + 2\overline{r_2r_2'} + \ldots + 2\overline{r_kr_k'} + \ldots)}, \qquad 11.24$$

in which $\overline{r_1r_1'}$, $\overline{r_2r_2'}$, ... are average products of corresponding serial correlation coefficients of n series. Other expressions may be used to account for the serial correlation. The averaging approach of r_kr_k' may be the simplest procedure once n correlograms are computed, but may be biased in several ways. In conclusion, both the regional correlation of hydrologic data at n-points, and the serial correlation of each series of size N, reduce the total size nN of all data to the effective number n_e of points and the effective sample size N_e, for the purposes of making the unbiased statistical inference. The total data of station-year method (or point-time interval method for smaller time units than a year) is now n_eN_e instead of nN.

Example 11.5. The correlation coefficient of annual runoff for water years (Oct. 1-Sept. 30) of the St. Lawrence River at Ogdensburg, and the Mississippi River at St. Louis, for the period of simultaneous observations, 1861–1957, is $r = 0.17$, with $N = 96$. For testing the hypothesis whether this r is not significantly different from zero, the Student t-statistic of 1.67 is used. It is greater than $t = 1.64$ at the 10 percent level, but smaller than $t = 1.96$ at the 5 percent level. Only the first serial correlation coefficient of the Mississippi River is significantly different from zero, and is $r_1' = 0.294$. Therefore, though several serial correlation coefficients of the St. Lawrence River are significantly different from zero, only $r_1 = 0.705$ has an effect on the sample effective size, because r_2', r_3', ... may be taken as zeros. By using Eq. 11.23, $N_e = N/(1 + 2r_1 r_1') = 68$. With this new sample size, the Student t is 1.42, which is smaller than $t = 1.64$, and $t = 1.96$ on both 10 and 5 percent level. Therefore, the value $r = 0.17$ is not significantly different from zero, and the annual flows of these two rivers may be considered as linearly uncorrelated.

11.4. Linear Bivariate Regression.

Definition of linear regression lines, and estimation of their parameters. If the correlation shows the two variables x and y to be linearly associated, the linear relations allow one variable to be predicted from the other. The linear regression relation for the variable y versus the variable x is defined by a straight line which gives the best estimate of y for a given value of x. Similarly, the best estimate of x for a given y is given by the regression line of x versus y.

The line connecting the mean values y_0 of all y_i, for given x_i, or more precisely, for the interval Δx around x_i, is for all practical purposes the regression line. In the simple linear regression analysis, as shown by Fig. 11.6, a straight line is fitted through these values.

The straight regression line is generally fitted analytically by the least-squares method of deviations from the line. The individual mean values y_0 for given x_i, or x_0 for given y_i, are not usually computed. For the regression line y versus x, the deviations are e_y along the ordinate. For the regression line x versus y, the deviations are e_x along the abscissa.

The two variables x and y for their population linear regressions are denoted as ξ and η, so that these two equations are

$$\eta = \alpha_1 + \beta_1 \xi \text{ , and } \xi = \alpha_2 + \beta_2 \eta. \tag{11.25}$$

The estimated regression lines from a sample are

$$y = a_1 + b_1 x, \text{ and } x = a_2 + b_2 y, \tag{11.26}$$

in which α_1, β_1, α_2, and β_2 are estimated by the sample values a_1, b_1, a_2, and b_2 by the least-squares method. These parameters are called regression coefficients, and particularly b_1 is the regression coefficient of y versus x, b_2 of x versus y_1, and a_1 and a_2 are intercepts. The regression line y versus x is different from the regression line x versus y. The variable on

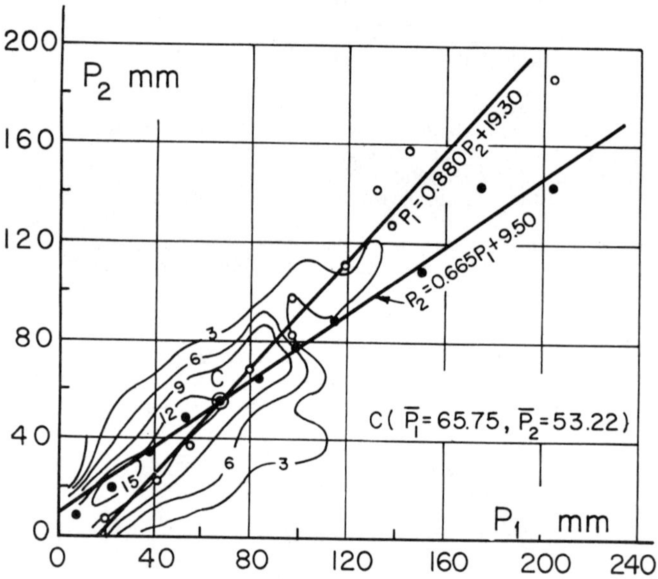

Figure 11.6. Frequency isolines and two regression lines, with the average values of one variable for class-interval of the other variable.

the right side of Eqs. 11.25 and 11.26 is called the independent variable, and on the left side it is called the dependent variable. In Eq. 11.26, y is the estimate of η, x of ξ, and e_y is an estimate of ε_y, and similarly e_x is an estimate of ε_x, with ε_y and ε_x residuals of population regression lines.

The least squares method is good for estimating α and β if the deviations $e_y = y_i - y$, with y_i observed and y predicted values of η by Eq. 11.26, have these properties: e_y are mutually independent, and the variance of e_y is independent of x. The estimates of α_1 and β_1 from the sample of size N are given by

$$b_1 = \frac{s_{xy}}{s_x^2} = \frac{\sum_{i=1}^{N} x_i y_i - N\bar{x}\,\bar{y}}{\sum_{i=1}^{N} x_i^2 - N\bar{x}^2}, \qquad 11.27$$

and

$$a_1 = \bar{y} - b_1\bar{x}, \qquad 11.28$$

because the regression lines must pass through the center $(\bar{x}, \bar{y})$ of the points.

Figure 11.7 shows the population (η versus ξ) and the sample (y versus x) regression straight lines, with their population, ε_y, and the sample deviations, e_y. If one regression is linear, it does not follow that the other is linear also. If both x and y are normally distributed, and one regression is linear, the other is also linear.

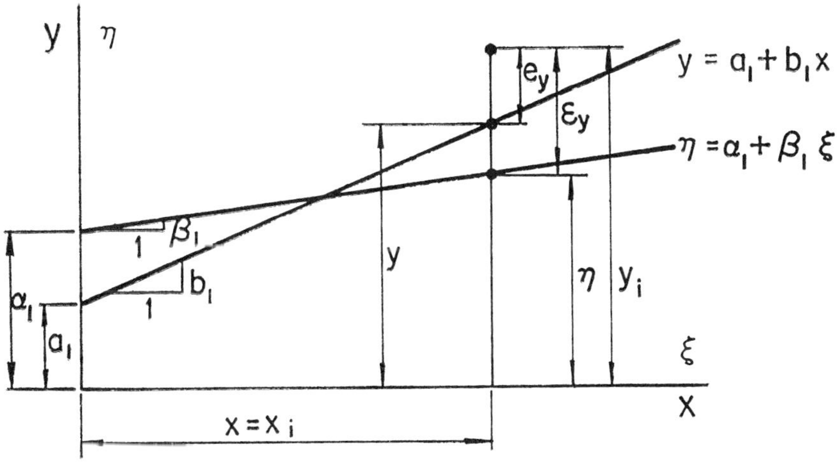

Figure 11.7. Schematic representation of true linear regression of η versus ξ, and estimated linear regression of y versus x.

For the regression x versus y, the estimates of parameters α_2 and β_2 are obtained similarly as in Eqs. 11.27 and 11.28, as

$$b_2 = \frac{S_{xy}}{S_y^2} = \frac{\sum_{i=1}^{N} x_i y_i - N\,\overline{xy}}{\sum_{i=1}^{N} y_i^2 - N\,\bar{y}^2}, \qquad 11.29$$

and

$$a_2 = \bar{x} - b_2 \bar{y}. \qquad 11.30$$

The estimates of Eqs. 11.27 through 11.30 are obtained by minimizing $\Sigma e_y^2 = \Sigma (y_i - a_1 - b_1 x_i)^2$, or $\Sigma e_x^2 = \Sigma (x_i - a_2 - b_2 y_i)^2$, $i = 1, 2, \ldots, N$, and by putting the partial derivatives $\partial \Sigma e_y^2/\partial a_1$ and $\partial \Sigma e_y^2/\partial b_1$, and $\partial \Sigma e_x^2/\partial a_2$ and $\partial \Sigma e_x^2/\partial b_2$, equal to zero, respectively, for the two regression lines.

In the case of grouped data in a bivariate distribution, the regression coefficients β_1 and β_2 are estimated by b_1 and b_2 as

$$b_1 = \frac{\sum_{i=1}^{m} \sum_{j=1}^{n} f_{ij}\, \Delta x_i\, \Delta y_j}{\sum_{i=1}^{m} f_{xi} (\Delta x_i)^2}, \text{ and } b_2 = \frac{\sum_{i=1}^{m} \sum_{j=1}^{n} f_{ij}\, \Delta x_i\, \Delta y_j}{\sum_{j=1}^{n} f_{yj} (\Delta y_j)^2}, \qquad 11.31$$

in which the symbols on the right side are the same as for Eq. 11.4. As

$$N\,\bar{x} = \sum_{i=1}^{m} f_{xi}\, x_i, \text{ and } N\,\bar{y} = \sum_{j=1}^{n} f_{yj}\, y_i, \qquad 11.32$$

$\bar{x}$ and $\bar{y}$ of Eq. 11.32 should be used in Eqs. 11.28 and 11.30 to estimate α_1 and α_2.

Example 11.6. In the correlation analysis of ungrouped data in Table 11.1 of the Flagstaff and Fort Valley rainfall, one is regarded as the dependent and the other as the independent random variable. The parameters of summer, winter, and annual regression straight lines are estimated by using the values $\bar{x}$, $\bar{y}$, s_x, s_y, and s_{xy} of Table 11.2, and Eqs. 11.27 through 11.30. The regression coefficients a_1, b_1, a_2, and b_2 are given in Table 11.3.

Table 11.3
Summary of Regression Coefficients of Fort Valley versus Flagstaff Precipitations and the Converse

Precipitation Variable	Regression Fort Valley versus Flagstaff		Regression Flagstaff versus Fort Valley	
	Intercept a_1	Slope b_1	Intercept a_2	Slope b_2
Summer	3.05	0.87	1.71	0.69
Winter	1.68	1.11	1.00	0.69
Annual	5.03	0.93	1.31	0.75

Example 11.7. Table 11.4 gives the monthly precipitations ($P_1 = x_i$ and $P_2 = x_j$) for two rainfall stations (Kruševac and Vrnjačka Banja, Yugoslavia), with their bivariate frequencies f_{ij} and their marginal frequencies f_{xi} and f_{yj} of grouped data. It also shows a practical procedure for computing the sums necessary to be used in Eqs. 11.31 and 11.32. In order to avoid computation with large absolute values of P_1 and P_2, the class intervals are coded. For coding or indexing class intervals, the new dimensionless variables u and v are introduced: $u = (113 - P_1)/\Delta P_1$, and $v = (83 - P_2)/\Delta P_2$, in which the length of class intervals is $\Delta P_1 = \Delta P_2 = 15$ mm (1 mm = 0.03937 in.). The new variables u and v are selected with an arbitrary central class-interval introduced, such that the central point is equal to zero. The variables u and v have positive and negative integers as marks of successive class intervals. Their means are in this example $\bar{u} = 3.150$ and $\bar{v} = 1.985$. By decoding class intervals, $\bar{P}_1 = 62.25$ mm, and $\bar{P}_2 = 53.32$ mm. The covariance of u and v is $s_{uv} = \frac{1}{N}\Sigma v \Sigma f_{ij}\, u - \bar{u}\,\bar{v} = 2921/279 - 3.150 \cdot 1.985 = 4.25$; their variances are $s_u^2 = \frac{1}{N}\Sigma f_{xi}\, u^2 - u^2 = 4546/279 - 3.150^2 = 6.40$; $s_v^2 = \frac{1}{N}\Sigma f_{yj}\, v^2 - \bar{v}^2 = 2456/279 - 1.985^2 = 4.84$. The regression coefficients are $b_1 = s_{uv}/s_u^2 = 0.665$, and $b_2 = s_{uv}/s_v^2 = 0.880$, and the two regression lines are $P_2 = 9.50 + 0.665\, P_1$, and $P_1 = 19.30 + 0.880\, P_2$. Figure 11.6 shows the relation between monthly precipitations P_1 and P_2 for the two stations, with lines of equal bivariate frequencies. The points of mean values y_0 of all y_j for given x_i (solid circles), the mean values x_0 of all x_i for given y_j of class intervals (open circles), and the two regression lines with means $\bar{P}_1$ and $\bar{P}_2$, are also given.

Table 11.4

Computation of Sums for Estimation of Parameters in Linear Regression and Correlation Analysis of Precipitations

$v\Sigma f_{ij}u$	$\Sigma f_{ij}u$	$f_{yj}v^2$	$f_{yj}v$	v	f_{yj}	Monthly precipitation Vrnjačka Banja, P_1 (mm) $= x_i$														
						8	23	38	53	68	83	98	113	128	143	158	173	188	203	Monthly Precipitation Kruševac, P_2 (mm) $= y_j$
30	−6	75	−15	−5	3							1		1					1	158
16	−4	64	−16	−4	4							1	1	1			1			143
15	−5	27	−9	−3	3							1	1						1	128
20	−10	48	−24	−2	12							3	2	5			2			113
−16	16	14	−14	−1	14				2	1	2	3	4	2						98
0	28	0	0	0	27			1	1	3	12	2	2	2	2		1	1		83
90	90	39	39	1	39			2	7	9	9	9	1	2						68
312	156	200	100	2	50		3	7	14	12	6	3	2	2	1					53
534	178	414	138	3	46		6	12	12	7	4	5								38
930	240	800	200	4	50		20	14	7	6	2		1							23
960	192	775	155	5	31	14	11	5		1										8
2921		2456	554		279	14	40	41	43	39	35	28	14	15	3	0	4	1	2	f_{xi}
						7	6	5	4	3	2	1	0	−1	−2	−3	−4	−5	−6	u
					876	98	240	205	172	117	70	28	0	−15	−6	0	−16	−5	−12	$f_{xi}u$
					4546	6861	440	1025	688	351	140	28	0	15	12	0	64	25	72	$f_{xi}u^2$
						70	159	133	97	82	39	9	−6	−15	2	0	−8	0	−8	$\Sigma f_{ij}v$
					2921	490	954	665	388	246	78	9	0	15	−4	0	32	0	48	$u\Sigma f_{ij}v$

NOTATION

$u\Sigma f_{ij}v - v\Sigma_{ij}u$

1 mm = 0.03937 in. Entries corresponding to x_i and y_j in the table are f_{ij} values.

Standard deviation of residuals. The residuals (departures, deviations) of the straight-line regression η versus ξ, given as $\varepsilon_y = y_i - \eta$ for the population, have the variance σ^2_{ey} as

$$\sigma^2_{ey} = \lim_{N \to \infty} \frac{1}{N} \sum_{i=1}^{N} \varepsilon^2_{yi} = (1 - \rho^2)\sigma^2_y . \qquad 11.33$$

It is estimated for the regression y on x in the case $e_{yi} = y_i - y$, with y_i observed and y from the line, by

$$s^2_{ey} = \frac{1}{N} \sum_{i=1}^{N} e^2_{yi} = (1 - r^2)\, s^2_y \,, \qquad 11.34$$

with $s^2_{ey} \leq s^2_y$. The standard deviation of residuals, e_{yi}, is

$$s_{ey} = s_y \sqrt{1 - r^2}, \qquad 11.35$$

where s_y is the estimate of σ_y and r is the estimate of ρ.

Considering the number of degrees of freedom in the case of small samples, the unbiased standard deviation of residuals, $\hat{s}_{ey}$, is given by

$$\hat{s}_{ey} = \hat{s}_y \sqrt{\frac{(N - 1)(1 - r^2)}{N - 2}} . \qquad 11.36$$

The greater s_{ey} or s_{ex}, the wider the spread of the points around the regression line, and the less accurate the values predicted from the regression lines.

Referring to the example in Table 11.4, the standard deviations for the variables u and v are $s_u = 2.52$ and $s_v = 2.20$. By retransforming u and v to P_1 and P_2, respectively, the standard deviations for the variables P_1 and P_2 are $s_x = 37.80$ mm and $s_y = 33.00$ mm. Since N is large, Eq. 11.35 gives $s_{ey} = 21.3$ mm and $s_{ex} = 24.4$ mm.

The variance, S^2_y, defined as

$$S^2_y = s^2_y - s^2_{ey} = r^2 s^2_y, \qquad 11.37$$

is the explained variance of y by the regression of y versus x, while s^2_{ey} is often called the unexplained variance of y by this regression. For s_{ex}, s^2_{ex}, or S^2_x, Eqs. 11.33–11.37 have y symbol replaced by x.

For $r^2 = 0$ and $S^2_y = 0$, none of the variance s^2_y of y is explained by the regression of y on x. For $r^2 = 1$ and $S^2_y = s^2_y$, all the variance s^2_y is explained by this regression. Since r^2 is defined as the coefficient of determination, its value is the proportion of the variance s^2_y explained by the regression, while $1 - r^2$ is the proportion of the variance unexplained by the regression, or $(1 - r^2)$ can be called the coefficient of non-determination.

Example 11.8. For the Flagstaff rainfall as the independent variable and the Fort Valley rainfall as the dependent variable, the values of s^2_y, s^2_{ey}, S^2_y, r^2, and $(1 - r^2)$

are given in Table 11.5. The explained variances range from 59 percent to 76 percent and the unexplained variances range from 24 percent to 41 percent of the variances for the Flagstaff rainfall data.

Table 11.5

Variances s_y, s_{ey}^2, and S_y^2, and r^2, and $(1 - r^2)$, for the Regression Analysis of Fort Valley Precipitation on Flagstaff Precipitation

	s_y^2	s_{ey}^2	S_y^2	r^2	$1 - r^2$
Summer	9.55	3.92	5.63	0.59	0.41
Winter	10.11	2.92	7.68	0.76	0.24
Annual	18.49	5.36	13.13	0.71	0.29

For the example of grouped data in Table 11.4, $D = r^2 = 0.581$, and $r = 0.762$. The coefficient D indicates that 58.1 percent of the variance is accounted for by linear regression in this example.

Relations and interpretation of parameters. Using Eqs. 11.3, 11.27, and 11.29, or Eqs. 11.4 and 11.31, the regression coefficients may be expressed in terms of the correlation coefficient r and the standard deviations s_x and s_y:

$$b_1 = r\frac{s_y}{s_x}, \text{ or } b_2 = r\frac{s_x}{s_y}. \qquad 11.38$$

Furthermore,

$$r^2 = b_1 b_2 = \frac{s_{xy}^2}{s_x^2 s_y^2}. \qquad 11.39$$

If Eq. 11.39 is used to compute r, r has the sign of s_{xy}, while both b_1 and b_2 also have the sign of s_{xy}. If $r^2 = 1$, b_1 and b_2 are reciprocal. If $r^2 = 0$, then both b_1 and b_2 must be zeros, so that the two regression lines parallel the two coordinate axes.

To interpret properly the practical aspects of linear regression, all three parameters b_1, r, and s_{ey} (or b_2, r, and s_{ex}) are useful and necessary, with b_1 indicating how quickly the dependent variable increases or decreases with a change of the other variable, r measuring the degree of the association, and s_{ey} measuring the spread of points around the straight line.

In the example of Table 11.4, $b_1 = 0.665$, $r = 0.762$, and $s_{ey} = 21.3$ mm. Thus P_2 increases rapidly with a corresponding increase of P_1, the association is sufficiently high, and despite the high association, the spread of points is also large.

Graphical estimate of regression parameters. The graphical method of estimating regression parameters is based on the principle that the graphically fitted straight line, by minimizing deviations from the line, leaves the same amount of scattered points on both sides of the line, and these points are nearly homogeneously distributed along both sides.

The parameters r, a_1, b_1, a_2, and b_2, estimated graphically, are not usually accurate. Assuming that x and y are distributed normally about $\bar{x}$ and $\bar{y}$, and ε-values are also normally distributed about the regression straight lines, then Fig. 11.8 shows the method of graphical estimation. The two

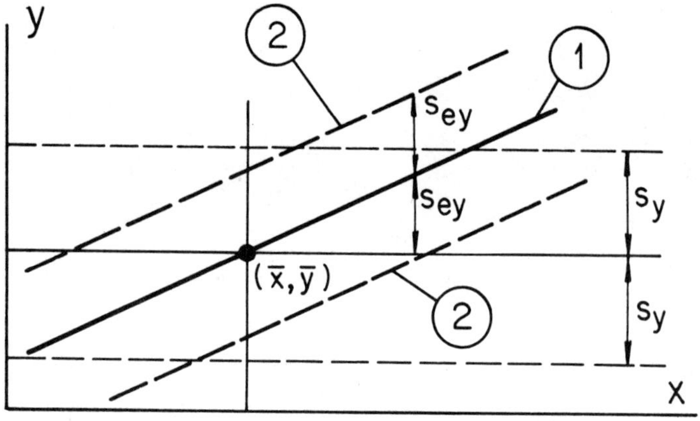

Figure 11.8. Graphical estimation of correlation and regression coefficients: (1) visual fit of a straight line through the center ($\bar{x}$, $\bar{y}$), and (2) two parallel and equidistant lines to the visually fitted line, which include between them about 68.26 percent of points (x_i, y_i).

horizontal lines equally distant from y, with half of points (x_i,y_i) between these two lines, produce $4s_y/3$. The two vertical lines equally distant from x, with half of points (x_i,y_i) between these two lines, produce $4s_x/3$. They produce estimates of s_y and s_x. For two straight equidistant lines parallel to $y = a_1 + b_1x$, with half of the y_i-points included between them, gives $4s_{ey}/3$, from which s_{ey} is estimated. From Eq. 11.35, $r^2 = 1 - s_{ey}^2/s_y^2$. Similarly, if s_{ex} is determined by drawing parallel lines to $x = a_2 + b_2y$, then $r^2 = 1 - s_{ey}^2/s_y^2$. From Eq 11.38 and r, s_x, and s_y, the regression coefficient estimates b_1 and b_2 are computed. The intercepts α_1 and α_2 are estimated by Eq. 11.28 as a_1, and by Eq. 11.30 as a_2.

Example 11.9. Figure 11.9 gives the graphical estimates of correlation and regression coefficients for the linear correlation of annual flows (water years) of the Kern River near Bakersfield and the Kings River at Piedra, California, for 58 years (1896—1953). Following the above procedure, $s_{ey} = 200$ cfs, $s_y = 520$ cfs, and $s_x = 980$ cfs. By using Eqs. 11.36 and 11.39, the coefficients are $r = 0.92$, $b_1 = 0.49$, and $b_2 = 1.99$. By using the analytical procedure of the least squares, the same coefficients are $r = 0.97$, $b_1 = 0.53$, and $b_2 = 1.83$.

Distribution of regression coefficients. If x and y follow a bivariate normal probability distribution, and if the regression of y on x is linear, then the least squares estimates a and b are unbiased estimates of α and β, respectively. Fisher has shown that for $\beta_1 = 0$, the statistic $b_1^2 \sigma_x^2/\sigma_y^2$ has a Beta-distribution of the second kind, with $B\left[\frac{1}{2}, \frac{1}{2}(N-1)\right]$.

The expected value of b_1^2 is $E(b_1^2) = \sigma_y^2/(N-1)\sigma_x^2$, and the distribution of b_1 is

$$f(b_1) = \frac{\sigma_x \sigma_y^{N-1}}{B\left[\frac{1}{2}, \frac{1}{2}(N-1)\right](\sigma_y^2 + b_1^2 \sigma_x^2)^{N/2}}, \qquad 11.40$$

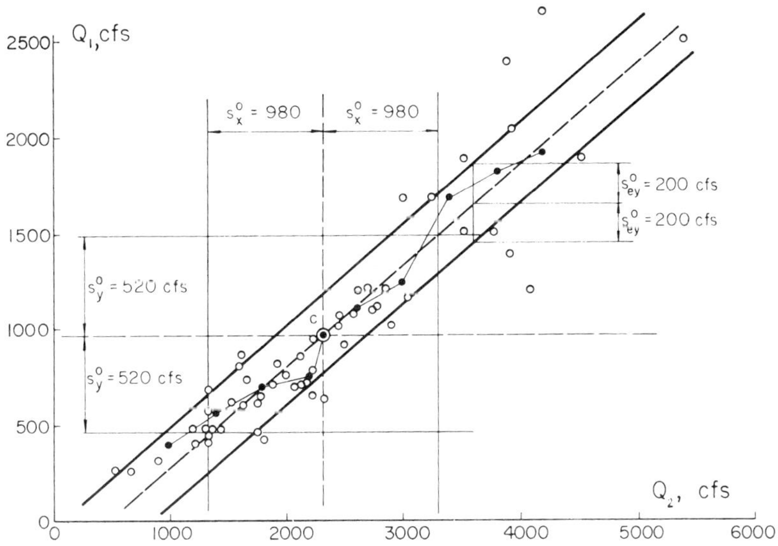

Figure 11.9. Graphical estimation of correlation and regression coefficients of the mean annual flows (1896–1953), of the Kern River near Bakersfield, California, as Q_1, and the mean annual flows (1896–1953), of the Kings River at Piedra, California, as Q_2, with the graphical estimates: $r^0 = 0.92$, $b^0_1 = 0.49$, and $b^0_2 = 1.88$.

with the population variances σ_x^2 and σ_y^2 of x and y estimated by $\hat{s}_x^2$ and $\hat{s}_y^2$. Equation 11.40 implies that

$$t = \frac{b_1 \sigma_x}{\sigma_y}\sqrt{N-1} \qquad 11.41$$

conforms to the Student t-distribution with N − 1 degrees of freedom. As σ_x and σ_y are unknown, the statistic $b_1^2 s_x^2/s_{ey}^2$ has the numerator and denominator which have the χ^2-distribution with 1 and N − 2 degrees of freedom, respectively. Their ratio has B-distribution of the second type, with $B\left[\frac{1}{2}, \frac{1}{2}(N-1),\right]$ or the statistic

$$t = b_1\sqrt{\frac{(N-2)s_x^2}{s_{ey}^2}} = r\sqrt{\frac{N-2}{1-r^2}} \qquad 11.42$$

has the Student t-distribution with N — 2 degrees of freedom. For the correlated x and y variables, with $\beta_1 \neq 0$, b_1 is replaced by $b_1 - \beta_1$, so that

$$t = \frac{(b_1 - \beta_1)s_x}{s_y}\sqrt{\frac{N-2}{1-r^2}} = \frac{b_1 - \beta_1}{b_1}\sqrt{\frac{(N-2)r^2}{1-r^2}} \tag{11.43}$$

follows the Student-t distribution with NDF = N − 2. For b_2 and β_2, s_x and s_y are interchanged. If β_1 or β_2 are assumed to be zero, Eq. 11.42 should be used, if not, Eq. 11.43 should be used.

The confidence limits are given then by

$$(b_1)_{1,2} = b_1 \pm \frac{b_1}{r} t_p \sqrt{\frac{1-r^2}{N-2}}, \tag{11.44}$$

in which t_p is the Student-distribution t-value for a given probability level.

The reduced value t of a_1 (intercept) is

$$t = \frac{r(a_1 - \alpha_1)}{b_1}\sqrt{\frac{N-2}{(1-r^2)(\hat{s}_x^2 + \bar{x}^2)}}, \tag{11.45}$$

with t following the Student distribution with (N — 2)-degrees of freedom. The confidence limits of the intercept are

$$(a_1)_{1,2} = a_1 \pm t_p \frac{b_1}{r}\sqrt{\frac{(1-r^2)(\hat{s}_x^2 + \bar{x}^2)}{N-2}}. \tag{11.46}$$

A similar equation can be derived for a_2, with corresponding interchanges. In hydrology, the usual test is whether a_1 and a_2 are not significantly different from zero, or whether the population regression straight line passes through the origin. This can be tested by putting $\alpha_1 = 0$ in Eq. 11.45, and testing whether t of that equation is smaller than Student t_p for a given probability level.

Example 11.10. In Table 11.6 results are given of the b_1-coefficient test for the Flagstaff-Fort Valley rainfall analysis, using a two-tail test and 5 percent level, with t computed by Eq. 11.42 for testing the hypotheses whether $\beta_1 = 0$, and with confidence limits $(b_1)_1$ and $(b_1)_2$ computed by Eq. 11.44.

Table 11.6

Test of Significance of b_1-Coefficient for the Regression of Fort Valley Precipitation on Flagstaff Precipitation.

	b_1	t	t_p	$(b_1)_1$	$(b_1)_2$
Summer	0.87	6.37	2.05	1.14	0.60
Winter	1.11	9.39	2.05	1.43	0.79
Annual	0.93	8.23	2.05	1.20	0.66

Since t is greater than t_p for summer, winter, and annual precipitations, all three b_1-coefficients are significantly different from zero. Since all three b_1-values are inside either of the three pairs of confidence limits, there is no significant difference at the 5 percent level between the slopes of the three regression lines. The test of hypothesis that $\beta = 1.0$, using Eq. 11.43, gives t-values for summer, winter, and annual rainfall: -0.96, 0.93, and -0.63. All three absolute values are much smaller than $t_p = 2.05$ for P = 5 percent. The hypothesis of $\beta = 1.0$ for all three cases should be accepted. The three a_1-values are 3.05, 1.68, and 5.03. Assuming $\alpha_1 = 0$, the three corresponding t-values computed by Eq. 11.45 are 2.25, 1.48, and 2.40. The t_p value for P = 5 percent is 2.05. Two a_1-values (summer and annual precipitations) are significantly different from zero at the 5 percent level.

Example 11.11. For the example of Table 11.4, Eq. 11.43 gives t = 19.5 for r = 0.762 and N = 279, assuming $\beta_1 = 0$. The t-table shows that the level of significance is very close to 100 percent for t = 19.5, so that $b_1 = 0.665$ is significantly different from zero at either the 90 or 95 percent level. For other values of β_1, the test that b_1 does not depart from β_1 significantly on the selected level is made by using Eq. 11.43. If $\beta_1 = 0.50$ in the above example, and the level of significance is selected to be 90 percent, Eq. 11.43 for $t = \pm 1.97$ gives the confidence limits $(b_1)_1 = 0.555$ and $(b_1)_2 = 0.455$. So at this level, $b_1 = 0.665$ is significantly different from $\beta_1 = 0.50$, because it is outside the confidence interval. The variables P_1 and P_2 are assumed serially uncorrelated.

For the same example, with $a_1 = 9.50$ mm, $b_1 = 0.665$, r = 0.762, N = 279, $s_x^2 = 752$, and $\bar{x}^2 = 4{,}330$, and for the 90 percent level of significance, the intercept confidence limits are $(a_1)_1 = 7.20$ mm and $(a_1)_2 = -7.20$ mm. At this level, the value $a_1 = 9.50$ mm is significantly different from zero. At the 98 percent level, $(a_1)_{1,2} = \pm 9.90$ mm and $a_1 = 9.50$ mm is not significantly different from zero.

Errors in estimated values by regression equations. The value y as defined for a given x by Eq. 11.26 is normally distributed about η given by Eq. 11.25 for $\xi = x$, with the variance $\sigma^2(y)$ being $\sigma_{ey}^2 N^{-1} [1 + (x - \mu)^2/\sigma_x^2]$. Replacing σ_{ey}^2 by its unbiased estimate, $\hat{s}_{ey}^2$, μ by $\bar{x}$, and σ_x^2 by $\hat{s}_x^2$, then

$$\hat{s}(y) = \frac{\hat{s}_{ey}}{\sqrt{N}} \sqrt{1 + \frac{(x - \bar{x})^2}{\hat{s}_x^2}}, \qquad 11.47$$

which is the standard error of an estimate, y. From the regression y on x, y is the estimate of η for given $\xi = x$.

The quantity $[(y - \eta)/\hat{s}(y)]$ has t-distribution.NDF = N − 2. The confidence limits $\eta_{1,2}$ (η_1 the upper limit, and η_2 the lower limit) for predicted value y are given by

$$\eta_{1,2} = \bar{y} + b(x_i - \bar{x}) \pm t_p \frac{\hat{s}_{ey}}{\sqrt{N}} \sqrt{1 + \frac{(x_i - \bar{x})^2}{\hat{s}_x^2}}. \qquad 11.48$$

Because the standard error of y increases as $(x_i - \bar{x})^2$ increases, the curves bounding the confidence intervals for different values of x_i are hyperbolas. The width of the confidence interval is a minimum for y corresponding to $x = \bar{x}$.

One may be more interested in the difference $(y_i - y)$ than in y itself. Since y_i are normally distributed about η with the variance σ_{ey}^2, independently of y, then the unbiased estimate $\hat{s}^2(y_i - y)$ of $\sigma^2(y_i - y)$ is

$$\hat{s}^2(y_i - y) = \hat{s}_{ey}^2 \left[1 + \frac{1}{N} + \frac{(x_i - \bar{x})^2}{N\hat{s}_x^2}\right]. \qquad 11.49$$

The quantity $(y_i - y)$ has Student t-distribution with the zero mean and the variance given by Eq. 11.49 with $(N - 2)$ degrees of freedom. The confidence limits $y_{1,2}$ (y_1 the upper limit and y_2 the lower limit) for an actual value of y are given by

$$y_{1,2} = \bar{y} + b(x_i - \bar{x}) \pm t_p \hat{s}_{ey} \sqrt{1 + \frac{1}{N} + \frac{(x_i - \bar{x})^2}{N \hat{s}_x^2}}. \qquad 11.50$$

The confidence interval of Eq. 11.50 increases as $(x - \bar{x})^2$ increases, and is a minimum at $x = \bar{x}$.

Equations 11.48 and 11.50 indicate that the predictions corresponding to values of x deviating greatly from $\bar{x}$ are unreliable. This is so even though s_{ey} may be small. The confidence limits can be computed also for the values of x outside the observed range, but this is in general not recommended. Predictions of y corresponding to those values of x are likely to have such wide confidence limits that the estimates would have little practical value. Also, the true regression may be approximately linear only within the observed range of x.

Example 11.12. An example of confidence limits $\eta_{1,2}$ and $y_{1,2}$ as a function of x is given in Fig. 11.10 for the linear regression of Fort Valley on Flagstaff, for summer precipitation, and for $t_p = 2.05$ at the $P = 5$ percent level.

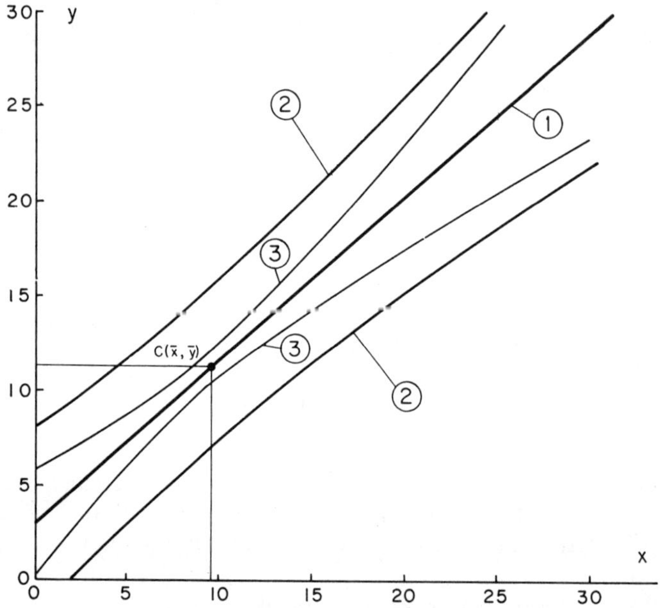

Figure 11.10. Confidence limits at the 95 percent level for the regression line $y = 11.34 - 0.87(x - 9.53)$, where y is the Fort Valley, and x is the Flagstaff summer precipitation in inches: (1) regression straight line; (2) the confidence limits for actual y-values, estimated by Eq. 11.50, and (3) confidence limits for predicted y-values by the above linear regression, obtained by Eq. 11.48.

Application of bivariate linear regression in hydrology. The regression and correlation analysis is one of the oldest statistical techniques used in hydrology. It was first applied to filling in missing data and extending short records at one hydrologic station by relating the available data at this station with those at adjacent stations. Equations 11.26 are used for predictive purposes, namely when one needs the best value of y for a given value of x, or the converse. Because Eqs. 11.26 do not include errors e_y or e_x, they fill in the missing data of a lesser variance than the values would have, if they were actually observed.

To predict y from x, the regression y on x is used, and to predict x from y, the regression x on y is applicable. One is free to choose which variable shall be taken as independent and dependent in filling in the missing data. For example, if one variable has N_1 observations, and the other $N_1 + N_2$, then it seems reasonable to use the longer record as the independent variable to obtain estimates of corresponding values for the shorter record as the dependent variable. However, it may be that one variable has N_1 observations, the other N_2, with N_3 simultaneous observations. By using both regression lines, and by estimating mutually the missing data of one variable when there is data for the other variable, the new equal sample sizes would be N_4. The new value N_4 is greater than either N_1 and N_2, but $N_4 < (N_1 + N_2)$. From N_3 simultaneous observations there will be N_4 new concurrent values of both variables.

Figure 11.11 shows the synoptic representation of the available data, with y and x having concurrent data for the period of k time units, and x for additional N − k (or a total of N) time units, so that N is the observed

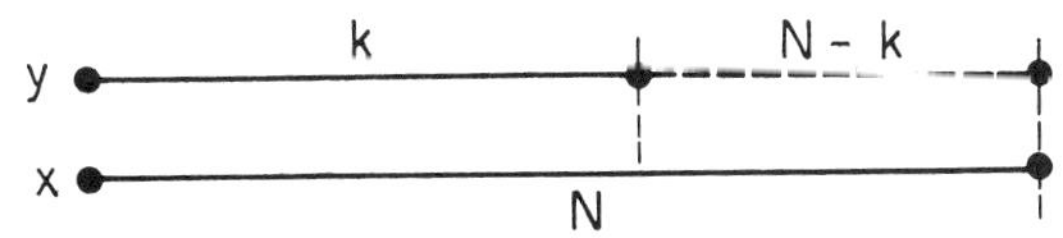

Figure 11.11. Filling in the missing data of the y series by the data of the x series.

length of x, k is the observed length of y, and N − k is the period to be filled in for y. The completion of data at y should be done by

$$y = a_1 + b_1x + e_y \ , \tag{11.51}$$

and not by Eqs. 11.26, with e_y generated from random numbers and having the same properties as $e_y = y_i - y$ of the period k. If Eq. 11.26 completes the data, the variance of y in the sample of N values will be somewhat smaller, on the average, than the variance for the period k.

If the mean and standard deviation of y for the period N are needed it is not necessary to introduce e_y (or e_x) of Eq. 11.51. Denote the means and the variances for the period k and N respectively as $\bar{x}_k$, $\bar{y}_k$, $\bar{x}_n$, $\bar{y}_n$,

s_{xk}^2, s_{yk}^2, s_{xn}^2, and s_{yn}^2, and let r_k and r_n be the corresponding correlation coefficients, then

$$\bar{y}_n = \bar{y}_k - r_k \frac{s_{yk}}{s_{xk}} (\bar{x}_k - \bar{x}_n), \quad 11.52$$

$$s_{yn}^2 = s_{yk}^2 - r_k^2 \frac{s_{yk}^2}{s_{xk}^2} (s_{xk}^2 - s_{xn}^2), \text{ and} \quad 11.53$$

$$r_n = r_k \frac{s_{yk}}{s_{xk}} \cdot \frac{s_{xn}}{s_{yn}}. \quad 11.54$$

For completing missing data, in order to use the sample of size N for the study of probability distribution of y variable, or the dependence of its series, Eq. 11.51 should be used with e_y, generated by the Monte Carlo method (data generation method), as an independent stochastic variable of given properties, usually as a normal variable with approximately $E(e_y) = 0$, and s_{ey} given by Eq. 11.35, or $N(0, s_{ey})$ distribution.

Before using Eqs. 11.52 through 11.54, it should be proven that the estimated values $\bar{y}_n$, s_{yn}^2, and r_n produce an increase of information by showing that the distributions of $\bar{y}_n$ and s_{yn}^2 have smaller variances than $\bar{y}_k$ and s_{yk}^2. R. Veron, see reference [5, p. 48], gives the ratio of variances of $\bar{y}_k$ and $\bar{y}_n$ as

$$E = \frac{\text{var } \bar{y}_n}{\text{var } \bar{y}_k} = 1 + \left(1 - \frac{k}{N}\right) \left[\frac{1 - (k-2) r_n^2}{k-3}\right], \quad 11.55$$

If var $\bar{y}_n <$ var $\bar{y}_k$, then $E < 1$, or the information is increased, $(1/\text{var } \bar{y}_n) > (1/\text{var } \bar{y}_k)$, and the estimate of Eq. 11.52 is acceptable. If $E > 1$, the estimate of Eq. 11.52 should be rejected. The development of the expression for $E_1 = \text{var } s_{yn}^2/\text{var } s_{yk}^2$ is more complex.

M. Roche [5] recommends using $k\, s_{yk}^2/(k-1)$ as an unbiased estimate of s_y^2 for the period k, except if s_{yn}^2 of Eq. 11.53 turns out to be greater than this unbiased variance.

Example 11.13. M. Roche [5] gives an example for the application of Eq. 11.55, with $\rho = 0.9$, $k = 10$, and $N = 40$, which give $E = 0.415$, and var $\bar{y}_n = 0.415$. var $y_k/10$. So, the new sample of N for the y series has an effective sample length N_e, computed from $s_{yn}^2/N_e = 0.415\, s_{yk}^2/10$, as $N_e = 10/0.415 = 24.1$ for $s_{yn} = s_{yk} = 1$. Therefore, filling in missing data does not effectively produce a length N for the y-series, but does produce an effective length N_e somewhere between k and N. The value N_e should be used for statistical inferences.

11.5. Nonlinear Correlation and Regression.

Nonlinear regression. For any nonlinear function $y = f(x)$ a regression may be obtained in one of three basic ways.

(1) Using the direct least-squares fit, which does not usually produce linear equations in a, b, c, ... parameters, as in the case of linear regression. But, by using digital computers, and an iterative procedure

in solving m equations with m unknown parameters, this method may apply in several cases.

(2) Transforming x and y in such a way that the new transformed variables u and v are linearly related. Often, $u = \log x$ and $v = \log y$ are used. The use of least squares of transformed variables has a bias. In case of log x and log y, the greater weight is given to the smaller values than to the larger values.

(3) The function $y = f(x)$ is developed in the power series form as

$$y = a_0 + a_1x + a_2x^2 + a_3x^3 + \ldots + a_mx^m, \qquad 11.56$$

and only the first three, four, or ... m terms of the polynomial are retained and fitted to data by the least-squares method. An equation similar to Eq. 11.56 is used for x versus y. The coefficients $a_0, a_1, \ldots, a_m$ in Eq. 11.56 are the regression coefficients to be determined by the least-squares method of estimation. The power order m is chosen so as to minimize the sum of the squares of deviations from the line. This power m should be much lower than the sample size N, to have a sufficient number of degrees of freedom $(N - m)$, or to have a reliable estimate of the standard deviation, s_{ey}. Also, to keep the arithmetical work reasonably simple, m should be small. Generally, m is between 2 and 4.

To select m, either a graphical plot will suggest the order of magnitude of m, or two or three different m values may be selected, and the unbiased standard error of deviations from the regression line for each case is taken as a measure of the most suitable m to be selected for purposes of prediction.

When a polynomial of Eq. 11.56 is selected for nonlinear regression, two methods of treatment may be used: compute the regression coefficients by the least-square method directly, or consider $x, x^2, x^3, \ldots, x^m$, as variables $u, v, z, \ldots$, and apply the multiple linear regression. The least-square method in the first approach involves the solution of $(m + 1)$ linear equations with $(m + 1)$ moments of x and y, and all cross products of powers of x and y, with the highest sum of powers being $m + 1$.

Correlation index. The nonlinear correlation is measured by two correlation indices

$$r_x = 1 - \frac{s_{ex}^2}{s_x^2}, \text{ and } r_y = 1 - \frac{s_{ey}^2}{s_y^2}, \qquad 11.57$$

with r_x the correlation index of x being a nonlinear correlation on y, and r_y the correlation index of y being a nonlinear correlation on x, where s_{ex}^2 and s_{ey}^2 are variances of e_x and e_y, respectively.

Transformations to linearize the curvilinear regressions. The transformation of variables has three basic objectives, which may or may not all occur in a given transformation: (1) a simple linear relation is obtained among the transformed variables, (2) the marginal distributions of transformed variables approach more closely normal distribution than do those of untransformed variables, and (3) the variation of points along the regression line is more homogeneous, or the variance is stabilized.

Table 11.7. Transformations for Linearization of Different Functions

	Type of function	Straight-line coordinate: Abscissa	Straight-line coordinate: Ordinate	Equation in linear form
1	$y = a + bx$	x	y	$[y] = a + b\,[x]$
2	$y = be^{ax}$	x	$\log y$	$[\log y] = \log b + (a \log c)\,[x]$
3	$y = ax^{b}$	$\log x$	$\log y$	$[\log y] = \log a + b\,[\log x]$
4	$y = a_0 + a_1 x + a_2 x^2$	$x - x_0$	$\dfrac{y - y_0}{x - x_0}$	$\left[\dfrac{y - y_0}{x - x_0}\right] = a_1 + 2a_1 x_0 + a_2\,[(x - x_0)]$
5	$y = a + b/x$	$1/x$	y	$[y] = a + b\,[1/x]$
6	$y = x/(a + bx)$	x	x/y	$[x/y] = a + b\,[x]$
7	$y = a/(b + cx)$	x	$1/y$	$[1/y] = (b/a) + (c/a)\,[x]$
8	$y = c + be^{ax}$	x	$\log \dfrac{\Delta y}{\Delta x}$	$\left[\log \dfrac{dy}{dx}\right] = \log (ab) + (a \log e)\,[x]$
9	$y = c + ax^{b}$	$\log x$	$\log \dfrac{\Delta y}{\Delta x}$	$\left[\log \dfrac{dy}{dx}\right] = \log (ab) + (b - 1)\,[\log x]$
10	$y = c + \dfrac{b}{x - a}$	$x - x_0$	$\dfrac{x - x_0}{y - y_0}$	$\left[\dfrac{x - x_0}{y - y_0}\right] = -\dfrac{a - x_0}{c - y_0} + \dfrac{1}{c - y_0}\,[x - x_0]$
11	$y = c + \dfrac{x}{a + bx}$	x	$\dfrac{x - x_0}{y - y_0}$	$\left[\dfrac{x - x_0}{y - y_0}\right] = a + bx_0) + \dfrac{b\,(a + bx_0)}{a}\,[x]$
12	$y = d + cx + be^{ax}$	x	$\log \dfrac{\Delta^2 y}{\Delta x^2}$	$\left[\log \dfrac{d^2 y}{dx^2}\right] = \log (a^2 b) + (a \log e)\,[x]$ $[y - be^{ax}] = d + c\,[x]$
13	$y = dc^{x} b^{m}$, where $m = a^{x}$	x	$\log \dfrac{\Delta^2 (\log y)}{\Delta x^2}$	$\left[\dfrac{\log d^2 (\log y)}{dx^2}\right] = \log \left[\dfrac{(\log b)(\log a)^2}{(\log e)^2}\right] +$ $+ (\log a)\,[x]$
14	$y = de^{cx} + be^{ax}$	$\dfrac{y_{k+1}}{y_k}$	$\dfrac{y_{k+2}}{y_k}$	$[\log y - a^{x} \log b] = \log d + (\log e)\,[x]$ $\left[\dfrac{y_{k+2}}{y_k}\right] = -e^{(a+c)\Delta x} + (e^{a\Delta x} + e^{c\Delta x})\left[\dfrac{y_{k+2}}{y_k}\right]$ $[ye^{-cx}] = d + b\,[e^{(a-c)x}]$
15	$y = e^{ax}(d \cos bx +$ $+ c \sin bx)$	$\dfrac{y_{k+1}}{y_k}$	$\dfrac{y_{k+2}}{y_k}$	$\left[\dfrac{y_{k+2}}{y_k}\right] = -e^{2a\Delta x} = (2e^{a\Delta x} \cos b\Delta x)\left[\dfrac{y_{k+2}}{y_k}\right]$ $\left[\dfrac{yc^{-ax}}{\cos bx}\right] = d + c\,[\tan bx]$

Remark: In types 14 and 15, y_k, y_{k+1}, and y_{k+2} are consecutive values for an increment Δx.

Table 11.7 gives the transformations for linearizing two-variable relations. In hydrology a logarithmic transformation is generally used more frequently than others.

Regression and correlation inference. When polynomial regression lines are used, the testing of the significance of the computed regression coefficients is less simple than is the case for the multiple linear regression. The general hypothesis tested is whether a selected power m of the polynomial represents the best regression, or whether a term of still higher power contributes significantly to the relation. In applying polynomial regression, it is not advisable to omit any intermediate term between the intercept a_0 and the highest power term.

Two tests are generally applicable. The first is the test of whether the intercept differs from zero. The second is the test of the highest significant term. Thus, the tests are made at the two extremes. From a given power term, the test may be carried out successively from lower to higher or from higher to lower terms [6].

11.6. Multiple Linear Correlation and Regression Analysis.

Multiple regression. The association of three or more variables can be investigated by multiple regression and correlation analysis. The general multiple regression relation may be expressed in the form

$$x_1 = f(x_2, x_3, \ldots, x_m) + \varepsilon, \qquad 11.58$$

in which $x_1, x_2, \ldots, x_m$ are m variables and ε are the residuals. This equation gives the estimate of x_1 for given values of all other variables.

If Eq. 11.58 is linear, the regression is referred to as the multiple linear regression, and the association is the multiple linear correlation. Because the linear equations are easier to treat than nonlinear multiple relations, the variables of nonlinear relations in hydrology are often transformed to linear relations for the multiple regression analysis.

If a regression equation with m parameters is fitted to a set of N data points of variables $x_1, x_2, \ldots, x_m$, and if N is the sample size, the number of degrees of freedom is $N - m$. Thus every parameter takes away one degree of freedom in the analysis of data. When $m = N$, the regression equation passes through all points, so that there are no degrees of freedom. Since the errors of the parameters are inversely related to the degrees of freedom, this regression equation cannot be used for prediction, otherwise, the errors would become infinite. For small-size samples, the sample size N, used in related formulas, should be replaced by the degrees of freedom $(N - m)$. This correction is necessary for avoiding a bias in the measure of reliability of the computed statistical parameters. If no such correction is made, the computed standard deviations of parameters are significantly smaller in very small samples than their true values.

In designing the multiple linear or nonlinear regression relations a sequence of the subjective and the objective approach is followed. First, the variable x_1 as the dependent variable is selected for purposes of prediction. It is defined by the problem to be solved. The next step is the selection of the so-called independent variables, $x_2, x_3, \ldots, x_m$ by which x_1 should be predictable. These $(m - 1)$-variables are often selected by two criteria:

(1) they have been observed in the past concurrently with x_1, so that the prediction regression equation may be established, and they will continue to be observed in the future, so that x_1 may be predicted from them when necessary, and

(2) the variables are selected for investigation by a subjective method on the grounds that the analysis of physical phenomena indicates an effect on x_1 by any of variables $x_2, x_3, \ldots, x_m$. These criteria are necessarily subjective. When the services or surveys have been established for gathering the hydrologic data, variables to be observed are decided on by some criteria, often of subjective nature.

Many variables $x_{m+1}, x_{m+2}, \ldots$, are either neglected, are not observed, are not known to affect x_1, or are found by analysis to insignificantly affect x_1. If known to affect x_1 but if they have not been observed, and if all variables $x_1, x_2, \ldots, x_m$ contain errors of random nature, it is impossible in hydrologic practice to arrive at a relation of the type of Eq. 11.58 in which var ε would be zero. The random variable, ε, of residuals between the predicted x_1 and the observed x_1' encompasses all neglected variables and all errors.

Linear regression of three variables. The linear regression of three variables is

$$x_1 = \beta_1 + \beta_2 x_2 + \beta_3 x_3, \qquad 11.59$$

in which x_1 is the dependent variable and x_2 and x_3 are independent variables. Three parameters must be estimated in this equation. The estimate of β_1 is the intercept b_1. The parameter β_2 is the multiple regression coefficient of x_1 on x_2 when x_3 is kept constant, and is estimated by b_2. The parameter β_3 is the multiple regression coefficient of x_1 on x_3 when x_2 is kept constant, and is estimated by b_3.

The residuals of x_1, or the difference between the observed value x_1' and the estimated value x_1 by Eq. 11.59 are $e_1 = x_1' - x_1 = x_1' - b_1 - b_2x_2 - b_3x_3$. The least squares method minimizes the sum of e_1^2. The three partial differential equations in b_1, b_2, and b_3 give the three linear equations of b_1, b_2, and b_3 for the fitting of a plane:

$$\begin{aligned} b_2 \Sigma (\Delta x_2)^2 + b_3 \Sigma \Delta x_2 \Delta x_3 &= \Sigma \Delta x_1 \Delta x_2, \\ b_2 \Sigma \Delta x_2 \Delta x_3 + b_3 \Sigma (\Delta x_3)^2 &= \Sigma \Delta x_1 \Delta x_3, \\ b_1 &= \bar{x}_1 - b_2 \bar{x}_2 - b_3 \bar{x}_3, \end{aligned} \qquad 11.60$$

with all sums taken from 1 to N, where $\bar{x}_i = \Sigma x_i/N$ and $\Delta x_i = x_i - \bar{x}_i$, with i taken from 1 to 3, and N is the sample size. The sums in the equations may be expressed and computed as $\Sigma (\Delta x_i)^2 = \Sigma x_i^2 - N\bar{x}_i^2$, and

$\Sigma (\Delta x_i \Delta x_j) = \Sigma x_i x_j - N\bar{x}_i\bar{x}_j$, with i and j having the corresponding values 1, 2, and 3 in the equations. Because the sum $\Sigma(\Delta x_1)^2$ is necessary for the computation of other parameters in the multiple correlation analysis, nine sums of the sample must be computed altogether for the three variables — for the variable values, their squares, and their three cross products. The solution of these three linear equations gives the three estimates b_1, b_2, and b_3 of the coefficients β_1, β_2, and β_3.

Linear regression of several variables. If there are m variables to correlate, including one dependent and m − 1 independent, the general equation for the multiple linear regression is

$$x_1 = b_1 + b_2x_2 + \ldots + b_ix_i + \ldots + b_mx_m, \qquad 11.61$$

in which b_1 is the intercept, and b_i is the multiple regression coefficient of the dependent variable x_i on the independent variable x_i, with all other variables kept constant. They are estimated by sample values. The b_i — coefficients are considered in this case both as the population parameters and their sample estimates.

Applying the least-squares method for the sum of residuals, $e_1 = x_1 - b_1 - b_2x_2 - \ldots - b_mx_m$, the m partial differential equations in b_1, $b_2, \ldots$, and b_m give m linear equations:

$$\begin{aligned} b_2 \Sigma (\Delta x_2)^2 + b_3 \Sigma \Delta x_2 \Delta x_3 + \ldots + b_m \Sigma \Delta x_2 \Delta x_m &= \Sigma \Delta x_1 \Delta x_2, \\ b_2 \Sigma \Delta x_2 \Delta x_3 + b_3 \Sigma (\Delta x_3)^2 + \ldots + b_m \Sigma \Delta x_3 \Delta x_m &= \Sigma \Delta x_1 \Delta x_3, \\ \ldots\ldots\ldots\ldots\ldots\ldots\ldots\ldots\ldots\ldots\ldots\ldots & \qquad 11.62 \\ b_2 \Sigma \Delta x_2 \Delta x_m + b_3 \Sigma \Delta x_3 \Delta x_m + \ldots + b_m \Sigma (\Delta x_m)^2 &= \Sigma \Delta x_1 \Delta x_m, \text{ and} \end{aligned}$$

$$b_1 = \bar{x}_1 - b_2\bar{x}_2 - b_2\bar{x}_3 - \ldots - b_m\bar{x}_m,$$

in which $\Delta x_i = x_i - \bar{x}_i$, with i = 1 to m. These equations facilitate determining m parameters. The computational work increases with the increase of either the number, m, of variables, or of the sample size, N.

When there are a small number of variables (say, three or four), either ungrouped or grouped data may be used for a multiple regression and correlation analysis. For more than four variables, the use of multivariate distributions in the form of grouped data becomes cumbersome because of difficulties in obtaining a suitable representation of multivariate distributions.

The selection of a feasible method for solving the linear equations depends on the number of variables and the available computational devices. When the number of variables and the sample size are relatively small, desk calculators can be used for a solution. In the past, the number of variables was restricted. Now, digital computers can make use of any number of variables that are statistically significant. Digital computers perform any operation in solving multiple linear regression problems by following a suitable set of instructions.

Example 11.14. In a study of floods in New England [3], a multiple linear regression and correlation analysis was carried out. In this analysis, the annual peak discharge, in cubic feet per second, for a recurrence interval of T years, was found to be

$$Q = aA^bS^cS_t^dI^et^f0^g, \tag{11.63}$$

in which A is the drainage area in sq. mi., S is the slope of main channel in ft/mile, S_t is the percent of water surface storage area plus 0.5 percent, I is the 24-hr rainfall in inches of a recurrence interval of T years, t is the average temperature in January in °F below freezing, 0 is an orographic factor, and ln a, b, c, d, e, f, g, are the multiple linear regression coefficients for a logarithmic transformation applicable to all variables.

By this logarithmic transformation for all seven variables, the general relation expressed by Eq. 11.63 is transformed to a multiple linear regression of the type of Eq. 11.61. Table 11.8 shows the results of computation of coefficients a, b, c, ..., for three recurrence intervals, T. The results given in this table show how the additional variables affect the values of the multiple regression coefficients previously computed.

This example demonstrates that regardless of recurrence interval (return period), the first independent variable in the step-wise selection of variables with the largest successive partial correlation coefficients is the area A. This variable explains about 75 percent of the variation of Q_{max}. The other variables add only another 20 percent of variation of Q_{max}, and the main rainfall parameter, I, turns to be the fifth or the sixth variable in importance. The geometric basin variables (A, S, O), in all three cases of Table 11.8, appeared to be most important. If the specific water yield during the flood peaks, in the form $q_{max} = Q_{max}/A$ in cfs/sq. mile or in m³/sec. km², is used as the dependent variable, the effect of area A will be nearly eliminated because its regression coefficient, b, is close to unity. Then q_{max} will be related to the remaining factors with a lesser explained variance.

Example 11.15. In studying the floods from small watersheds, Reich [7] used 47 flood hydrographs fitted by the Gamma function with three parameters: q_0 the hydrograph peak specific yield (Q_{max}/A), W the hydrograph volume in inches per square mile, and G the difference between the time of hydrograph center of gravity and the time of the peak. The multiple linear correlation gave the prediction equation $q_0 = 0.2917 + 0.4600\ I - 0.00040\ L_1 + 0.00018\ L_2$, with I the 30-minute rainfall intensity (in/hour), L_1 the distance in feet from the station of flood investigation to a point of the river near the center of gravity, and L_2 the length in feet of the largest collector carried to the divide. Similar results are obtained for W and G.

Measures of multiple linear correlation. The degree of correlation of a dependent variable x_1 to many independent variables in a multiple linear association is measured by any of five parameters: the standard deviation of residuals, the multiple correlation coefficient, the coefficient of multiple determination, the partial correlation coefficients, and the beta coefficients.

The standard deviation of residuals. This measure is determined as the standard deviation of differences between the observed values x_1' and the values of x_1 estimated by Eq. 11.61. This parameter is a measure of the closeness with which the estimated values approach the observed values. It is designated by S_1 and should be distinguished from the standard deviation s_1 of the marginal distribution of x_1.

If the total number N of points of the multivariate ($x_1, x_2, x_3, \ldots, x_m$) is small in comparison with the number m of variables involved, the standard deviation of residuals computed by the usual procedure is biased, since the degrees of freedom, N − m, may be much smaller than the sample size N.

Table 11.8. Multiple Linear Regression and Correlation Analysis of Floods in New England

Recurrence interval T	No. stations N	Independent variables included	a	b	c	d	e	f	g	R_1	Standard deviation*	D_1
											164.4	
		A	113.06	.7858						.883	63.6	.780
		A, S	5.9808	1.0182	.5720					.942	44.4	.888
10	164	A, S, 0	3.4623	1.0804	.6119				.8482	.965	34,5	.932
		A, S, 0, S_t	6.1854	1.0508	.5153	.2645			.8543	.972	30.4	.945
		A, S, 0, S_t, t	5.1680	.9707	.4306	.2524		.3549	1.0815	.978	27.2	.959
		A, S, 0, S_t, t, I	3.6365	.9726	.4255	.2592	.1915	.3963	1.0668	.978	27.2	.959
											145.4	
		A	342.32	.6764						.875	59.9	.765
		A, S	17.892	.9197	.5704					.931	44.6	.865
50	116	A, S, 0	9.7302	1.0006	.5727				.9894	.962	33.0	.925
		A, S, 0, S_t	17.115	.9675	.4978	.2520			.9288	.969	29.9	.940
		A, S, 0, S_t, t	13.433	.9354	.4602	2.554		.2165	1.1089	.971	29.0	.940
		A, S, 0, S_t, t, I	1.8401	.9458	.4349	.2757	.8773	.4504	1.0866	.975	26.8	.951
											134.6	
		A	395.75	.6922						.862	62.8	.743
		A, S	25.504	.9095	.5438					.910	50.6	.827
100	100	A, S, 0	13.937	.9922	.5332				1.0627	.950	37.7	.902
		A, S, 0, t	10.964	.9480	.4667			.2837	1.2930	.952	36.8	.905
		A, S, 0, t, I	1.1194	.9524	.4409		.9507	.4409	1.2730	.958	34.6	.920
		A, S, 0, t, I, S_t	1.4203	.9226	.3823	−.2308	1.0685	.5632	1.2129	.965	32.2	.935

From Benson [3]. * Standard deviation in percent of the mean of original peak discharges of recurrence interval T.

Taking into account m parameters and N − m degrees of freedom, then

$$\hat{S}_1 = \sqrt{\frac{NS_1^2}{N-m}} = \sqrt{\frac{\Sigma e_1^2}{N-m}} =$$

$$= \sqrt{\frac{1}{N-m}[\Sigma(\Delta x_1)^2 - b_2\Sigma\Delta x_1\Delta x_2 - \ldots - b_m\Sigma\Delta x_1\Delta x_m]}, \quad 11.64$$

in which S_1 is the biased and $\hat{S}_1$ the unbiased standard deviation of residuals. If, however, $\hat{S}_1$ exceeds $\hat{s}_1$ (the unbiased standard deviation of the marginal distribution of x_1), $\hat{s}_1$ would be used for the standard deviation of residuals instead of $\hat{S}_1$. This can occur only for a very small number of degrees of freedom.

The addition of a new independent variable x_{m+1} in a multiple linear regression will decrease the standard deviation of residuals if this variable influences the dependent variable x_1. Whether this variable should be included in the multiple regression depends on how much the standard deviation of residuals is decreased.

For the study of floods in New England Table 11.8 shows how the standard deviation of residuals, in percentage of the mean flood discharges, changes with a change of the number and the selection of independent variables.

Multiple correlation coefficient. The correlation coefficient is a measure of the linear association of two variables. In this case, the two variables used are the estimated and observed values. The correlation coefficient is the ratio of the standard deviation of estimated values to the standard deviation of observed values,

$$r = \frac{s_{e1}}{s_1} = \sqrt{1 - \frac{S_1^2}{s_1^2}}, \quad 11.65$$

in which s_{e1} is the standard deviation of estimated values, s_1 is the standard deviation of observed values of x_1, and S_1 is the standard deviation of residuals.

In a similar manner, the measure of linear association of many variables, including one dependent and the others mutually independent, is expressed by the ratio of the standard deviation of estimated values to the standard deviation of observed values of the dependent variable. This is the multiple correlation coefficient, R, or the coefficient of multiple correlation. It measures the combined influence of all independent variables in terms of the difference between the observed and the estimated values of the dependent variable. The coefficient is expressed as

$$R_1 = \frac{s_{e1}}{s_1} = \sqrt{1 - \frac{S_1^2}{s_1^2}} =$$

$$= \sqrt{\frac{b_2\Sigma\Delta x_1\Delta x_2 + b_3\Sigma\Delta x_1\Delta x_3 + \ldots + b_m\Sigma\Delta x_1\Delta x_m}{\Sigma(\Delta x_1)^2}}. \quad 11.66$$

The unbiased value $\hat{R}_1$ is obtained by substituting into Eq. 11.66 the unbiased values $\hat{S}_1$ and $\hat{s}_1$ for S_1 and s_1, respectively.

Table 11.8 gives the values of R_1 for the example of floods in New England, showing the change of R_1 with the change of the number and the selection of independent variables.

The multiple correlation coefficient is a dimensionless measure of linear association, while the standard deviation of residuals is a measure having the dimension of the variable x_1. Because s_1 is always the same, regardless of the number of other variables, a decrease of the standard deviation of residuals will cause an increase of the multiple correlation coefficient.

The multiple correlation coefficient can be defined as the correlation coefficient of linear association of two variables, one being the observed values of the dependent variable and the other the estimated values of the dependent variable, based on all other independent variables.

Coefficient of determination. The square of the multiple correlation coefficient, R_1^2, is referred to as the coefficient of multiple determination, D_1, and $1 - D_1 = 1 - R_1^2$ as the coefficient of multiple non-determination.

The square of the multiple correlation coefficient indicates the part of the variance of the dependent variable, s_1^2, that has been mathematically accounted for, whereas $1 - D_1 = 1 - R_1^2$ indicates the part of the variance that has not been accounted for by the multiple linear correlation of the variable x_1 on $x_2, \ldots, x_m$.

Example 11.16. Table 11.8 of Example 11.14 gives values of D_1 in percentages for different numbers and different selections of variables in the example of floods in New England. This table also shows that $D_1 = R_1^2$ varies from 0.743 to 0.959 for the cases analyzed. In Examples 11.15 the unbiased $\hat{R}^2$ (replacing N in the above equations by $N - m$) is from 0.51 to 0.80. The expressions for q_0, W, and G of the type of Eq. 11.63 did not produce a significant increase in $\hat{R}^2$, though the increase existed in several cases. Because the sample size was small ($N = 47$), the degrees of freedom $N - m$ decreased rapidly with the number of independent variables included. Figure 11.12 gives both the biased and unbiased coefficient of determination, R^2 and $\hat{R}^2$respectively, as they change with m in this example, and for the specific volume of flood hydrograph, W. The coefficient R^2 always increases toward unity, while $\hat{R}^2$ comes to a maximum, in this case for $m = 10$. No information can be gained for $m > 10$, nor does adding a new independent variable compensate for the loss of one more degree of freedom, going from $m = 10$ to $m = 11$.

Partial correlation coefficients. The partial correlation coefficients measure the association of the dependent variable x_1 with any given independent variable x_i. Thus, they measure the variation of x_1, which is explained by, and only by, the given x_i. A simple method of estimating the partial correlation coefficient r_{1-i} involves the determination of (1) the multiple correlation coefficient R_1 between x_1 and all the independent variables, and (2) the multiple correlation coefficient R_{1-i} between x_1 and all the independent variables except the chosen x_i. The partial correlation coefficient r_{1-i} is then determined by [6]

$$r^2_{1-i} = \frac{(1 - R_{1-i}^2) - (1 - R_1^2)}{1 - R_{1-i}^2} = 1 - \frac{1 - R_1^2}{1 - R_{1-i}^2}. \qquad 11.67$$

Since R^2_{1-i} is smaller than R^2_1, $r_{1-i} < 1.0$. The greater the difference between R_1 and R_{1-i}, the smaller the ratio on the right side of the equation and the greater r_{1-i}. It is seen from Eq. 11.67 that considerable computation is necessary in the case of many variables if all partial correlation coefficients are to be determined.

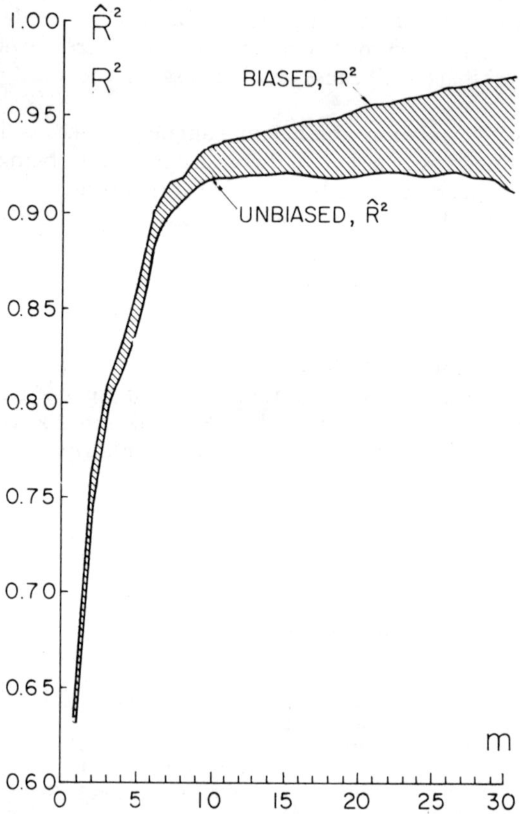

Figure 11.12. Typical difference between biased, R^2, and unbiased, $\hat{R}^2$, coefficients of determination for a sample size of 47, for the multiple linear correlation of the flood hydrograph specific volumes (given in inches per square mile), as expressed to topographic, soil, vegetation, and rainfall parameters, as independent variables ($m = 1, 2, \ldots, 30$).

It is important to stress that partial correlation coefficients should have the sign of their corresponding regression coefficients. If b_i is negative, r_{1-i} is also negative. For a given number of variables and their multiple correlation coefficient R_1, the numerator of the fraction in the right term of Eq. 11.67 is a constant, and only the denominator changes from one partial correlation coefficient to another. This agrees with the definition of the coefficient, which measures the reduction of variation of the dependent variable when all other variables except the considered one have been taken into account.

If a new variable x_{m+1} is added, all the partial correlation coefficients computed for the m variables will change.

The partial correlation coefficients are important parameters, since they measure the association of each independent variable with the dependent one, after the influence of certain related variables has been accounted for. It is possible that the simple correlation coefficient between the dependent and the new independent variable is very small, but this variable may still have a significant influence on the variation of the dependent variable. It may be remembered that the converse can also occur. The partial correlation coefficient is the measure of the association, after the relation of the dependent variable to other independent variables has been considered and included in the analysis. Therefore, the simple correlation coefficient is not as true a measure of association between two variables in multiple linear correlation as is the partial correlation coefficient.

An *independent* variable in Eq. 11.61 may be highly correlated with another *independent* variable, when simple linear correlation is applied to the two variables, but it may not be highly correlated with the dependent variable in a multiple correlation. The partial correlation coefficient may give a reliable answer as to the real effect of an independent variable on the variation of the dependent variable. In hydrology, one may find many examples of this kind, (such as the example given in Table 11.8).

Beta coefficients. The effect of individual independent variables on the dependent variable may be measured by a dimensionless form of the regression coefficient. Expressing each variable in ratio to its standard deviation, Eq. 11.61 may be rewritten as

$$\frac{x_1}{s_1} = \beta_1 + \beta_2 \frac{x_2}{s_2} + \beta_3 \frac{x_3}{s_3} + \ldots + \beta_m \frac{x_m}{s_m}, \qquad 11.68$$

in which

$$\beta_1 = \frac{b_1}{s_1};\ \beta_2 = b_2 \frac{s_2}{s_1};\ \beta_3 = b_3 \frac{s_3}{s_1};\ \ldots;\ \beta_m = b_m \frac{s_m}{s_1}. \qquad 11.69$$

These are beta coefficients, the dimensionless parameters that measure the effect of the individual independent variables on the variation of the dependent variable.

Comparing beta coefficients with partial correlation coefficients shows, in general, that the rank or the order of magnitude for the effect of the independent variables is the same for both these measures. This does not always hold true, since they are two different sets of measures. It is easier to compute beta coefficients than partial correlation coefficients, since b_i values and s_i values in Eq. 11.69 must always be computed for regression analysis. For this reason, there is a general preference for beta coefficients, although the partial correlation coefficients are better measures of the influence of individual *independent* variables.

Statistical inference for regression coefficients. The same general conclusions derived for the correlation coefficient in a simple linear association can be applied to measures of association in multiple linear regression.

When only the extreme low and extreme high values of variables in multiple linear regression are available, the multiple correlation coefficient, the partial correlation coefficients, and the beta coefficients are likely to be higher. Thus, if the sample values of variables are available only for a narrow range, these coefficients are likely to be smaller than the true coefficients for the population.

The unbiased standard deviation of any multiple regression coefficient b_i is [6]

$$\hat{S}_{bi} = \sqrt{\frac{\hat{S}_1^2}{N s_i^2 (1 - R_i^2)}} = \frac{S_1}{s_i}\sqrt{\frac{1}{(N - m)(1 - R_i^2)}}, \qquad 11.70$$

in which s_i is the standard deviation of marginal distribution of variable x_i, $\hat{S}_1$ is the unbiased standard deviation of residuals for variable x_1, S_1 is the biased standard deviation of residuals for variable x_1, $N - m$ is the number of degrees of freedom, and R_i is the multiple correlation coefficient of x_i with respect to all variables except the variable x_1, with i varying from 2 to m. If the variables $x_2, x_3, \ldots, x_m$ are independent among themselves, $R_i = 0$, and Eq. 11.70 becomes

$$\hat{S}_{bi} = \frac{S_1}{s_i\sqrt{N - m}}. \qquad 11.71$$

It may be noted that the more the variables $x_2, x_3, \ldots, x_m$ are related among themselves, the greater R_i. Consequently, $\hat{S}_{bi}$ is larger and the regression coefficient is less reliable.

The standard deviation of individual values of b_i can be very high. Some of the values are more accurate and some are less accurate than others. If x_1 is estimated using all b_i values, the error in one value of b_i may be compensated for by an error in another value of b_i, so that the error of estimated x_1 may be smaller than that which may appear from the errors in b_i. If the extreme values of x_1 are estimated by the multiple regression linear equation, the error may be large because b_i values have much greater error effects for the extremes.

For the test whether R is or is not significantly different from zero, the distribution of R is

$$f(R) = \frac{2\Gamma\left(\frac{N-1}{2}\right)}{\Gamma\left(\frac{m-1}{2}\right)\Gamma\left(\frac{N-m}{2}\right)} R^{m-2} (1 - R^2)^{\frac{N-m}{2}-1} \qquad 11.72$$

which is F-distribution with $(N - m)$ and $(m - 2)$ degrees of freedom.

Multiple nonlinear regression and correlation. The use of any function in multiple regression and correlation analysis other than a linear function is referred to as multiple nonlinear regression and correlation. Three methods are used in hydrology to determine a multiple nonlinear regression equation:

(a) least-squares fit by using a computer approach, usually solving equations by iterative procedures;

(b) expansion in power series form, and by neglecting the higher order terms, which is an increase of the number of variables, or a decrease of degrees of freedom, and

(c) linearization of relations such as

$$x_1 = b_1 x_2^{b_2} x_3^{b_3} \ldots x_m^{b_m}, \tag{11.73}$$

with the logarithms giving

$$\ln x_1 = \ln b_1 + b_2 \ln x_1 + b_3 \ln x_3 + \ldots + b_m \ln x_m. \tag{11.74}$$

This latter equation has the negative aspect that large and small values of x_i have different weights.

The statistical inference about the reliability of the b_i-coefficients of Eq. 11.73 is not simple. The measure of nonlinear multiple correlative association is difficult to determine, though the variance S_1^2 of e_1, is the best measure available. The data generation method (Monte Carlo method), but only in a complex way, may be used to determine the distributions of the b_i-coefficients of Eq. 11.73.

Use of correlation and regression analysis in hydrology. Two main objectives of the use of correlation and regression analysis in hydrology are as follows.

(1) The prediction of values of a variable x_1 from given values of other variable(s), once the regression equation is available. This prediction gives the most probable value of x_1 for given values of other variables, but not the real fluctuations of x_1 at the predicted times. Because x_1-values obtained by a regression equation fluctuate less than the observed values, a smaller variance of x_1 is obtained if the random variable unaccounted for, ε, is not added. Equation 11.61 should be extended to

$$x_1 = b_1 + b_2 x_2 + \ldots + b_m x_m + \varepsilon, \tag{11.75}$$

so that the independent random values of ε are added. However, the predication of individual values of x_1 should be made by Eq. 11.61.

(2) The investigation of hydrologic physical relations is limited by the correlation and regression analysis, mainly because the *so called independent variables* $x_2, \ldots, x_m$ on the right side of Eqs. 11.61 and 11.75 are rarely independent in hydrology. Besides, they are very often also serially correlated. If x_1 is physically dependent on x_i and x_j, but if x_i and x_j are also interdependent, the correlation analysis may show only

a significant effect of x_j, which indirectly takes into account the effect of x_i. This and the requirement that the functional form (type of correlation) must be assumed in advance act as a limitation for the use of regression analysis in the investigation of physical hydrologic relations. Despite this limitation, the correlation and regression analysis may provide good insight into many hydrologic connections, if critically evaluated.

Experience shows that the correlation analysis is a major technique of hydrologic statistical work. All techniques presently available in correlation analysis are also used in hydrology. One technique is the step-wise correlation and regression. In applying Eq. 11.61 to a large number of variables, $x_1, x_2, \ldots, x_m$, and looking for a multiple linear regression equation of x_1 as the dependent variable, with all others to be so-called independent variables, standard computer programs for the step-wise correlation and regression computational method are available. Usually, the correlation matrix of m variables (symmetrical matrix of simple correlation coefficients between any pair of variables) is first computed:

$$[r_{ij}] = \begin{vmatrix} 1 & r_{1,2} & \ldots & r_{1,m} \\ r_{2,1} & 1 & \ldots & r_{2,m} \\ \ldots & \ldots & \ldots & \ldots \\ r_{m,1} & r_{m,2} & \ldots & 1 \end{vmatrix} . \qquad 11.76$$

The largest value $r_{1,i}$ of the matrix gives the first variable on the right side of Eq. 11.61 as

$$x_1 = b_1 + b_2 x_2. \qquad 11.77$$

Then, keeping this $(x_2 =)x_i$ in the equation, the next variable $(x_3 =)x_j$ is selected so that it adds most to the explained variance of x_1 (or the second variable on the right side with the largest partial correlation coefficient), and so on. This approach has two basic biases. First, after x_i one obtains x_j, which adds most to x_1 in the $x_1 = b_1 + b_2x_i + b_3x_j$ equation. However, there may be x_p and x_q variables, which give a larger R^2-coefficient than x_i and x_j selected by the step-wise method. Second, since this method always selects $x_i, x_j, \ldots$ with the largest value of $r_{1-i}, r_{1-j}, \ldots$, it may be by pure sampling fluctuations that r_{1-i}, r_{1-j}, are much greater than zero, though x_i and x_j may not be related to x_1. A careful interpretation should be made, and one should try to have $(N - m)$, the sample size minus the number of variables, as large as possible. The statistical inference whether b_j-coefficients are or are not significantly different from zero should be continuously performed at every step of this step-wise method. Despite these difficulties, the application of step-wise correlation analysis in hydrology is extensive and beneficial.

11.7. Problems for Discussion and Solution.

11.1. The annual flows of two rivers are given below in modular coefficients, with the water years (October 1 – September 30) as designated.

Umpqua River, near Elkton, Oregon, 1905–1957, and the Rogue River near Raygold, Oregon, 1905–1957, designated respectively by U and R:

	U	R	U	R	U	R	U	R	
(1906)	0.928	0.943	1.196	0.971	1.015	0.923	0.823	0.879	
	1.322	1.463	0.914	0.706	1.023	0.954	1.165	1.137	
	1.050	1.081	1.631	1.415	0.443	0.524	0.854	0.785	
	1.128	1.222	1.201	0.992	1.002	0.940	1.192	1.103	
	1.035	1.264	0.868	0.771	0.912	0.920	0.973	0.982	
	0.852	1.071	0.532	0.575	0.798	0.842	1.109	1.059	
	1.073	1.215	1.294	1.098	1.436	1.421	1.483	1.470	
	0.970	1.050	0.544	0.523	0.763	0.810	1.295	1.465	
	0.902	0.981	1.175	1.184	0.749	0.824	1.242	1.361	
	0.646	0.658	0.906	0.957	0.615	0.672	1.336	1.399	
	1.283	0.978	0.576	0.627	0.894	0.902	0.810	0.729	
	1.152	1.071	0.596	0.613	1.512	1.510	1.800	1.693	
	0.950	0.837	0.425	0.399	0.591	0.719	1.015	1.308	(1957)

Umpqua River		**Rogue River**	
$N_1 =$	52	$N_2 =$	52
$C_v =$	0.303	$C_v =$	0.294
$C_s =$	0.251	$C_s =$	0.231
$r_1 =$	-0.004	$r_1 =$	0.152

with N the sample size, C_v the coefficient of variation, C_s the skewness coefficient and r_1 the first serial correlation coefficient.

Compute the (product-moment) correlation coefficient r as the measure of the linear association of the two above variables of annual flows. Test the hypothesis that r is significantly different from zero at the 99 percent level by using the three approximate tests.

11.2. For the two variables selected above in Problem 11.1:

(a) estimate the regression coefficients for the linear regression equation by the least squares method, $y = a + bx$, where y are modular coefficients of the Umpqua River and x of the Rogue River;

(b) determine the standard error of estimate for y, and

(c) determine the explained and unexplained variance for the y-variable in the case of the fitted regression line.

11.3. For $N_1 = 52$ years of the Umpqua River, estimate y and determine 52 values of $e_i = y - y_i$, with y the estimated and y_i the observed modular coefficients. Determine also the distribution properties of e_i (mean, standard deviation, skewness coefficient, and the first serial correlation coefficient). See if the variance of e_i is constant along the values of x by dividing the values of e_i into two groups for $x > \bar{x}$ and $x < \bar{x}$, and test whether the means and the variances of these two groups are significantly different.

11.4. Assume that the period 1951–1957 is missing at the Rogue River. Fill in the missing data of the Rogue River by taking into account e_i from the excess data of the Umpqua River, and show how much the information of the mean of the Rogue River is increased by that extension of data (use for the Rogue River only the data of the period 1905–1950).

11.5. Because one of the two variables of Problem 11.1 is not serially correlated, the effective length, N, of the sample, is equal to $N = 52$. Assume that $r_1 = 0.250$ of the Upqua River, however, and then determine N_e, and test the same hypotheses of Problem 11.1 by using this N_e.

11.6. Plot the points of the example in Problem 11.1 for 52 pairs, and estimate r, a, and b by the graphical method. Plot the regression line obtained analytically, in Problem 11.2, and the line graphically obtained in this Problem, and compare them.

11.7. For the linear regression line established in Problem 11.2 determine the confidence limits about this line on the 90 percent level by using the standard error of estimate, and its confidence limits, and the limits for $e_i = y - y_i$, also at the 90 percent level.

11.8. Assume that the product-moment correlation coefficient, $r = 0.55$, obtained for k simultaneous values of x and y (i. e., the period of simultaneous record is k). Now consider that the N-k values of x are to be used in order to extend the data of the variable y from k to N values. If the ratio of N/k is (1) 1.25; (2) 1.50; (3) 1.75, and (4) 2.00, compute the ratio of the variance of y for N values and the variance of y for k values, provided the variances in Eqs. 11.52 through 11.54 are unities, and $N = 20$. Comment, in general, about the information of the mean. How long is the effective size of the new sample of size N of the random variable y ?

11.9. The straight line, which bisects the two linear regression lines $y = a_1 + b_1x$ and $x = a_2 + b_2y$, is called the structural line. Discuss the relation of these three straight lines, including when each may be used.

11.10. State and discuss the three procedures by which a nonlinear correlation and regression analysis may be performed on two variables x and y.

11.11. State the two usual hypotheses that should be tested for a nonlinear regression of y against x, if a nonlinear function is assumed to be well approximated by a polynomial of the m-th order. Discuss the number of degrees of freedom applicable.

11.12. Describe the tests necessary to determine whether the two indices of correlation are significantly different from zero.

11.13. Select three variables x_1, x_3, and x_2 (annual values of three adjacent precipitation stations), such that x_1 is linearly correlated with x_2 and x_3. Select a sample size of about 30 values in order to reduce computation, and compute: the multiple correlation coefficient, R; the partial correlation coefficient, r_{1-2} and r_{1-3}; the regression coefficients, b_1, b_2, and b_3; the beta-coefficients, and the explained and unexplained variance of x_1. Also, test the hypothesis that R is significantly different from zero at the 90 percent level, and compare the simple product-moment correlation coefficients r_{12} and r_{13} with the r_{1-2} and r_{1-3}.

11.14. Describe the methods for the estimation of regression coefficients of multiple nonlinear regression. Describe measures of the multiple nonlinear correlation association of m variables, and outline experimental methods for making statistical inferences about the estimated regression coefficients and the estimated measures of multiple nonlinear correlation.

CHAPTER 12

MULTIVARIATE ANALYSIS

The *multivariate analysis* as applied to various hydrologic problems is still in its development and trial states. The three types of analyses, *principal components*, *factor analysis*, and *canonical correlations*, show promise in many aspects for assisting in the discovery of basic hydrologic relations of complex problems with a large number of interrelated variables. This chapter is only a very brief introduction to these techniques; for a more exhaustive study the reader is referred to several pertinent references listed at the end of this book under references of Chapter 12.

12.1. Types of Multivariate Analyses.

Definition of multivariate analysis. *Multivariate analysis*, considered either in a broad or in a narrow sense, has no generally agreed definition. In this text multivariate analysis is defined as those statistical methods that investigate the relations among a set, or among several sets, of random variables. For n variables, each of the concurrent sample of size N, a matrix $n \times N$ of sample data is usually the background observational material on which the multivariate statistical analysis is based. Sometimes two matrices, $n \times N$ and $m \times N$, of two sets of n and m random variables, are used. Any additional information from physical considerations of the problem helps to postulate hypotheses in multivariate analysis and improves the obtained relation and the accuracy of the estimated parameters.

Multiple correlation and regression as part of multivariate analysis. Multiple correlation and regression is used to evaluate numerically the parameters of the assumed mathematical prediction model of a hydrologic random variable by n other variables, assuming these variables are not correlated. In this case only the dependent variable is considered as a random variable. Snyder [1] points out the so-called independent variables are usually mutually correlated in hydrologic analyses. For example, a multiple regression equation predicting river runoff does not provide a satisfactory estimate of the independent contributions of the intensities, total amount, and duration of rainfall to runoff, because a positive correlation usually exists between rainfall intensities, total amount, and its duration. In such cases, multiple regression may not appear to offer a workable and mathematically good solution in fitting the relation equation to observed hydrologic data.

Multivariate analysis is the study of sets of random variables. In such an analysis, a vector of means and a matrix of covariances of several random variables are used instead of the simple mean and the variance of a single dependent variable of the multiple regression, with this depenent variable used as the only random variable subject to errors ε. This concept of multivariate analysis makes it possible to associate errors in two or more random variables.

Principal components. The principal components represent an objective method of orthogonal transformation in multivariate analysis of a set of mutually correlated random variables to a new set of mutually uncorrelated random variables. The principal components in the multivariate analysis may be either an intermediate step in the investigation of relations among many random variables, or they may be the final investigation, provided that physical explanations of components are feasible.

Factor analysis. Factors represent either a subjective or a semi--objective method of selecting variables in multivariate analysis. The analysis consists of finding the factors in such a way that all original variables become linear functions of a smaller number of these factors. Then, the dependent variable in the eventual regression equation may be related to factors instead to the original variables. The transformation of principal components may help in identifying and selecting the most relevant factors.

Multivariate canonical analysis. This analysis investigates a set of mutually dependent variables as related to another set of also mutually dependent variables. These two sets of variables are transformed to the two new but smaller sets of highly pairwise correlated variables between sets.

12.2. Multivariate Analysis by Principal Components.

The concept of principal components. Let a set of $n + 1$ random variables be available, each with N concurrent observations. Let one variable be dependent, y, and all or the most of the other n variables, x_1, $x_2, \ldots, x_n$, be mutually correlated, though they are considered independent variables in multiple regression analysis. The multiple regression of Eq. 11.61 does not give the real effect of each of variables $x_1, \ldots, x_n$ on the dependent variable, though it produces the best prediction of y. One of the objectives in any multivariate analysis should be to make n as small and N as large as practically feasible, or the number of degrees of freedom $N - n$ to be as large as possible. This requires economy in the number of variables used in the analysis. The smaller a sample of size N the more important this objective. It is particularly important in hydrology because of the prevalence of small samples. Howerer, the concept of principal components is not necessarily related in hydrology to multiple regression and correlation. Simply the concept relates to the transformation of a set of n mutually dependent variables into a new set of n lineraly uncorrelated variables.

The linear transformation

$$\xi_i = \sum_{j=1}^{n} a_{ij} x_j \ , \qquad 12.1$$

in which $j = 1, 2, \ldots, n$, and $i = 1, 2, \ldots, n$, replace the set of n random variables x_j by the set of n random variables ξ_i in such a manner that ξ_i variables become mutually uncorrelated. Because some of these new variables often have very small variances, they may be neglected in practice. The m variables ξ_i of significant variances, with $m < n$, are mutually and linearly uncorrelated variables. The eventual relation of the type $y = f(x_1, x_2, \ldots, x_n)$ is transformed to a relation of $y = f(\xi_1, \xi_2, \ldots, \xi_m)$, with $m \ll n$, or $N - n$ is replaced by $N - m$, and the variables ξ_i are linearly uncorrelated. Similarly, if the multivariate distribution $F(x_1, x_2, \ldots, x_n)$ is studied, it may be complex because of mutual correlation among the pairs of x_j variables, $j = 1, 2, \ldots, n$. In replacing the set x_j by the set ξ_i, $i = 1, 2, \ldots, m$, the multivariate distribution $F(\xi_1, \xi_2, \ldots, \xi_m)$ may be much simpler to obtain because the components ξ_i are mutually uncorrelated. These new variables ξ_i are called the principal components. When the principal components are determined the first principal component has maximum variance, the second principal component has the largest variance among the remaining components, etc. The linear transformation of Eq. 12.1 is used because it is simple. The objective of using the analysis of principal components is to save on the number of independent variables in a multiple regression, and reduce the number of variables to be observed, by also making them linearly uncorrelated. The method consists of sorting ξ_i variables into descending order of magnitude of their variances, σ_ξ^2, so that ξ_1 has the largest variance, ξ_2 the second largest, and so forth. It happens often in hydrology that $\xi_1, \xi_2, \ldots, \xi_m$ may account for 85—95 percent of the variation of the dependent variable y in a multiple correlation and regression analysis, while the remaining $n - m$ variables of ξ_i give only a small contribution to the explanation of variation of y, so they may be neglected. The mathematical basis for the linear (orthogonal) transformation of Eq. 12.1 is the rotation of the coordinate system of $x_1, x_2, \ldots, x_n$ to a new coordinate system of $\xi_1, \xi_2, \ldots, \xi_n$, so that the vector X, with the correlated projection components $x_1, x_2, \ldots, x_n$, has the uncorrelated principal components $\xi_1, \xi_2, \ldots, \xi_n$ as the projections on the axes of the new system.

The main characteristics of principal components are: (a) they are obtained by a simple objective mathematical operation; (b) they are often reduced to a smaller number of significant principal components in comparison with the original set of variables, and (c) it is necessary in practice to give physical meanings to selected significant ξ_i variables.

Example 12.1. A simple example of the application of principal components in hydrology is the runoff-rainfall relation. To predict runoff from the rainfall variables $x_1, \ldots, x_k$, the antecedent river basin characteristics $x_{k+1}, \ldots, x_p$, the river basin geometric variables $x_{p+1}, \ldots, x_q$, the variables of climatic conditions $x_{q+1}, \ldots, x_r$, the variables of soil and vegetal conditions $x_{r+1}, \ldots, x_s$, and all other variables $x_{s+1}, \ldots, x_n$, a large number of the so-called independent variables must be used in Eq. 11.61. However, if the principal components $\xi_1, \xi_2, \ldots$, are computed, and m of them selected as significant, m may be much smaller than n. The new prediction equation is

$$y = c_0 + c_1 \xi_1 + c_2 \xi_2 + \ldots + c_m \xi_m. \tag{12.2}$$

In practice the main problem is finding physical meaning for significant variables. If one identifies, for this example of runoff-rainfall relation, ξ_1 with the river basin water yield, ξ_2 with the storage of water, ξ_3 with the time distribution of runoff, ξ_4 with another runoff parameter, and similar, principal components would be very useful in hydrology. The present application of principal components in hydrology is, however, far from this ideal.

Computation of principal components from the correlation matrix. Let a sample of size N of n + 1 variables be given as a N × (n + + 1) matrix of original data in a multiple correlation and regression analysis for purposes of predicition. By singling out the dependent variable y, the remaining n so-called independent variables of multiple regression analysis make a N × n matrix from the original data. The objective is to transform these n variables x_j into n uncorrelated principal components ξ_i, and by neglecting n − m variables ξ_i of very small variances, σ_ξ^2 to maximize the parameter N − m in the multivariate analysis. For this computation the concept of n + 1 variables and the multiple correlation and regression analysis is not necessary.

Standardize each variable x_j by $\varepsilon_j = (x_j - m_j)/s_j$, where m_j is the mean and s_j^2 the variance of any x_j variable, and compute the correlation matrix $[r_{jk}]$ of Eq. 11.76. A new coordinate system is obtained for ξ_i components by rotating the n-dimensional x_j coordinate system so that ξ_i become mutually and linearly uncorrelated variables. A line through the end point of the vector of means $(m_1, m_2, \ldots, m_n)$ of the x_j coordinate system (this point is also the origin of the ε_j coordinate system) is drawn in such a way as to obtain the least-squares sum of distances of N point of x_j variables from that line. It can be shown that this line passes through the origin of the x_j coordinate system. The line has the cosines of directions b_j, so that

$$\frac{x_1 - m_1}{b_1} = \frac{x_2 - m_2}{b_2} = \ldots = \frac{x_n - m_n}{b_n} \tag{12.3}$$

represents the equation of the straight line in the x_j coordinate system, with $\Sigma b_j^2 = 1$, $j = 1, 2, \ldots, n$. The sum of squares of distances of N points of N × n matrix of the data of standardized variables ε_j on to the line of Eq. 12.3 is

$$S = \frac{1}{N} \sum_{k=1}^{N} \left[\sum_{j=1}^{n} \varepsilon_{jk}^2 - \left(\sum_{j=1}^{n} b_j \varepsilon_{jk} \right)^2 \right]. \tag{12.4}$$

Because $\Sigma \varepsilon_{jk}^2 / n = 1$,

$$S = n - \frac{1}{N} \sum_{k=1}^{N} \left(\sum_{j=1}^{n} b_j \varepsilon_{jk} \right)^2. \tag{12.5}$$

The conditions for S to be the minimum, and also stationary for variations in b_j, with an undetermined multiplier λ, produce n equations [2, p. 14–15]:

$$\begin{aligned} b_1(1-\lambda) + b_2 r_{1,2} + \ldots + b_n r_{1,n} &= 0 \\ b_1 r_{2,1} + b_2(1-\lambda) + \ldots + b_n r_{2,n} &= 0 \\ \ldots\ldots\ldots\ldots\ldots\ldots\ldots\ldots \\ b_1 r_{n,1} + b_2 r_{n,2} + \ldots + b_n(1-\lambda) &= 0. \end{aligned} \qquad 12.6$$

By eliminating the b_j cosines from Eq. 12.6 the characteristic equation of the correlation matrix is obtained as $\lambda = f(r_{j,k})$, which produces n roots of λ as $\lambda_1, \lambda_2, \ldots, \lambda_n$, with $\Sigma\lambda_j = n$, $j = 1, 2, \ldots, n$. Each λ_j gives a set of n coefficients b_{ij} by Eq. 12.6, where i refers to ξ_i and j to x_j. The sum of Eq. 12.5 is $S_j = n - \lambda_j$ for each individual principal component, and λ_j is always positive. Equation 12.1 is expressed in terms of a_{ij}. It can be rewritten in terms of b_{ij} and ε_j as a set of new variables ζ_i,

$$\zeta_i = \sum_{j=1}^{n} b_{ij}\,\varepsilon_j\,. \qquad 12.7$$

By replacing $\varepsilon_j = (x_j - m_j)/s_j$ in Eq. 12.7, and comparing it with Eq. 12.1, the coefficients a_{ij} of Eq. 12.1 are $a_{ij} = b_{ij}/s_j$, and ξ_i of Eq. 12.1 is now $\xi_i = \zeta_i + \Sigma b_{ij}\,m_j/s_j$, $j = 1, 2, \ldots, n$, or ξ_i and ζ_i differ by a constant for each $i = 1, 2, \ldots, n$. The largest root of $\lambda = f(r_{j,k})$ gives λ_1, which is the variance of ζ_1. The second largest root is λ_2, the variance of ζ_2, and so forth. The results are that $\text{cov}(\zeta_p, \zeta_q) = 0$, or ζ_j components

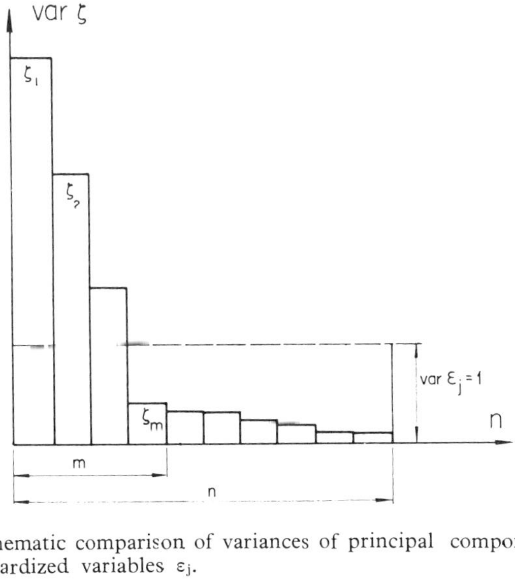

Figure 12.1. Schematic comparison of variances of principal components ζ_j, and variances of standardized variables ε_j.

are mutually uncorrelated, and their variances decrease more rapidly from the largest to the smallest while all ε_j have the variance unity, as shown schematically in Fig. 12.1. The coefficients a_{ij} of Eq. 12.1 may be computed from the covariance matrix without passing through the $[r_{j,k}]$ matrix and the standardized variables ε_i. In this case ξ_i are also computed as uncorrelated variables.

For a more detailed computational procedure or discussion of principal components see Kendall [2, p. 18—36], Anderson [3, p. 281—286], Rao [4], Roy [5], and others [6—12].

Example 12.2. In the application of multivariate analysis to the study of the chemistry of groundwater quality in the Mojave River Valley, California, Dawdy and Feth [13] computed a correlation matrix, Table 12.1, for eight chemical measures of quality by using 108 analyses of samples from 85 ground water wells. If a variable W_q of water quality is related to these eight variables, Table 12.1 shows that the so-called independent variables of Eq. 11.61 are highly linearly correlated except the last two variables, which do not show a significant correlation with the other six variables or among themselves. Rotation of the coordinate system gives the coefficients a_{ij} of eight principal components of Table 12.2, in terms of eight observed chemical variables. In terms of the percent of variance of W_q which would be explained by all eight original variables, the first four principal components would explain 50.2, 16.9, 13.2, and 10.5 percent of it, respectively, or a total of 90.8 percent The remaining four principal components would contribute only 9.2 percent. In practice then in a multiple regression analysis eight original variables may be replaced by four principal components.

Table 12.1

Correlation Matrix for the 8 Chemical Measures from 108 Analyses of Samples from 85 Groundwater Wells in the Mojave River Valley

	x_1 Ca	x_2 Mg	x_3 Na + k	x_4 HCO_3	x_5 SO_4	x_6 Cl	x_7 F	x_8 NO_3
x_1	1.000	0.898	0.424	0.622	0.681	0.576	−0.061	−0.042
x_2		1.000	0.543	0.438	0.760	0.691	−0.062	0.008
x_3			1.000	0.192	0.794	0.923	0.201	−0.080
x_4				1.000	0.330	0.129	−0.074	−0.114
x_5					1.000	0.719	0.226	−0.083
x_6						1.000	0.066	−0.053
x_7							1.000	−0.078
x_8								1.000

Interpretation of principal components. It is often difficult to assign physical meaning for computed principal components, though it is not impossible. If some mutually independent physical parameters could be identified with the principal components, an objective method of their computation would be practical and valuable. In the hydrologic future various investigators should attempt to assign physical meanings to the most significant principal components in the case of some already classical multivariate hydrologic problems.

Table 12.2
Coefficients of Principal Components ξ_i, i = 1, 2, ..., 8, Derived from the Correlation Matrix of Eight Original Chemical Variables, x_j, j = 1, 2, ..., 8.

	x_1 Ca	x_2 Mg	x_3 Na + k	x_4 HCO_3	x_5 SO_4	x_6 Cl	x_7 F	x_8 NO_3
ζ_1	−0.425	−0.442	−0.404	−0.245	−0.462	−0.425	−0.045	0.062
ζ_2	−0.346	−0.205	0.370	−0.514	0.159	0.258	0.582	−0.085
ζ_3	−0.042	0.102	0.089	−0.344	−0.011	0.186	−0.303	0.858
ζ_4	−0.145	−0.020	0.187	−0.353	−0.078	0.297	−0.704	−0.479
ζ_5	−0.343	−0.460	0.479	0.612	−0.079	0.124	−0.141	0.156
ζ_6	−0.235	−0.074	0.018	−0.053	0.802	−0.493	−0.223	0.012
ζ_7	−0.655	0.717	0.037	0.156	−0.149	−0.075	0.048	−0.027
ζ_8	−0.274	−0.128	−0.649	0.174	0.307	0.602	0.010	0.016

12.3. Multivariate Factor Analysis.

The concept of factors. Factors are a set of new variables that can replace the original observed variables. The selection and the introduction of factors in the multivariate analysis is less direct than in the case of principal components. In some ways factor analysis is either an opposite or a supplemental approach in comparison with analysis by principal components. In an opposite approach factors are postulated by various considerations. In a supplemental approach factors are obtained from the original data, or from further treating principal components, until an interpretation of identifiable factors is satisfactory.

Postulation of factors. A mathematical model for the factors is postulated, and its parameters estimated. If the results are not satisfactory, the factors and model are changed, the procedure repeated, and so forth. It usually takes a sequence of successive approximations until an acceptable relation between a variable and its postulated factors is obtained.

Basically the following model of factors replaces a variable x_j in the multivariate analysis

$$x_j = \sum_{k=1}^{p} a_{jk} f_k + \sum_{g=1}^{q} b_{jg} s_g + c_j \varepsilon_j, \qquad 12.8$$

j = 1, 2, ..., n, in which f_k are factors common to many or to all variables $x_1, x_2, \ldots, x_n$; s are factors, or usually only a factor, relating only to the x_j variable, or they are specific to x_j; ε_j is the error in x_j, or it accounts for the unreliability of factors in the x_j model, and a_{jk}, b_{jg}, and c_j are coefficients (factor loadings) to be estimated from data.

The new factors, f_k and s_g, may be independent among themselves, but are usually dependent; they are either standardized or nonstandardized, and p should be as small as feasible, while usually q = 1. The factors

f_k may be divided into general factors, which are in all variables, and common factors which are in two or more but not in all variables. Basic difficulties in applying factor analysis in this manner are the arbitrary and subjective procedures in both the postulation of the number of factors and the selection of the mathematical model similar to Eq. 12.8.

For two variables designated by x_i and x_j, and with the variables f_k, s_g, and ε_j standardized, the covariance of x_i and x_j is

$$\text{cov}(x_i, x_j) = \sum_{k=1}^{p} a_{ik} a_{jk} + \sum_{g=1}^{q} b_{ig} b_{jg} \text{cov}(s_{ig}, s_{jg}) + c_i c_j \text{cov}(\varepsilon_i, \varepsilon_j). \quad 12.9$$

If variables s_{ig} and s_{jg}, and ε_i and ε_j of x_i and x_j are mutually independent, then

$$\text{cov}(x_i, x_j) = \sum_{k=1}^{p} a_{ik} a_{jk}, \text{ for } i \neq j, \quad 12.10$$

and

$$\text{var } x_j = \sum_{k=1}^{p} a_{jk}^2 + \sum_{g=1}^{q} b_{jg}^2 + c_j^2, \text{ for } i = j. \quad 12.11$$

For x_j of Eq. 12.8 standardized, or var $x_j = 1$, the term Σa_{jk}^2 is called the communality, Σb_{jg}^2 the specificity, and c_j^2 the unreliability of factor analysis. Similarly, $\Sigma a_{jk}^2 + \Sigma b_{jg}^2$ is called the reliability, and $1 - \Sigma a_{jk}^2$ the uniqueness of factor analysis [2].

Example 12.3. A large number of geomorphological river basin parameters are either in use, or proposed for use, in hydrologic investigations, such as the parameters of river basin shape, slope, stream density, surface roughness, and similar. They are highly mutually dependent random variables of river basin morphology. Introducing them into the multiple regression analysis usually means some parameters will account for the effects in variation of other parameters. What hydrology needs are geomorphological parameters in the form of factors relevant to hydrology, such as those measuring water retention, surface water losses (either as evaporation or infiltration), time delays of surface runoff, and similar. The problem of geomorphological basin factors of particular importance for various responses of hydrologic environments to known inputs has yet to obtain practical solutions.

Direct computation of factors. It is possible to derive factors from principal components ζ_i, or directly from the original variables x_j, by further rotating the ζ_i coordinate system, or by applying a different rotation to the x_j coordinate system than in the case of principal components.

For various aspects of factor analysis, the reader is referred to abundant literature on the subject of multivariate analysis [1 — 18].

Application of factor analysis in hydrology. Opinions among hydrologists are far from unanimous about the merits of factor analysis in hydrology. The Tennessee Valley Authority used factor analysis in the design of the hydrologic condition survey [19]. Wallis [20] used a known prediction equation of a hollow cylinder example and investigated the accuracy of prediction by various techniques of multivariate analysis including the multiple regression. Matalas and Reiher [21] subjected the application of factor analysis in hydrology to a critical review, considering it technically underdeveloped and questionable in its usefulness for studying interrelated hydrologic variables. Wallis [22] suggested that factor analysis will never be of great value to hydrology if used in the classical manner, but he does consider it useful as a numerical procedure in building effective multiple regression equations. Other authors [23—27] have discussed various aspects of the application of multivariate (mainly factor) analysis in hydrology. The general conclusion is that factor analysis has many undesired properties, but it may have promise in some of its aspects if it is properly applied to hydrologic multivariate problems.

12.4. Multivariate Canonical Analysis.

The general aspects of canonical analysis. Instead of investigating the relation of a dependent random variable y to a set of mutually correlated variables x_j, a set of mutually correlated random variables y_i is related to the set x_j. When interest is in the relation of the set y_i regarded as the dependent set, to the set x_j considered as the fixed or independent set, rather than in the relations of the x_j's among themselves or of the y_i's among themselves, the analysis is called multivariate canonical. This type of relation may be considered either as a correlation analysis (symmetrical analysis of the two sets of variables), or as a regression analysis, with a depedent and an independent or fixed set of variables.

One approach in the analysis may be to transform the y_i set to the principal components η_i, and the x_j set to the principal components ξ_j, as the two individually uncorrelated sets, and to relate them. However, in the canonical analysis the y_i and x_j sets are transformed to two new orthogonal (uncorrelated) sets η_i and ξ_j but not so as to maximize the variance in any particular direction. The approach used is to maximize the correlations (covariances) between certain pairs of variables of the two sets while minimizing or reducing to zeros the correlations (covariances) between the remaining pairs of variables of the two sets.

For details and examples of canonical analysis see Kendall [2, p. 75—85], Morrison [7, p. 207—220], and others [27—32].

Example 12.4. An example of the use of multivariate canonical analysis in hydrology is the relation of monthly average precipitation of several coastal areas (as the dependent y_i set of random variables) to monthly average water temperatures of ocean surface of several areas adjacent to the coast (as the dependent x_j set of random variables). Both sets are mutually and individually correlated, so the mutiple canonical analysis may reduce the sets y_i and x_j to some pairs highly correlated while the remaining pairs are uncorrelated.

12.5. Problems for Disscusion.

12.1. Discuss the multivariate principal components analysis as follows.

(1) Define the principal components analysis; compare it with the multiple regression analysis of interdependent variables.

(2) Describe the use of the correlation matrix.

(3) State the objectives of principal components analysis.

(4) State the basic difficulty in applying principal components in hydrology, once they have been computed.

(5) Describe briefly two examples of potential application of principal components analysis in hydrology, indicating the possible physical meanings that may be given to some expected components.

12.2. Discuss the multivariate factor analysis as follows.

(1) Define the concept of factor analysis.

(2) Express a known hydrologic variable in terms of different factors and errors, in a practical example, similar to Eq. 12.8.

(3) State the difference between the factor and principal components analyses.

REFERENCES

Chapter 1.

1. *Chow, Ven Te, Hydrology and its Developments, Section 1, Handbook of Applied Hydrology, Ven Te Chow, ed., McGraw-Hill, New York, 1964.

2. Biswas, A. K., History of Hydrology, American Elsevier, New York, 1970.

3. Yevjevich, V., Fluctuation of Wet and Dry Years, Part II, Analysis by Serial Correlation, Colorado State University Hydrology Paper No. 4, June 1964, Fort Collins, Colorado.

4. Yevjevich, V., and Jeng, R. I., Properties of Non-Homogeneous Hydrologic Series, Colorado State University Hydrology Paper No. 32, April 1969, Fort Collins, Colorado.

5. Hurst, H. E., Black, R. P., and Simaika, Y. M., The Nile River, The Future Conservation of the Nile, Vol. 7, S. O. P. Press, Cairo, Egypt, 1946.

Chapter 2.

1. Kolmogorov, A. N., Foundations of the Theory of Probability, 2nd ed., Chelsea, New York, 1956.

2. Kolmogorov, A. N., Grundbegriffe der Wahrscheinlichkeitsrechnung, Berlin, 1933.

3. Mises, R. von, Wahrscheinlichkeitsrechnung, Leipzig and Vienna, 1931.

4. Mises, R. von, Probability, Statistics, and Truth, MacMillan, New York, 1939.

5. Gnedenko, B. V., The Theory of Probability, 4th ed., Chelsea, New York, 1968.

6. Wilks, Samuel S., Mathematical Statistics, Wiley, New York, 1962.

---

* This source contains 137 general references related to hydrology and its development. The reader unfamiliar with hydrology can find among these references those which he may need to understand the basic principles of hydrology.

7. Pinkayan, S., Conditional Probabilities of Occurrence of Wet and Dry Years Over a Large Continental Area, Colorado State University Hydrology Paper No. 12, April 1966, Fort Collins, Colorado.

8. Todorović, P., and Yevjevich, V., Stochastic Process of Precipitation, Colorado State University Hydrology Paper No. 35, September 1969, Fort Collins, Colorado.

9. Kendal, M. G., and Stuart, Alan, The Advanced Theory of Statistics, 2nd Ed., Vol. 1, Griffin, London, 1963.

10. Fisz Marek, Probability Theory and Mathematical Statistics, 3rd ed., Wiley, New York, 1963.

Chapter 3.

1. Riordan, John, An Introduction to Combinatorial Analysis, Wiley, New York, 1958.

2. David, R. N., and Barton, D. E., Combinatorial Chance, Hafner, New York, 1962.

3. Parsen, Emanual, Modern Probability Theory and Its Applications, Wiley, New York, 1960.

4. Burington, R. S., and May, D. C., Handbook of Probability and Statistics with Tables, Handbook Publishers, Sandusky, Ohio, 1958.

5. Beckmann, Petr, Elements of Applied Probability Theory, Harcourt, Brace and World, New York, 1968.

6. Anderson, D. V., Review of Basic Statistical Concepts in Hydrology, Proceedings of Hydrology Symposium No. 5 on Statistical Methods in Hydrology, Held at McGill University, Feb. 1966, National Research Council of Canada.

7. Gnedenko, B. V., The Theory of Probability, 4th ed., Chelsea, New York, 1968.

8. Anis, A. A., and Lloyd, E. H., On the Range of Partial Series of a Finite Number of Independent Normal Variates, Biometrika, Vol. 22, 1955, pp. 96—101.

9. Feller, W., The Asymptotic Distributions of the Range of Series of Independent Random Variables, Ann. Math. Stat., Vol. 22, 1951, pp. 427—432.

10. Yevjevich, V., Mean Range of Linearly Dependent Normal Variables with Application to Storage Problems, AGU Water Resources Research Journal, Vol. 3, No. 3, 3rd quarter 1967, pp. 663—671.

Chapter 4.

1. Weatherburn, C. E., First Course in Mathematical Statistics, 2nd ed., Cambridge University Press, 1968.

2. Spiegel, Murray, Theory and Problems of Statistics, Schaum's Outline Series, Schaum Publishing, 1961.

3. Feller, W., An Introduction to Probability Theory and Its Application, 2nd ed., Wiley, 1960.

4. Parzen, E., Modern Probability Theory and its Applications, Wiley, 1960.

5. Cramér, Harold, Mathematical Methods of Statistics, 8th printing, Princeton University Press, 1958.

6. Fisz, Marek, Probability Theory and Mathematical Statistics, 3rd ed., Wiley, 1963.

7. Benson, M. A., Characteristics of Frequency Curves Based on a Theoretical 1,000-year Record, Flood Frequency Analysis, Manual of Hydrology, Part 3 of Flood-Flow Techniques, U. S. Geological Survey Water Supply Paper 1543—A, 1960.

8. Knisel, W. G., and Yevjevich, V., The Statistical Measure of Hydrologic Time Series, Proceedings of the International Hydrology Symposium, Sept. 6—8, 1967, Colorado State University, Fort Collins, Colorado, Vol. 1, Paper No. 40, pp. 306—313.

Chapter 5.

1. Sturges, H. A., The Choice of a Class Interval, P. Amer. Statistical Assoc., March 1926.

2. Gumbel, E. J., Statistics of Extremes, Columbia University Press, New York, 1958.

3. Weibull, W., A Statistical Theory of the Strength of Materials, Vol. 151, Ing. Vetenskaps Akad. Handl., Stockholm, 1939, pp. 15, (also, Vol. 153, p. 17, 1939).

4. Hazen, Allen, Flood Flows, A Study of Frequencies and Magnitudes, Wiley, New York, 1930.

5. Chow, Ven Te, ed., Handbook of Applied Hydrology, McGraw-Hill, New York, 1964.

Chapter 6.

1. Kendall, M. G., and Stuart, Alan, The Advanced Theory of Statistics, Vol, I, Distribution Theory, Hafner, New York, 1958.

2. Wilks, S. Samuel, Mathematical Statistics, Wiley, New York, 1962.

Chapter 7.

1. National Bureau of Standards, Tables of the Binomial Probability Distribution, Applied Mathematics Series 6, Washington, D. C., 1950.

2. Pearson, E. S., and Hartley, H. O., Biometrika Tables for Statisticians, 1954.

3. Molina, E. C., Poisson's Exponential Binomial Limit, D. Van Nostrand, New York, 1942.

4. Kendall, M. G., and Stuart, Alan, The Advanced Theory of Statistics, Vol. 1, Distribution Theory, Hafner, New York, 1958.

5. Greenwood, M., and Yule, G. U., An Inquiry into the Nature of Frequency Distributions of Multiple Happenings, etc., Journal Ray. Statistical Soc., Vol. 83, 1920, p. 255.

6. Marković, R. D., Probability Functions of Best Fit to Distributions of Annual Precipitation and Runoff, Colorado State University Hydrology Paper No. 8, August 1965, Fort Collins, Colorado.

7. Hazen, Allen, Storage to be Provided in Impounding Reservoirs for Municipal Water Supply, Transactions Amer. Soc. of Civil Engineers, Vol. 77, Paper 1308, 1914, p. 1539.

8. McAlister, D., The Law of Geometric Mean, Proc. Royal Soc., Vol. 29, (London), 1879, p. 367.

9. Galton F., Philos. Magazine, London, 1875, ser. 4. V. 49; also Galton, F., The Geometric Mean in Vital and Social Statistics, Proceedings Royal Soc., Vol. 29, (London), 1879, p. 365.

10. Aitchison, J., and Brown J. A. C., The Lognormal Distribution, Cambridge University Press, England, 1957, pp. 1—176.

11. Lane, E. W., and Kai Lei, Streamflow Variability, Transactions Amer. Soc. of Civil Engineers, Vol. 115, Paper No. 2418, 1950, p. 1084.

12. Gibrat, R., Une loi des répartitions économiques: l'effet proportionnel (A Law of Economic Distributions: the Proportionate Effect), Bull. Statistique Génerale (France), Vol. 19, 1930, p. 469; also Gibrat, R., Les inégalitiés économiques (The Economic Inegalities), Libraire du Recueil Sirey, Paris, 1931.

13. Wicksell, S. D., On the Genetic Theory of Frequency, Arkives Mat. Astr., Fys., Vol. 12, No. 20, 1917, p. 1.

14. Nydell, S., The Mean Error of the Characteristics in Logarithmic-Normal Distribution, Skand. Aktuar. Tidskr., Vol. 2, 1919, p. 134.

15. Finney, D. J., On the Distribution of a Variate whose Logarithm is Normally Distributed, Journal of Royal Statistical Soc., Vol. 7, 1941, p. 151.

16. Quensel, C. E., Studies of the Logarithmic-Normal Curve, Skand. Aktuar. Tidskr., Vol. 28, 1945, p. 141.

17. Cochran, W. G., Some Difficulties in Statistical Analysis of Replicated Experiments, Emp. J. Exp. Agric., Vol. 6, 1938, p. 157.

18. Curtiss, J. H., On Transformations Used in the Analysis of Variance, Am. Math. Statist., Vol. 14, 1943, p. 107.

19. Cohen, A. C., Estimating Parameters of Logarithmic-Normal Distribution by Maximum Likelihood, J. Amer. Statist. Assoc., Vol. 46, 1951, p. 206.

20. Kolmogorov, A. N., Ueber das logaritmisch normale Verteilungsgesetz der Dimensionen der Teilchen bei Zerstueckelung (On the Lognor-

mal Disbritution of Particle Dimension in Breakage), C. R. Acad. Sci., U.R.S.S. Vol. 31, 99, 1941.

21. Rasumovsky, N. K., Distribution of Metal Values in Ore Deposits, C. R. Akad. Sci., U.R.S.S., Vol. 28, 814, 1940.

22. Pearson, Karl, On the Curves which are Most Suitable for Describing the Frequency of Random Samples of a Population, Biometrika, Vol. 5, 1905, pp. 172—175.

23. Pearson, Karl, Tables of the Incomplete Gamma Function, Cambrid-dge University Press, England, 1921.

24. Gumbel, E. J., Statistics of Extremes, Columbia University Press, New York, 1958.

25. Fuller, W. E., Flood Flows, Trans. ASCE, Vol. 77, 1914, pp. 564—617.

26. Wadsworth, P. G., and Bryan, G. F., Introduction to Probability Theory and Random Variables, McGraw-Hill, New York, 1960.

27. Kibler, F. D., and Yevjevich, V., Effects of Sampling Interval, Periodicity, Dependence and Skewness on Extreme Values, Proc. Int. Hydrology Symposium, Sept. 6—8, 1967, Colorado State University, Fort Collins, Colorado, Vol. 1, Paper 67, pp. 536—545.

Chapter 8.

1. Cramér, H., Mathematical Methods of Statistics, 8th printing, Princeton University Press, 1958.

2. Davis, H. T., Tables of Higher Mathematical Functions, Vol. I and II, Principia Press, Bloomington, 1933.

3. Masuyama, M., and Kuroiwa, Y., Tables of the Likelihood Solutions of Gamma Distribution and its Medical Applications, Statistical Application Research, Vol. 1, No. 1, 1951.

4. Nörlund, N. E., Vorlesungen ueber Differenzenrechnung (Lectures on Difference Calculus), Springer, Berlin, 1924, pp. 101, 106.

5. Thom, H. C. S., A Note on the Gamma Distribution, Monthly Weather Review, Vol. 86, No. 4, April 1958, pp. 117—121.

Chapter 9.

1. Cramér, H., Mathematical Methods of Statistics, 8th printing, Princeton University Press, 1958

2. Siddiqui, M. M., Some Statistical Theory for the Analysis of Radio Propagation Data, Journal of Research of the National Bureau of Standards, D. Radio Propagation, Vol. 66D, No. 5, Sept.-Oct. 1962.

3. Brooks, C. E. P., and Carruthers, N., Handbook of Statistical Methods in Meterology, M. O. 538, Meteorological Office, London, 1953.

4. Matalas, N. C., and Langbein, W. B., Information Content of the Mean, Journal of Geophysical Reseearch, Vol. 67, No. 9, 1962.

5. Yevjevich, V., Fluctuations of Wet and Dry Years, Part I, Research Data Assembly and Mathematical Models, Colorado State University Hydrology Paper No. 1, July 1963, Fort Collins, Colorado.

6. Yevjevich, V., Fluctuations of Wet and Dry Years, Part II, Analysis by Serial Correlation, Colorado State University Hydrology Paper No. 4, June 1964, Fort Collins, Colorado.

7. Yevjevich, V., Some General Aspects of Fluctuations of Annual Runoff in the Upper Colorado River Basin, Colorado State University Publication, Report CER61VMY54, October 1961, Fort Collins, Colorado.

8. Todorović, P., On Some Problems Involving Random Number of Random Variables, Ann. of Math. Statistics, Vol. 41, No. 3, 1970.

9. Zelenhasić, E., Theoretical Probability Distributions for Flood Peaks, Colorado State University Hydrology Paper No. 42, December 1970, Fort Collins, Colorado.

10. Knisel, W. G., and Yevjevich, V., The Statistical Measure of Hydrologic Time Series, Proceedings of International Hydrology Symposium, Sept. 6—8, 1967, Colorado State University, Fort Collins, Colorado, Paper No. 40 , pp. 306—313.

Chaper 10.

1. Neyman, T., Basic Ideas and Some Recent Results of the Theory of Testing Statistical Hypotheses, Journal Roy. Stat. Soc., Vol. 105, p. 299, 1942.

2. Hoel, G. Paul, Introduction to Mathematical Statistics, 3rd ed., Wiley, New York, 1962.

3. Cramér, Harold, Mathematical Methods of Statistics, 8th printing, Princeton University Press, 1958.

4. Ostle, Bernard, Statistics in Research, Iowa State University Press, 1963.

Chapter 11.

1. Pearson, Karl, On a Form of Spurious Correlation Which May Arise When Indices are Used in the Measurements of Organs, Proc. Royal Soc. of London, Vol. 60, 1932, pp. 489—502.

2. Benson, A. M., Spurious Correlation in Hydraulics and Hydrology, ASCE Journal of Hydraulics Division, HY4, July 1965, pp. 35—42.

3. Benson, A. M., Factors Influencing the Occurrence of Floods in a Humid Region of Diverse Terrain, U. S. Geological Survey, Water-Supply Paper 1580-B, 1962.

4. Bartlett, M. S., Some Aspects of the Time-Correlation Problem in Regard to Tests of Significance, Journal Royal Statistical Soc., Vol. 98, 1935, pp. 536—543.

5. Roche, M., Hydrologie de surface (Surface Hydrology), Gauthier--Villars, Paris, 1963, p. 48.

6. Ezekiel, Mordecai, and Fox, K. A., Methods of Correlation and Regression Analysis: Linear and Curvilinear, 3rd ed., Wiley, New York, 1959.

7. Reich, B. M., Design of Hydrographs for Very Small Watersheds from Rainfall, Colorado State University Report CER62BMR41, July 1962, Fort Collins, Colorado.

Chapter 12.

1. Snyder, W. M., Some Possibilities for Multivariate Analysis in Hydrologic Studies, Journal of Geophysical Research, Vol. 67, No. 2, February 1962, pp. 721—729.

2. Kendall, M. G., A Course in Multivariate Analysis, Hafner, New York, 1957.

3. Anderson, T. W., An Introduction to Multivariate Statistical Analysis, Wiley, New York, 1958.

4. Rao, C. R., Advanced Statistical Methods in Biometrik Research, Wiley, New York, 1952.

5. Roy, S. N., Some Aspects of Multivariate Analysis, Wiley, New York, 1958.

6. Barlett, M. S., Multivariate Analysis, Journal Roy. Statistical Soc., B, 9, 1947, p. 176.

7. Morrison, D. F., Multivariate Statistical Methods, Mc-Graw-Hill, New York, 1967.

8. Whittle, P., On Principal Components and Least Square Methods of Factor Analysis, Skand. Akt., Vol. 35, 1953, p. 223.

9. Lawley, D. N., A Statistical Examination of the Centroid Method, Proc. Royal Soc. Ed., A. 64, 1955, p. 175.

10. Holzinger, K. J., and Harman, H. H., Factor Analysis, University of Chicago Press, 1941.

11. Burt, Sir Cyril, Alternative Methods of Factor Analysis, British Journal of Psychology (Stat. Sect.), Vol. 2, 1949, p. 98.

12. Anderson, T. W., and Rubin, H., Statistical Inference in Factor Analysis, Proc. 3rd Berkeley Symposium, Vol. 5, 1956, p. 111.

13. Dawdy, D. R., and Feth, J. H., Application of Factor Analysis in Study of Chemistry of Ground Water Quality, Mojave River Valley, California, AGU Journal of Water Resources Research, Vol. 3, No. 2, 1967, pp. 505—510.

14. Kaiser, H. F., The Varimax Criterion for Analytic Rotation in Factor Analysis, Psychometrica, Vol. 23 (3), 1958, pp. 187—200.

15. Thurstone, L. L., Multiple Factor Analysis, University of Chicago Press, 1947.

16. Lawley, D. N., and Maxwell, A. E., Factor Analysis as a Statistical Method, Butterworths, London ,1963.

17. Horst, Paul, Factor Analysis of Data Matrices, Holt, Rinehart and Winston, New York, 1965.

18. Harman, H. H., Modern Factor Analysis, University of Chicago Press, 1960.

19. Design of a Hydrologic Condition Survey Using Factor Analysis, Tennessee Valley Authority Div. Water Control Planning Res. Paper 5, 1965, 71 pp.

20. Wallis, J. R., Multivariate Statistical Methods in Hydrology — A Comparison Using Data of Known Functional Relationship, Water Resources Research, Vol. 1, No. 4, 1965, pp. 447—461.

21. Matalas, N. C., and Reiher, J. Barbara, Some Comments on the Use of Factor Analysis, Water Resources Research, Vol. 3, No. 1, 1967, pp. 213—223.

22. Wallis, J. R., Factor Analysis in Hydrology — An Agnostic View, Water Resources Research, Vol. 4, No, 3, 1968, pp. 521—527.

23. Wong, S. T., A Multivariate Statistical Model for Predicting Mean Annual Flood in New England, Ann. Assoc. Am. Geog., Vol. 52 (3), 1963, pp. 298—311.

24. Fiering, M. B., Multivariate Technique for Synthetic Hydrology, Proc. Amer. Soc. of Civil Engineers, 1964, pp. 43—60.

25. Wallis, J. R., A Factor Analysis of Soil Erosion and Stream Sedimentation in Northern California, Ph. D. Dissertation, University of California, Berkeley, 1965.

26. Wallis, J. R., When is it Safe to Extend a Prediction Equation? — An Answer Based upon Factor and Discriminant Function Analysis, Water Resources Research, Vol. 3, No. 2, 1967, pp. 375—384.

27. Hotteling, H., Relations Between Two Sets of Variates, Biometrica Vol. 28, 1936, pp. 321—377.

28. Waugh, F. V., Regressions Between Sets of Variables, Econometrica, Vol. 10, 1942, pp. 290—310.

29. Tinter, G., Some Applications of Multivariate Analysis in Economic Data, Journal of the Amer. Statistical Assoc., Vol. 41, 1946, pp. 472—500.

30. Meredith, W., Canonical Correlations with Fallible Data, Psychometrika, Vol. 29, 1961, pp. 55—66.

31. Horst, P., Relations Among m Sets of Measures, Psychometrika, Vol. 26, 1961, pp. 1929—149.

31. Horst, P., Generalized Canonical Correlations and Their Applications to Experimental Data, Journal of Clinical Psychology (monograph supplement), No. 14, Oct. 1961, pp. 331—347.

INDEX